TM 9-2815-237-34
CUCV
Commercial Utility Cargo Vehicle
Direct Support / General Support
Engine Maintenance Manual
January 1996 w/ Change 1&2

The CUCV or Commercial Utility Cargo Vehicle is a US Military vehicle based on readily available commercial trucks. Originally intended to augment the purpose-built, but expensive GAMA Goat 6x6 and older Jeeps. The first generation was based on Dodge / Chrysler trucks.

This book is focused on the M1008 series second generation CUCV which was General Motor's first major light-truck military vehicle production since World War II. They began production in 1984 and ended production in 1996 with most units being produced as 1984 model year units. Later production was focused on replacements for existing CUCV's. The majority of units were built from existing heavy duty light truck commercial parts. The M1009 was an upgraded/up-rated Chevy K5 Blazer with a 3/4 ton capacity. The M1008 series trucks were a 1-1/4 ton or 5/4 ton rated truck. In all 70,000 units were produced with three power trains.

This manual is the Direct Support and General Support repair instructions for the diesel engines used in these vehicles. Including the 6.5 liter and other models. It is published as a convenience to enthusiasts who may wish to have a quality professionally printed copy of the manual.

This publisher has also printed other manuals for this series of vehicles. Should you have suggestions or feedback on ways to improve this book please send email to Books@OcotilloPress.com

---

Edited 2021 Ocotillo Press
ISBN 978-1-954285-74-3

Ocotillo Press
Houston, TX 77017
Books@OcotilloPress.com

ARMY TM 9-2815-237-34
AIR FORCE TO 38V1-6-2
MARINE CORPS TM 2815-34/3 A

(SUPERSEDES TM 9-2815-237-34, 25 June 1985)

TECHNICAL MANUAL

DIRECT SUPPORT AND GENERAL SUPPORT MAINTENANCE

ENGINE, DIESEL: 8 CYLINDER, NATURALLY ASPIRATED, FUEL-INJECTED, WATER COOLED, DDA MODEL 6.2 LITER (PRE-1990) (NSN 2815-01-231-3672), (POST-1990) (NSN 2815-01-314-7940); DDA MODEL 6.5 LITER (NSN 2815-01-410-9710) (1995,1996), (NSN 2815-01-439-6665) (1997), (NSN 2815-01-453-7403) (1998);

ENGINE, DIESEL: 8 CYLINDER, NATURALLY ASPIRATED, DETUNED FUEL-INJECTED, WATER COOLED, DDA MODEL 6.5 LITER (NSN 2815-01-406-6675) (1996), (NSN 2815-01-439-6664) (1997), (NSN 2815-01-453-7404) (1998);

ENGINE, DIESEL: 8 CYLINDER, TURBOCHARGED, FUEL-INJECTED, WATER COOLED, (NSN 2815-01-420-4180) (1996), (NSN 2815-01-439-8164) (1997), (NSN 2815-01-453-7402) (1998).

HEADQUARTERS, DEPARTMENTS OF THE ARMY, THE AIR FORCE, AND MARINE CORPS

JANUARY 1996

WARNING

EXHAUST GASES CAN KILL

Brain damage or death can result from heavy exposure. Precautions must be followed to ensure personnel safety when the engine is operated for any purpose.

1. DO NOT operate your vehicle engine in enclosed area.
2. BE ALERT at all times for exhaust odors.
3. BE ALERT for exhaust poisoning symptoms. They are:
    - Headache
    - Dizziness
    - Sleepiness
    - Loss of muscular control
4. If YOU SEE another person with exhaust poisoning symptoms:
    - Remove person from area
    - Expose to open air
    - Keep person warm
    - Do not permit person to move
    - Administer artificial respiration, if necessary*
    - Notify a medic

    * For artificial respiration, refer to FM 21-11
5. BE AWARE, the field protective mask for nuclear-biological-chemical (NBC) protection will not protect you from exhaust poisoning.

    THE BEST DEFENSE AGAINST EXHAUST POISONING IS ADEQUATE VENTILATION.

WARNING SUMMARY

- Always wear eyeshields when using compressed air. Failure to wear eyeshields may result in eye injury.
- Compressed air used for cleaning purposes will not exceed 30 psi (207 kPa). Use only with effective chip guarding and personal protective equipment (goggles/shield, gloves, etc.).
- Do not use compressed air to dry bearings. Spinning a dry bearing with compressed air may cause injury to personnel or damage to equipment.
- Improper cleaning methods and use of unauthorized cleaning solutions will injure personnel and damage equipment. See TM 9-247 for correct information.
- Drycleaning solvent is flammable and will not be used near an open flame. A fire extinguisher will be kept nearby when the solvent is used. Use only in well-ventilated places. Failure to do this may result in injury to personnel and/or damage to equipment.
- When steam cleaning, protective clothing must be used. Failure to use protective clothing may result in serous injury.
- Diesel fuel is highly flammable. Do not perform this procedure near fire, flame, or sparks. Severe injury or death may result.
- Cylinder block must be supported during removal and installation from engine stand. Failure to support cylinder block may cause injury to personnel or damage to equipment.
- Crankshaft must be supported during removal and installation. Failure to support crankshaft may cause injury to personnel or damage to equipment.
- Do not place hands or arms near nozzle during testing. Penetrating force of oil may cause serious injury or death.
- Direct personnel to stand clear during hoisting operation. Failure to do this may cause injury to personnel.
- Gaskets installed on some 6.2L engines assembled prior to 1991 may contain asbestos. Gaskets should be removed with a scraper or putty knife and then disposed of IAW current directives. Inhalation of asbestos fibers can cause respiratory ailments.

ARMY TM 9-2815-237-34
AIR FORCE TO 38V1-6-2
MARINE CORPS TM 2815-34/3 A

CHANGE

NO. 2

HEADQUARTERS,
DEPARTMENTS OF THE ARMY,
THE AIR FORCE, AND MARINE CORPS
• Washington, D.C., 15 July *2004*

TECHNICAL MANUAL

DIRECT SUPPORT AND GENERAL SUPPORT MAINTENANCE

ENGINE, DIESEL: 8 CYLINDER, NATURALLY ASPIRATED, FUEL-INJECTED, WATER COOLED, DDA MODEL 6.2 LITER (PRE-1990) (NSN 2815-01-231-3672), (POST-1990) (NSN 2815-01-314-7940); DDA MODEL 6.5 LITER (NSN 2815-01-410-9710) (1995,1996), (NSN 2815-01-439-6665) (1997), (NSN 2815-01-453-7403) (1998); ENGINE, DIESEL: 8 CYLINDER, NATURALLY ASPIRATED, DETUNED, FUEL-INJECTED, WATER COOLED, DDA MODEL 6.5 LITER (NSN 2815-01-406-6675) (1996), (NSN 2815-01-439-6664) (1997), (NSN 2815-01-453-7404) (1998); ENGINE, DIESEL: 8 CYLINDER, TURBOCHARGED, FUEL-INJECTED, WATER COOLED, (NSN 2815-01-420-4180) (1996), (NSN 2815-01-439-8164) (1997), (NSN 2815-01-453-7402) (1998)

TM 9-2815-237-34, 31 January 1996, is changed as follows:

1. Remove old pages and insert new pages as indicated below.
2. New or changed material is indicated by a vertical bar in the margin of the page.

Remove pages	**Insert pages**
A/(B blank)	A/(B blank)
i and ii	i and ii
1-1 and 1-2	1-1 and 1-2
1-5/(1-6 blank)	1-5/(1-6 blank)
2-23 and 2-24	2-23 and 2-24
2-85 through 2-90	2-85 through 2-90
2-101 and 2-102	2-101 and 2-102
2-113 and 2-114	2-113 and 2-114
2-117 through 2-120	2-117 through 2-120
2-125 and 2-126	2-125 and 2-126
2-155 and 2-156	2-155 and 2-156
2-173 and 2-174	2-173 and 2-174
A-1 and A-2	A-1 and A-2
Index 1 through Index 4	Index 1 through Index 4
DA Form 2028	DA Form 2028
Cover	Cover

3. File this change sheet in front of the publication for reference purposes.

By Order of the Secretary of the Army:

Peter J. Schoomaker
General, United States Army
Chief of Staff

Official:

JOEL B. HUDSON
Administrative Assistant to the
Secretary of the Army
0223410

By Order of the Secretary of the Air Force:

RONALD R. FOGLEMAN
General, United States Air Force
Chief of Staff

Official:

HENRY VICCELLIO, JR.
General, United States Air Force
Commander, Air Force Materiel Command

By Order of the Marine Corps:

R. P. SHOCKEY
Director, Program Support
Marine Corps Systems Command

Distribution:

To be distributed in accordance with the initial distribution number (IDN) 380623, requirements for TM 9-2815-237-34.

ARMY TM 9-2815-237-34
AIR FORCE TO 38V1-6-2
MARINE CORPS TM 2815-34/3 A

CHANGE

NO. 1

HEADQUARTERS,
DEPARTMENTS OF THE ARMY,
THE AIR FORCE, AND MARINE CORPS
Washington, D.C., *31 May 1999*

TECHNICAL MANUAL

DIRECT SUPPORT AND GENERAL SUPPORT MAINTENANCE

ENGINE, DIESEL: 8 CYLINDER, NATURALLY ASPIRATED, FUEL-INJECTED, WATER COOLED, DDA MODEL 6.2 LITER (SERIAL NUMBER 1-99,999) (NSN 2815-01-231-3672), (SERIAL NUMBER 100,000 AND ABOVE) (NSN 2815-01-314-7940) (1990); DDA MODEL 6.5 LITER (NSN 2815-01-410-9710) (1995,1996), (NSN 2815-01-439-6665) (1997), (NSN 2815-01-453-7403) (1998); ENGINE, DIESEL: 8 CYLINDER, NATURALLY ASPIRATED, DETUNED, FUEL-INJECTED, WATER COOLED, DDA MODEL 6.5 LITER (NSN 2815-01-406-6675) (1996), (NSN 2815-01-439-6664) (1997), (NSN 2815-01-453-7404) (1998); ENGINE, DIESEL: 8 CYLINDER, TURBOCHARGED, FUEL-INJECTED, WATER COOLED, (NSN 2815-01-420-4180) (1996), (NSN 2815-01-439-8164) (1997), (NSN 2815-01-453-7402) (1998).

TM 9-2815-237-34, 31 January 1996, is changed as follows:

1. New engine models have been added to the front cover. The new cover, located at the end of the change package, replaces the front cover.
2. Remove old pages and insert new pages as indicated below.
3. New or changed material is indicated by a vertical bar in the margin of the page.
4. File this change sheet in front of the publication for reference purposes.

Remove pages	Insert pages	Remove pages	Insert pages
Warning a and b	Warning a and b	2-153 through 2-156	2-153 through 2-156
None	A (After Warning b)	2-169 and 2-170	2-169 and 2-170
i and ii	i and ii	2-183 through 2-186	2-183 through 2-186
1-3 and 1-4	1-3 and 1-4	2-189 and 2-190	2-189 and 2-190
None	2-6.1 and 2-6.2	None	2-193/(2-194 blank)
2-7 through 2-18	2-7 through 2-18	D-1 through D-3/(D-4 blank)	D-1 through D-3/(D-4 blank)
2-23 through 2-26	2-23 through 2-26	E-3/(E-4 blank)	E-3 and E-4
2-31 through 2-34	2-31 through 2-34	F-3/(F-4 blank)	F-3/(F-4 blank)
2-37 through 2-48	2-37 through 2-48	Index 1 through Index 4	Index 1 through Index 4
2-53 through 2-56	2-53 through 2-56	Cover	Cover
2-83 and 2-84	2-83 and 2-84		
None	2-140.1 through 2-140.25/(2-140.26 blank)		

By Order of the Secretary of the Army:

ERIC K. SHINSEKI
General, United States Army
Chief of Staff

Official:

JOEL B. HUDSON
Administrative Assistant to the
Secretary of the Army
05694

By Order of the Secretary of the Air Force:

RONALD R. FOGLEMAN
General, United States Air Force
Chief of Staff

Official:

HENRY VICCELLIO, JR.
General, United States Air Force
Commander, Air Force Materiel Command

By Order of the Marine Corps:

D.R. BLOOMER
Colonel, USMC
Director, Program Support
Marine Corps Systems Command

Distribution:

To be distributed in accordance with the initial distribution number (IDN) 380623, requirements for TM 9-2815-237-34.

LIST OF EFFECTIVE PAGES

INSERT LATEST CHANGED PAGES. DESTROY SUPERSEDED PAGES.

NOTE: The portion of the text affected by the changes is indicated by a vertical line in the outer margins of the page.

Dates of issue for original and changed pages are:
Original ..... 0 .... 31 January 1996
Change ...... 1 .... 31 May 1999
Change ...... 2 .... 15 July 2004

TOTAL NUMBER OF PAGES IN THIS PUBLICATION IS 232, CONSISTING OF THE FOLLOWING:

Page No.	*Change No.
Warning a and b	1
A	2
B Blank	1
i - ii	2
iii	0
iv Blank	0
1-1	2
1-2 - 1-3	0
1-4	1
1-5	2
1-6 Blank	0
2-1 - 2-6	0
2-6.1 - 2-6.2 Added	1
2-7 - 2-8	1
2-9	0
2-10 - 2-11	1
2-12	0
2-12.1 - 2-12.2 Added	1
2-13 - 2-15	1
2-16 - 2-17	0
2-18	1
2-19 - 2-22	0
2-23	2
2-24	0
2-25 - 2-26	1
2-27 - 2-31	0
2-32 - 2-33	1
2-34 - 2-37	0
2-38 - 2-39	1
2-40 - 2-41	0
2-42 - 2-47	1
2-48 - 2-53	0
2-54 - 2-56	1
2-57 - 2-82	0
2-83 - 2-84	1
2-85	0
2-86	2
2-86.1 - 2-86.2 Added	2
2-87 - 2-90	2
2-91 - 2-101	0
2-102	2
2-103 - 2-112	0
2-113	2
2-114 - 2-117	0
2-118 - 2-120	2
2-121 - 2-124	0
2-125	2
2-126-2-140	0
2-140.1 - 2-140.25 Added	1
2-140.26 Blank	1
2-141 - 2-152	0
2-153	1
2-154	0
2-155	1
2-156	2
2-157 - 2-168	0
2-169	1
2-170 - 2-172	0
2-173	2
2-174 - 2-182	0
2-183 - 2-184	1
2-184.1 - 2-184.2 Added	1
2-185	0
2-186	1
2-187 - 2-188	0
2-189	1
2-190 - 2-192	0
2-193 Added	1
2-194 Blank	1
A-1	0
A-2	2
B-1 - B-3	0
B-4 Blank	0
C-1 - C-3	0
C-4 Blank	0
D-1	0
D-2 - D-3	1
D-4 Blank	0
E-1 - E-2	0
E-3	1
E4 Added	1
F1 - F2	0
F3	1
F4 Blank	0
Index 1	1
Index 2 - Index 4	2

*Zero in this column indicates original page.

*ARMY TM 9-2815-237-34
AIR FORCE TO 38V1-6-2
MARINE CORPS TM 2815-34/3 A

TECHNICAL MANUAL
NO. 9-2815-237-34
NO. 2815-34/3 A
TECHNICAL ORDER
NO. 38V1-6-2

HEADQUARTERS,
DEPARTMENTS OF THE ARMY,
THE AIR FORCE, AND MARINE CORPS
Washington, D.C. *31 January 1996*

TECHNICAL MANUAL

DIRECT SUPPORT AND GENERAL SUPPORT MAINTENANCE

ENGINE, DIESEL: 8 CYLINDER, NATURALLY ASPIRATED, FUEL-INJECTED, WATER COOLED, DDA MODEL 6.2 LITER (PRE-1990) (NSN 2815-01-231-3672), (POST-1990) (NSN 2815-01-314-7940); DDA MODEL 6.5 LITER (NSN 2815-01-410-9710) (1995,1996), (NSN 2815-01-439-6665) (1997), (NSN 2815-01-453-7403) (1998); ENGINE, DIESEL: 8 CYLINDER, NATURALLY ASPIRATED, DETUNED, FUEL-INJECTED, WATER COOLED, DDA MODEL 6.5 LITER (NSN 2815-01-406-6675) (1996), (NSN 2815-01-439-6664) (1997), (NSN 2815-01-453-7404) (1998); ENGINE, DIESEL: 8 CYLINDER, TURBOCHARGED, FUEL-INJECTED, WATER COOLED, (NSN 2815-01-420-4180) (1996), (NSN 2815-01-439-8164) (1997), (NSN 2815-01-453-7402) (1998)

REPORTING ERRORS AND RECOMMENDING IMPROVEMENTS

You can help improve this publication. If you find any mistakes or if you know of a way to improve the procedures, please let us know. Submit your DA Form 2028 (Recommended Changes to Publications and Blank Forms), through the Internet, on the Army Electronic Product Support (AEPS) website. The Internet address is http://aeps.ria.army.mil . If you need a password, scroll down and click on "ACCESS REQUEST FORM." The DA Form 2028 is located in the ONLINE FORMS PROCESSING section of the AEPS. Fill out the form and click on SUBMIT. Using this form on the AEPS will enable us to respond quicker to your comments and better manage the DA Form 2028 program. You may also mail, fax or E-mail your letter or DA Form 2028 direct to: AMSTA-LC-CI Tech Pubs, TACOM-RI, 1 Rock Island Arsenal, Rock Island, IL 61299-7630. The E-mail address is TACOM-TECH-PUBS@ria.army.mil. The fax number is DSN 793-0726 or Commercial (309) 782-0726. (Marine Corps) Submit NAVMC 10722 to Commander Code 835-2, Marine Corps Logistic Base, 814 Radford Boulevard, Albany, GA 31704-1128.

* This publication supersedes TM 9-2815-237-34 dated 25 June 1985 and all changes.

CHAPTER 1
INTRODUCTION

Section I. GENERAL INFORMATION

1-1. SCOPE

This technical manual contains instructions for direct support and general support maintenance of the 6.2 liter naturally aspirated, 6.5 liter naturally aspirated detuned, and 6.5 liter turbo V-8 diesel engines.

1-2. MAINTENANCE FORMS, RECORDS, AND REPORTS

(Army) Department of the Army forms and procedures used for equipment maintenance will be those prescribed by DA Pam 738-750, The Army Maintenance Management System (TAMMS). (Marine Corps) Refer to MCO 4855-10.

1-3. DESTRUCTION OF ARMY EQUIPMENT TO PREVENT ENEMY USE

Procedures for destruction of Army tank-automotive equipment to prevent enemy use are found in TM 750-244-6.

1-4. PREPARATION FOR SHIPMENT

(Army) Refer to TM 746-10, Marking, Packaging, and Shipment of Supplies and Equipment: General Packaging Instructions for Field Use. (Marine Corps) Refer to MCO 4450-7.

1-5. EQUIPMENT REQUIRING CALIBRATION

Calibration requirements in this manual cover the fuel injection pump and can be found in paras. 2-33, 2-34, and 2-35 of this manual.

1-6. REPORTING EQUIPMENT IMPROVEMENT RECOMMENDATIONS (EIRs)

If your vehicle needs improvement, let us know. Send us an EIR. You, the user, are the only one who can tell us what you don't like about your equipment. Let us know why you don't like the design or performance. The preferred method for submitting QDRs is through the Army Electronic Product Support (AEPS) website under the Electronic Deficiency Reporting System (EDRS). The web address is: https://aeps.ria.army.mil. This is a secured site requiring a password that can be applied for on the front page of the website. If the above method is not available to you, put it on an SF 368, Product Quality Deficiency Report (PQDR), and mail it to us at: Department of the Army, U.S. Army Tank-automotive and Armaments Command, ATTN: AMSTA-TR-E/PQDR MS 267, 6501 E. 11 Mile Road, Warren, MI 48397-500. We'll send you a reply. (Marine Corps) Submit QDR's in accordance with MCO 4855-10.

1.7. EQUIPMENT IMPROVEMENT REPORT AND MAINTENANCE DIGEST (EIR MD)

The quarterly Equipment Improvement Report and Maintenance Digest, TB 43-0001-62 series, contains valuable field information on the equipment covered in this manual. The information in the TB 43-0001-62 series is compiled from some of the Equipment Improvement Reports that you prepared on the vehicles covered in this manual. Many of these articles result from comments, suggestions, and improvement recommendations that you submitted to the EIR program. The TB 43-0001-62 series contains information on equipment improvements, minor alterations, proposed Modification Work Orders (MWOs), warranties (if applicable), actions taken on some of your DA Form 2028s (Recommended Changes to Publications and Blank Forms), and advance information on proposed changes that may affect this manual. The information will help you in doing your job better and will help in keeping you advised of the latest changes to this manual. Also refer to DA Pam 25-30, Consolidated Index of Army Publications and Blank Forms, and appendix A, References, of this manual. (Marine Corps) Submit QDR's in accordance with MCO 4855-10. For those with access to the World Wide Web (WWW), the EIR MD can be viewed through the Army Electronic Product Support. The site is http://aeps.ria.army.mil.

1-8. METRIC SYSTEM

The engines described herein contain metric components and require metric common and special tools; therefore, metric units in addition to standard units will be used throughout this publication. In addition, a metric conversion table is located on the inside back cover of this publication.

1-9. MANDATORY REPLACEMENT PARTS

The maintenance instructions contained herein make reference to removing and discarding piece parts such as: gaskets, lockwashers, cotter pins, O-rings, seals, etc.; these items should be considered mandatory replacement items and replaced with new parts during assembly/installation.

Section II. EQUIPMENT DESCRIPTION AND DATA

1-10. DESCRIPTION

The engine used in the M998 series vehicles is an eight-cylinder, four-cycle, liquid-cooled, naturally aspirated diesel engine. The 6.2L engine has a compression ratio of 21.5:1 and develops approximately 150 horsepower (112 kW) at 3600 rpm. The 6.5L engine has a compression ratio of 21.5:1 and develops approximately 160 horsepower (119 kW) at 3400 rpm. For proper orientation, the left and right sides of the engine are as viewed from the rear.

1990 and above 6.2L and 1994 6.5L engines have the following new and revised parts:

***Cylinder Heads.** The 1990 cylinder heads have been revised to meet 1990 emissions standards.

Precombustion Valves. The 1990 engine is equipped with reverse throat precombustion chambers.

Exhaust Valves. The 1990 exhaust valves are coated with chrome to prevent burning and to increase durability.

***Exhaust Valve Stem Seal.** The 1990 engine is equipped with a new exhaust valve stem seal that is not available for previous model engines.

Cylinder Head Gaskets. The 1990 cylinder head gaskets have new circular combustion seals and are made of a non-asbestos material.

Fuel Injection Pump. The 1994 fuel injection pump has a TP sensor and a revised calibration procedure.

Fuel Injection Pump. The 1990 fuel injection pump has a new governor and a revised calibration procedure.

***Fuel Injection Nozzles.** The 1990 fuel injection nozzles are shorter and are orientated ten degrees from the vertical.

***Fuel Injection Lines.** The 1990 engine is equipped with high pressure fuel injection lines.

Fuel Injection Line Brackets. The 1990 fuel injection line brackets have been revised to provide clearance for the injection nozzles.

***RPM Sensors.** The 1994 engine is equipped with two rpm sensors. One is for STE-ICE-R and one is for the engine.

***Fan Cut-Off Switch.** The 1994 engine is equipped with a fan cut-off switch which replaced the kick-down switch.

*These 1990 parts are not interchangeable with parts for previous model engines. Differences in maintenance procedures are identified where applicable. Refer to engine decal number on left rocker arm cover before ordering replacement parts.

The 1990 6.2L engine is identified by a decal which is adhered to the left rocker arm cover. The 6.5L engine is identified by a serial number bar code decal.

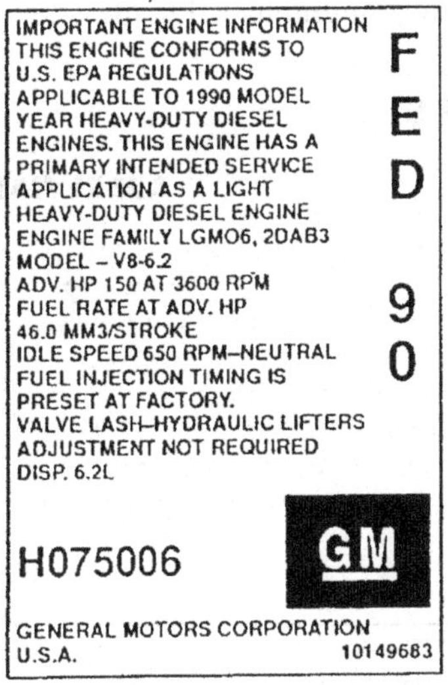

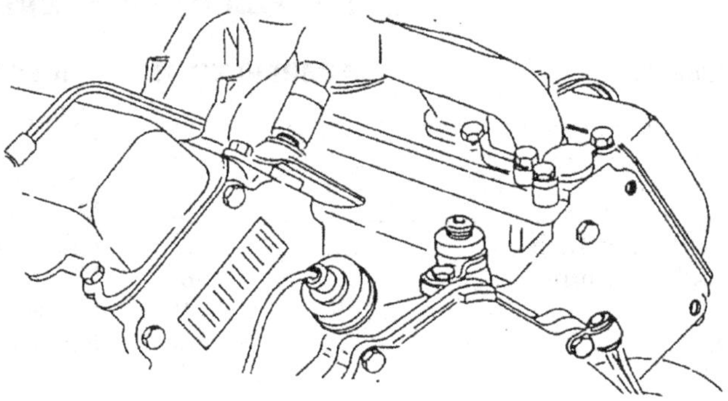

1-11. COMPONENT LOCATION

The locations of components described below are common to all engines in the M998 series vehicles. Special differences are described in para. 1-10.

(A) FAN CLUTCH
(B) ROCKER ARM COVERS (2)
(C) FUEL INJECTORS (8)
(D) OIL FILTER
(E) EXHAUST MANIFOLDS (2)
(F) OIL PAN
(G) TIMING GEAR COVER
(H) CRANKSHAFT PULLEY

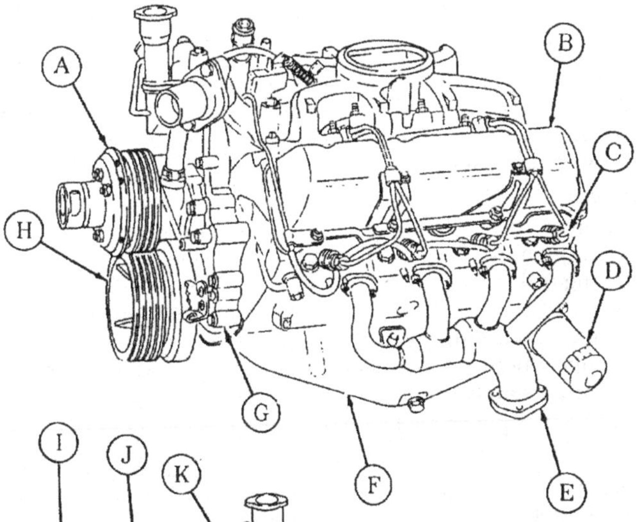

(I) INTAKE MANIFOLD
(J) FUEL INJECTION PUMP
(K) WATER CROSSOVER
(L) WATER PUMP
(M) FUEL PUMP
(N) GLOW PLUGS (8)

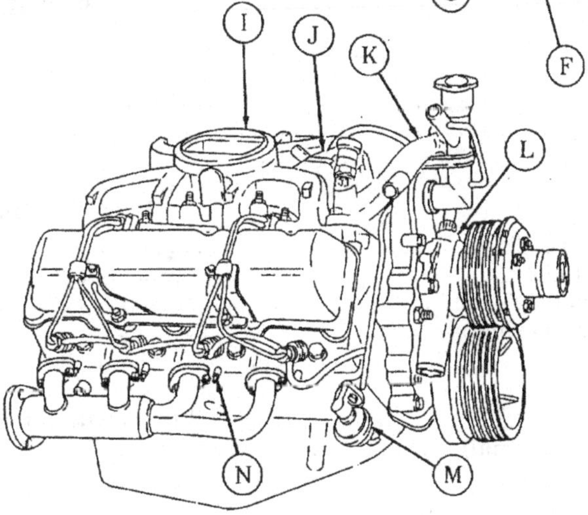

1-12 TABULATED DATA

Engine tabulated data is listed in tables 1-1 and 1-2. This information includes only those data applicable to direct support and general support maintenance. Information not covered can be found in TM 9-2320-280-20 and TM 9-2320-280-10.

TABULATED DATA ABBREVIATIONS

MEASUREMENT	ABBREVIATION	MEASUREMENT	ABBREVIATION
Quart	qt	Celsius	C
Kilogram	kg	Liter	L
Inch	in.	Centimeter	cm
Pound	lb	Millimeter	mm
Pounds Per Square Inch	psi	Kilopascal	kPa
Revolutions Per Minute	rpm	Newton Meter	N·m
Pound-Feet	lb-ft	Horsepower	hp
Cubic Inch	cu-in.	Kilowatt	kW
Fahrenheit	F		

Table 1-1. Tabulated Data (6.2 L)

NOTE

Standard and metric measurements will be used in this table.
A list of their abbreviations is provided above.

Manufacturer	Detroit Diesel Allison	
Model	6.2 L	
Type	Four-cycle, liquid-cooled, naturally aspirated	
	STANDARD	**METRIC**
Dimensions:		
Length	35 in.	89 cm
Width	28 in.	71 cm
Height	28 in.	71 cm
New weight, dry	650 lbs	295 kg
Cylinders:		
Number	Eight	
Arrangement	90° V	
Firing order	1-8-7-2-6-5-4-3 (clockwise)	
Bore	3.98 in.	10.1 cm
Stroke (nominal)	3.82 in.	9.7 cm
Displacement	392 cu-in.	6.2 L
Compression ratio	21.5:1	
Maximum Torque (gross)	260 lb-ft @ 2000 rpm	353 N·m @ 2000 rpm
Governed Speed:		
Full load	3600 rpm	
No load	4000 rpm	
Idle speed	650±25 rpm	
Lubrication System:		
Type	Pressure feed	
Operating pressure (normal)	40-50 psi	276-345 kPa
Operating pressure (at idle)	10 psi	69 kPa
System capacity (including filter)	8 qts	7.6 L
Operating temperature (normal)	180°-260°F	82°-126°C
Oil pump	Gear-driven	
Cooling system:		
Type	Liquid with fan and radiator	
Operating temperature (normal)	190°-230°F	88°-110°C

CHAPTER 2
ENGINE MAINTENANCE

Section I. REPAIR PARTS, SPECIAL TOOLS, TEST, MEASUREMENT, AND DIAGNOSTIC EQUIPMENT (TMDE), AND SUPPORT EQUIPMENT

2-1. COMMON TOOLS AND EQUIPMENT

Refer to Modified Table of Organization and Equipment (MTOE) for authorized common tools and equipment applicable to your unit.

2-2. SPECIAL TOOLS AND SUPPORT EQUIPMENT

Special tools and support equipment are listed in Appendix E of this manual and illustrated in TM 9-2815-237-34P.

2-3. TEST, MEASUREMENT, AND DIAGNOSTIC EQUIPMENT (TMDE)

Calibrate all measuring and test equipment used to determine equipment conformance in accordance with TB 43-180.

2-4. REPAIR PARTS

Repair parts are listed and illustrated in TM 9-2815-237-34P.

Section II. SERVICE UPON RECEIPT

2-5. ENGINE INSPECTION

a. Inspect for any damage done to engine during its removal from the vehicle. Refer to TM 9-2320-280-34 for engine removal procedure.

b. Upon receipt of engine, ensure that all components are present and that no engine subassemblies have been removed. Refer to TM 9-2320-280-34 for engine dress after removal.

c. If the engine has been transferred to you from another DS/GS maintenance facility, check the equipment against the packing slip to see if the shipment is complete. Report all discrepancies in accordance with the instructions of DA Pam 738-750.

Section III. TROUBLESHOOTING

2-6. GENERAL

Information in this section is for use of support maintenance personnel in conjunction with, and as a supplement to, troubleshooting procedures in TM 9-2320-280-20 and TM 9-2320-280-34.

2-7. ENGINE TROUBLESHOOTING

CAUTION

Operation of a deadlined engine without preliminary inspection will cause further damage.

Information to be used for troubleshooting the engine is contained in TM 9-2320-280-20 and TM 9-2320-280-34.

a. Excessive oil consumption is generally caused by leaky gaskets or loose line connections.

b. Always check the easiest and most obvious things first. This simple rule saves time and trouble.

c. Double check before disassembly. The source of most engine problems can be traced to more than one part in a system. For example:

(1) Excessive fuel consumption may not be caused by the fuel pump alone. Instead, the trouble could be a clogged air cleaner, or a restricted exhaust passage causing severe back pressure.

(2) Engines very often are disassembled in search of a complaint and the real evidence of the problem is destroyed. Check again to be sure an easier solution to the problem has not been overlooked.

d. Before correcting a problem, diagnose the cause of the problem. Do not allow the same failure to occur again.

Section IV. GENERAL MAINTENANCE INSTRUCTIONS

2-8. GENERAL MAINTENANCE INSTRUCTIONS TASK SUMMARY

2-9. CLEANING

a. General Instructions. Cleaning procedures will be the same for the majority of parts and components which make up engine subassemblies. General cleaning procedures are detailed in "b" through "i".

b. The Importance of Cleaning. Great care and effort are required in all cleaning operations. The presence of dirt and foreign material is a constant threat to satisfactory engine operation and maintenance. The following will apply to all cleaning operations:

(1) Hands must be kept free of any accumulation of grease which can collect dust and grit.

(2) Clean all parts before inspection, after repair, and before assembly.

(3) After cleaning, all parts must be covered or wrapped in plastic or paper to protect them from dust and/or dirt.

WARNING

Compressed air used for cleaning purposes will not exceed 30 psi (207 kPa). Use only with effective chip guarding and personal protective equipment (goggles/shield, gloves, etc.).

c. External Engine Cleaning. All electrical equipment and other parts that could be damaged by steam cleaning or moisture must be removed, and all openings covered before cleaning. Dry with compressed air.

d. Disassembled Parts Cleaning. Place all disassembled parts in wire baskets for cleaning.

(1) Dry and cover all cleaned parts.

(2) Place on or in "racks" and hold for inspection or repair.

(3) All parts subject to rusting must be lightly oiled and wrapped.

(4) Keep all related parts and components together. Do not mix parts.

WARNING

Improper cleaning methods and use of unauthorized cleaning solutions will injure personnel and damage equipment. See TM 9-247 for correct information.

e. Castings.

(1) Clean inner and outer surfaces of castings and all areas subject to grease and oil with cleaning solvents. Refer to TM 9-247.

(2) Use a stiff brush to remove sludge and gum deposits.

(3) Use compressed air to blow out all tapped capscrew holes and dry castings after cleaning.

f. Oil Passages. Particular attention must be given to all oil passages in castings and machined parts. Oil passages must be clean and free of any obstructions.

(1) Clean passages with wire probes to breakup any sludge or gum deposits.

(2) Wash passages by flushing with solvents. See TM 9-247,

(3) Dry passages with compressed air.

CAUTION

Do not allow drycleaning solvents to come in contact with seals, cables, or flexible hoses. These cleaners cause leather, rubber, and synthetic materials to dry out, rot, and lose pliability making them unserviceable.

g. Nonmetallic Parts. Clean hoses and other nonmetallic parts with soap and water.

h. Bearings.

WARNIN G

Do not use compressed air to dry bearings. Spinning a dry bearing with compressed air may cause injury to personnel or damage to equipment.

(1) Bearings require special cleaning. After removing surface oil and gum deposits, place bearings in hot oil, 140°F (60•C), to loosen congealed oil and grease. Wipe bearings dry with a Iint-free cloth; do not use compressed air.

(2) See TM 9-214 for information and care of bearings.

i. Electrical Components.

(1) Clean electrical components with clean cloth dampened with drycleaning solvent. Care must be taken not to damage protective insulation.

WARNING

Compressed air used for cleaning purposes will not exceed 30 psi (207 kPa). Use only with effective chip guarding and personal protective equipment (goggles/shield, gloves, etc.).

(2) Use compressed air oa dry electrical components.

2-10. INSPECTION

a. General Instructions. Procedures for inspections will be the same for many parts and components which make up the engine subassemblies. General procedures are detailed in "b" through "k", Dimensional standards for parts have been fixed at extremely close tolerances, so use specification tables. Use specified inspection equipment for inspection where cracks and other damage cannot be spotted visually. Exercise extreme care in all phases of inspection. Repair or replace all unserviceable components; refer to para. 2-11.

b. Castings.

(1) Inspect all ferrous and nonferrous castings for cracks. See MIL-I-6866, Inspection, Penetrant Methods, and MIL-I-6868, Inspection Process, Magnetic Particles. Particularly check areas around studs, pipe plugs, threaded inserts, and sharp comers. Replace cracked castings.

(2) Inspect machined surfaces for nicks, burrs, and raised metal. Mark damaged areas for repair or replacement.

(3) Inspect all pipe plugs, pipe plug openings, capscrews, and capscrew openings for damaged and stripped threads. Replace if damaged or threads are stripped.

(4) Check all gasket mating surfaces, flanges on housings, and supports for warpage with a straight edge or surface plate. Inspect mating flanges for discolorations which may indicate leakage. Replace if warped.

(5) Check all castings for conformance to applicable repair standards.

c. Bearings.

CAUTION

All engine connecting rod and main bearings will be replaced if one or more bearings fail.

Check all bearings for conformance to applicable repair standards.

d. Bushings and Bushing-Type Bearings.

(1) Check all bushings and bushing-type bearings for secure fit, evidence of heating, wear, burrs, nicks, and out-of-round conditions.

(2) Check for dirt in lubrication holes or grooves. Holes and grooves must be clean and free from damage.

e. **Machined Parts.**

(1) Check machined parts for cracks, distortion, and damage.

(2) Check all surfaces for nicks, burrs, and raised metal.

f. **Studs, Bolts, and Capscrews.** Replace if bent, loose, stretched, or threads are damaged.

g. **Gears.**

NOTE

When gear teeth wear limits are not established, good judgment is required to determine if gear replacement is necessary.

(1) Inspect all gears for cracks and missing teeth. Replace if cracked or teeth are missing.

(2) Inspect gear teeth for wear, sharp fins, burrs, and galled or pitted surfaces.

(3) Inspect splines for wear, burrs, and galled or pitted surfaces.

(4) Check keyway slots for wear and/or damage.

h. **Oil Seals.** Oil seals are mandatory replacement items.

i. **Casting Plugs.** Inspect for leakage. Replace plugs when leakage is present.

j. **Springs.** Inspect for damaged, distorted, and collapsed coils.

k. **Snaprings, Retaining Rings, and Washers.** Many of these parts are mandatory replacement items. Inspect all others for obvious damage.

2-11. REPAIR

a. General Instructions. Repair of most parts and components is limited to general procedures outlined in applicable maintenance instructions and the following detailed procedures "b" through "h".

CAUTION

Repaired items must be thoroughly cleaned to remove metal chips and abrasives to prevent them from entering working parts of the engine.

b. Castings.

(1) All cracked castings will be replaced.

(2) Only minor repairs to machined surfaces, flanges, and gasket mating surfaces are permitted. Remove minor nicks, burrs, and/or scratches with:

(a) Fine mill file.

(b) Abrasive crocus cloth dipped in cleaning solvent.

(c) Lapping across a surface plate.

(d) Remachining of machined surfaces to repair damage, warpage, or uneven surfaces is not permitted. Replace castings.

(3) Repair damaged threaded pipe plug and/or capscrew holes with a thread tap or repair oversize holes with threaded inserts.

c. Bearings. See TM 9-214.

d. Studs. Replace all bent and stretched studs. Repair minor thread damage with a thread restorer file. Replace studs having stripped or damaged threads as outlined below:

(1) Remove, using a stud remover. Back studs out slowly to avoid heat buildup and seizure which can cause stud to break off.

(2) If studs break off too short to use with a stud remover, use extractor to remove.

(3) Replacement studs have a special coating and must have a small amount of antiseize compound (Appendix B, Item 3) applied on threads before stud is installed. Install replacement stud slowly to prevent heat buildup and snapping off.

e. Gears.

(1) Remove gears using pullers, as required.

(2) Use the same methods described in paragraph 2-11.b.(2) for castings t o remove minor nicks, burrs, or scratches on gear teeth.

(3) If keyways are worn or enlarged, replace gear.

f. Bushings and BushingType Bearings. When bushings and bushing-type bearings seize to a shaft and spin in the bore, the associated part must also be inspected and replaced, as required.

g. Oil Seals.

(1) Remove oil seals, being careful not to damage casting or adapter bore.

(2) Always install new seal in bore using proper seal replacing tool.

h. Cylinder Block or Cylinder Head. Repair of cylinder block and cylinder head is limited to procedures outlined in paragraphs 2-16 and 2-21. Cylinder block or cylinder head must be replaced if measurements are outside tolerance limits listed in table 2-2 or table 2-3.

2-12. ASSEMBLY

a. Cleanliness is essential in all component assembly operations. Dirt and dust, even in minute quantities, are abrasive. Parts must be cleaned as specified, and kept clean. Wrap or cover parts and components when assembly procedures are not immediately completed.

b. Coat all bearings and contact surfaces with engine oil (MIL-L-2104) to ensure lubrication of parts during initial operation after repair.

c. Use new gaskets and preformed packings during assembly of all components.

Section V. ENGINE DISASSEMBLY, REPAIR, AND ASSEMBLY

2-13. ENGINE DISASSEMBLY, REPAIR, AND ASSEMBLY TASK SUMMARY

2-13. ENGINE REPLACEMENT IN SHIPPING/STORAGE CONTAINER

This task covers:

a. Removal   b. Installation

INITIAL SETUP:

Tools

General mechanic's tool kit:
automotive (Appendix E, Item 1)

Special Tools

Engine lifting sling (Appendix E, Item 3)

Materials/Parts

Eight lockwashers (Appendix D, Item 26.1)

Manual References

TM 9-2815-237-34P

Personnel Required

One mechanic
One assistant

General Safety Instructions

Direct personnel to stand clear during hoisting operation.

a. Removal

1. If engine container (5) is closed and sealed, press air-release button (1) located at center of breather valve (2) before opening engine container (5).
2. Remove twenty-six nuts (19), washers (18), capscrews (16), and upper container (4) from lower container (13).
3. Position engine lifting sling on engine assembly (6) and install sling on right cylinder head (25) with two capscrews (24).
4. Install sling bracket (21) to engine lifting sling with pin (22) and on left cylinder head (20) with two capscrews (23). Tighten capscrews (23).
5. Remove four capscrews (14), lockwashers (8), and washers (7) from two engine mounts (12). Discard lockwashers (8).
6. Remove four capscrews (9), lockwashers (8), and washers (7) from two rear engine mounts (10). Discard lockwashers (8).

WARNING

Direct personnel to stand clear during hoisting operation. Failure to do this may cause injury to personnel.

7. Attach hoist to engine lifting sling and remove engine assembly (6) from lower container (13).
8. Position engine on repair stand (para. 2-14).

b. Installation

1. Install engine assembly (6) in lower container (13).
2. Install engine assembly (6) on two rear engine mounts (10) with four washers (7), lockwashers (8), and capscrews (9).
3. Install engine assembly (6) on two engine mounts (12) with four washers (7), lockwashers (8), and capscrews (14).
4. Remove pin (22) from sling bracket (21).
5. Remove two capscrews (24) and lifting sling from right cylinder head (25).
6. Remove two capscrews (23) and sling bracket (21) from left cylinder head (20).
7. Visually check humidity indicator (17) for discolorization. If indicator (17) is dark purple, replace desiccant (3).

2-13. ENGINE REPLACEMENT IN SHIPPING/STORAGE CONTAINER (Cont'd)

NOTE

Ensure gasket is seated properly.

8. Using alignment pins (15), position upper container (4) on gasket (11) and lower container (13).
9. Install upper container (4) on lower container (13) with twenty-six capscrews (16), washers (18), and nuts (19).

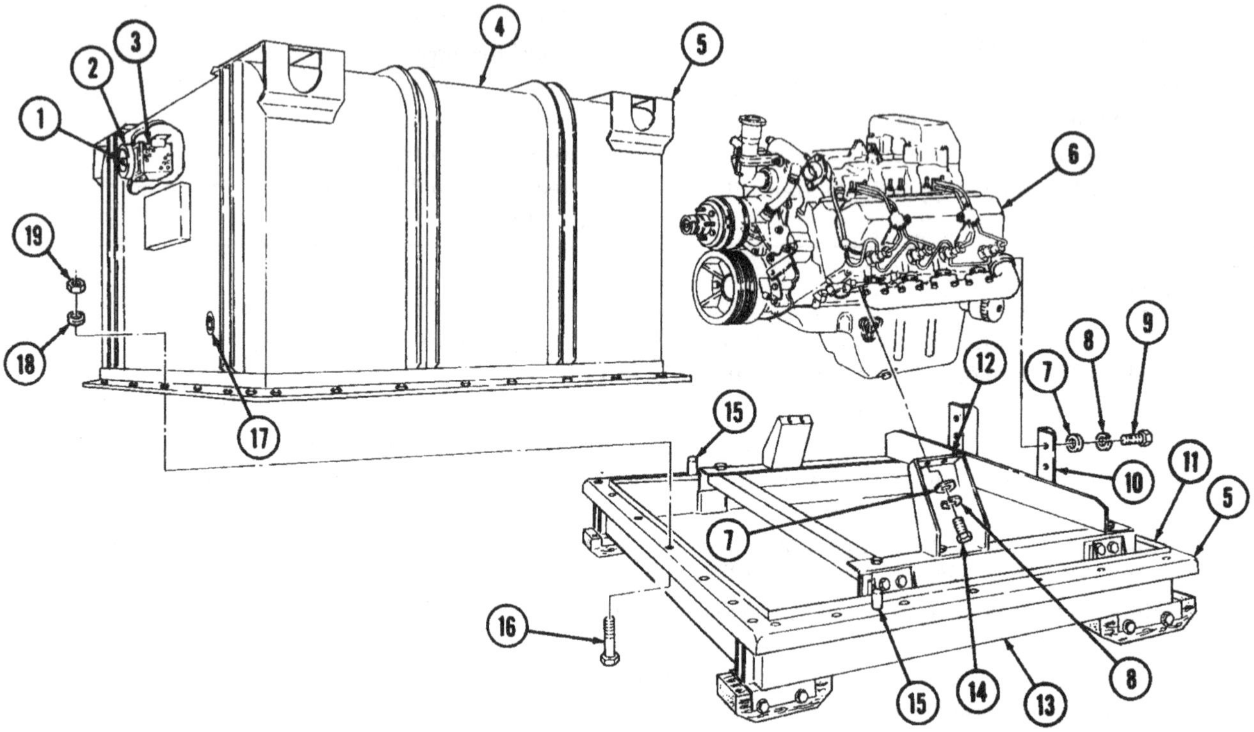

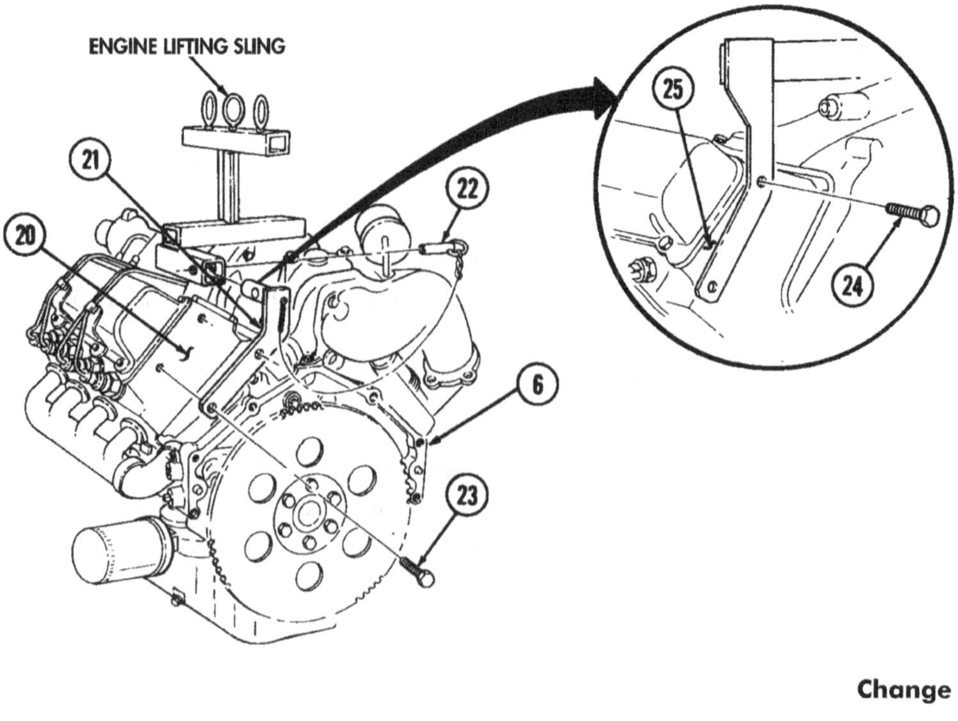

2-14 MOUNTING ENGINE ON REPAIR STAND

This task covers:

a. **Installation** b. **Removal**

INITIAL SETUP:

Tools
General mechanic's tool kit: automotive (Appendix E, Item 1)

Special Tools
Engine repair stand (Appendix E, Item 2)
Engine lifting sling (Appendix E, Item 3)

General Safety Instructions
Direct personnel to stand clear during hoisting operation.

a. Installation

NOTE

Do not perform step 1 if engine lifting sling has been previously installed.

1. Position engine lifting sling on engine (4) and secure to right cylinder head (10) with two bolts (11). Finger tighten bolts (11).
2. Install sling bracket (2) to engine lifting sling with pin (3) and on left cylinder head (1) with two bolts (9). Tighten bolts (9) and (11).

WARNING

Direct personnel to stand clear during hoisting operation. Failure to do this may cause injury to personnel.

3. Attach hoist to engine lifting sling and hoist engine (4) into position over engine repair stand.
4. Loosen four capscrews (5) securing repair stand arms (6) on engine repair stand and align arms (6) with holes in rear of engine (4).
5. Install repair stand arms (6) to engine (4) with four washers (8) and capscrews (7).
6. Tighten capscrews (5) and (7).
7. Disconnect hoist and remove two bolts (11) and (9), engine lifting sling, and bracket (2) from engine (4).

b. Removal

1. Position engine lifting sling on engine (4) and secure to right cylinder head (10) with two bolts (11). Finger tighten bolts (11).
2. Install sling bracket (2) to engine lifting sling with pin (3) and on left cylinder head (1) with two bolts (9). Tighten bolts (9) and (11).
3. Attach hoist to engine lifting sling and hoist engine (4) to relieve pressure on engine repair stand.
4. Remove four capscrews (7) and washers (8) from repair stand arms (6) and hoist engine away from engine repair stand.

2-14. MOUNTING ENGINE ON REPAIR STAND (Cont'd)

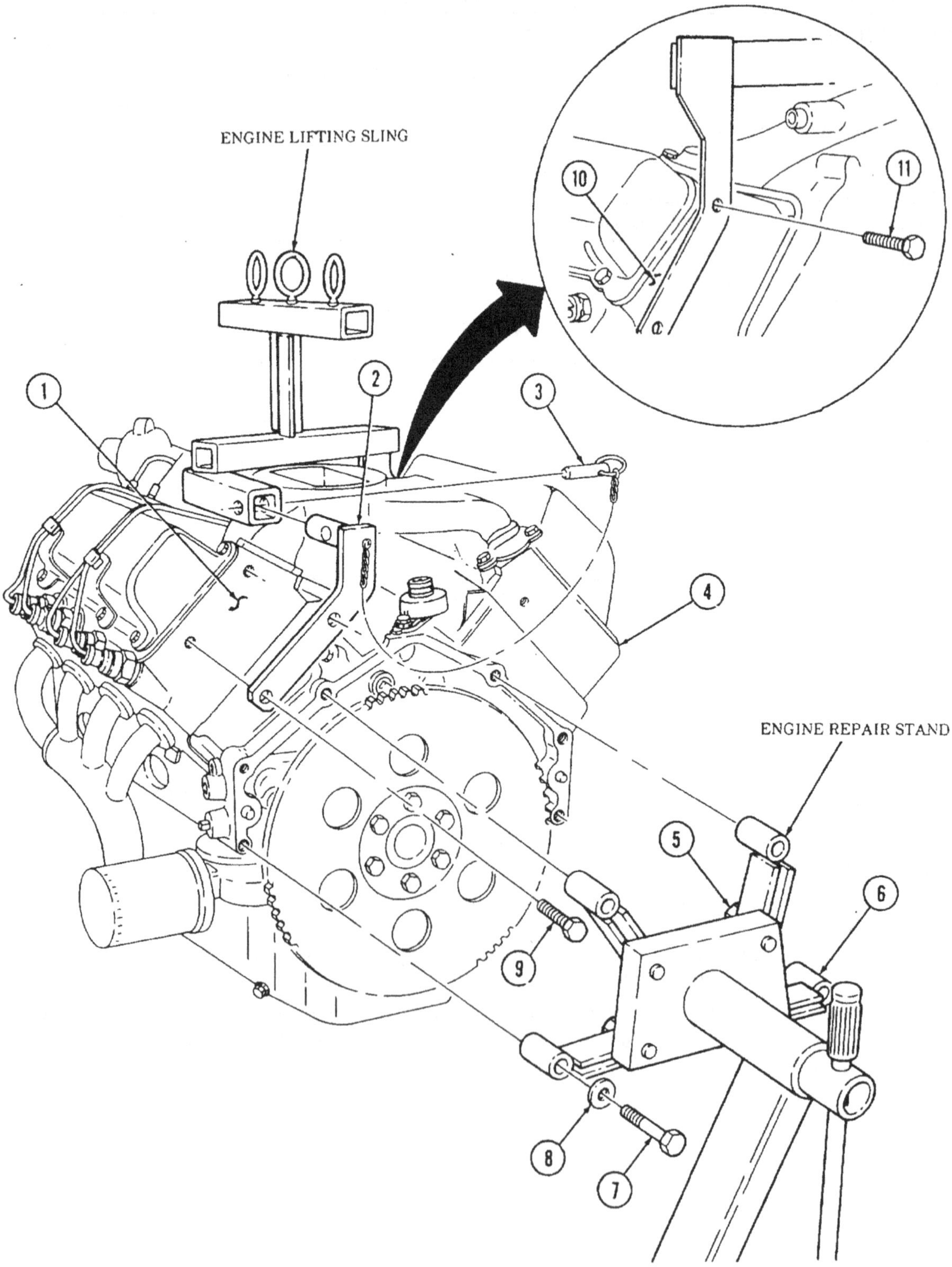

2 - 1 5ENGINE DISASSEMBLY INTO SUBASSEMBLIES

This task covers:

a. **Fan Drive and Water Pump Pulley**
b. **Crankshaft Pulley**
c. **Torsional Damper**
c.1. **Wastegate Actuator**
c.2. **Wastegate Housing**
c.3. **Manifold-to-Turbocharger Exhaust Pipe**
c.4. **Turbocharger**
d. **Exhaust Manifolds**
e. **Intake Manifold**
f. **Water Crossover**
g. **Fuel Injection Lines**
h. **Fuel Supply and Return Lines**
i. **Modulator Link**
j. **Rocker Arm Covers**
k. **Rocker Arm Shafts and Pushrods**
l. **Fuel Injection Nozzles**
m. **Glow Plugs**
n. **Cylinder Heads**
o. **Valve Lifters**
p. **Water Pump and Adapter Plate**
q. **Fuel Injection Pump**
r. **Timing Gear Cover**
s. **Timing Chain and Drive Sprockets**
t. **Oil Filter, Adapter, and Oil Pressure Sending Unit**
u. **Oil Pan**
v. **Oil Pump**
w. **Fuel Pump**
x. **Oil Pump Drive**
y. **Camshaft**
z. **Pistons and Connecting Rods**
aa. **Flywheel**
bb. **Crankshaft and Main Bearings**

INITIAL SETUP:

Tools

General mechanic's tool kit:
automotive (Appendix E, Item 1)
Puller (Appendix E, Item 19)
Cylinder ridge reamer (Appendix E, Item 11)
Dial indicator (Appendix E, Item 10)

Special Tools

Hydraulic valve lifter remover (Appendix E, Item 4)
Injection nozzle socket (Appendix E, Item 5)
Glow plug socket (Appendix E, Item 6)
Hex-head driver, 6 mm (Appendix E, Item 7)
Hex-head driver, 8 mm (Appendix E, Item 8)
Hex-head driver, 5/16-in. (Appendix E, Item 9)

Personnel Required

One mechanic
One assistant

Equipment Condition

Engine mounted on repair stand (para. 2-14).

General Safety Instructions

- Do not perform this procedure near fire, flame, or sparks.
- Gaskets installed on some 6.2L engines assembled prior to 1991 may contain asbestos. Gaskets should be disposed of IAW current directives.

NOTE

- Work area should be clean, well-ventilated, and free from blowing dirt and dust.
- In some cases, flanged head fasteners may be present instead of standard fasteners and washers. In all cases, washers should be used when replacing a flanged head fastener with a standard fastener.
- Disassembly procedures for pre-1990, 1990 and above 6.2L and 6.5L engines are basically the same. Any differences in procedures are noted. Refer to para. 1-10 to determine engine model before ordering replacement parts.

a. Fan Drive and Water Pump Pulley

NOTE

It may be necessary to apply compressed air to fan drive fitting. This disengages fan drive and allows access to socket head capscrews.

1. Position pry bar between water pump pulley (3) and crankshaft pulley (6) and apply pressure.

2-15 ENGINE DISASSEMBLY INTO SUBASSEMBLIES (Cont'd)

2. Using a 5/16-in. hex-head driver, remove four socket-head screws (1) and fan drive (2) from water pump pulley (3).
3. Using a 6-mm hex-head driver, remove four socket-head screws (5) and water pump pulley (3) from water pump (4).

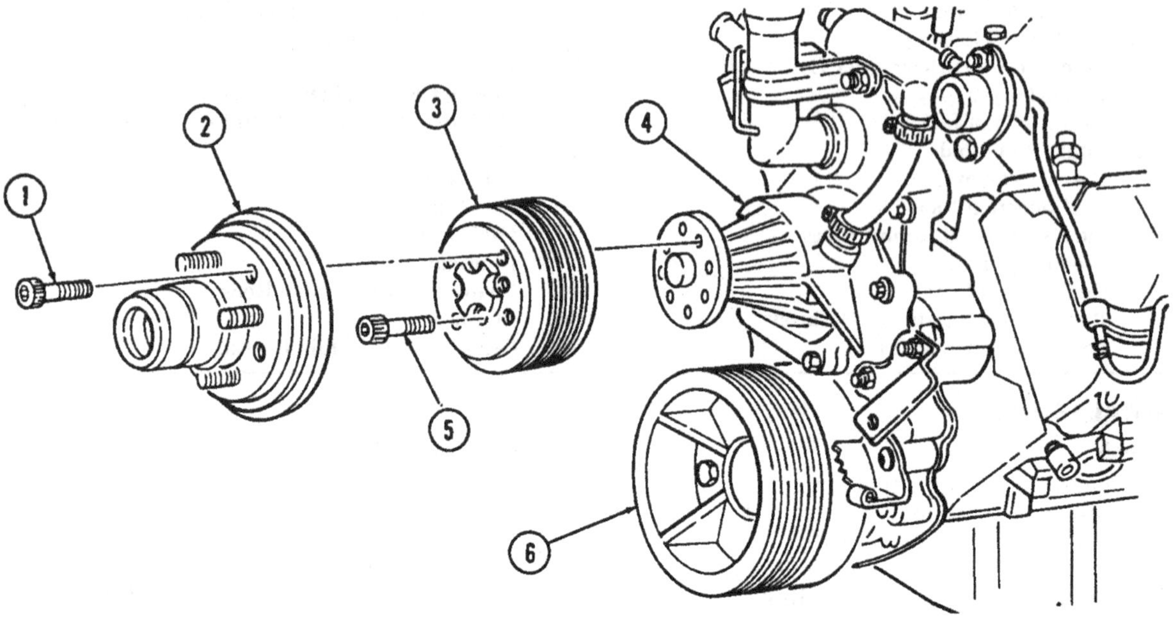

b. Crankshaft Pulley

Remove four capscrews (9), crankshaft pulley (6), and spacer (8) (if installed) from torsional damper (7).

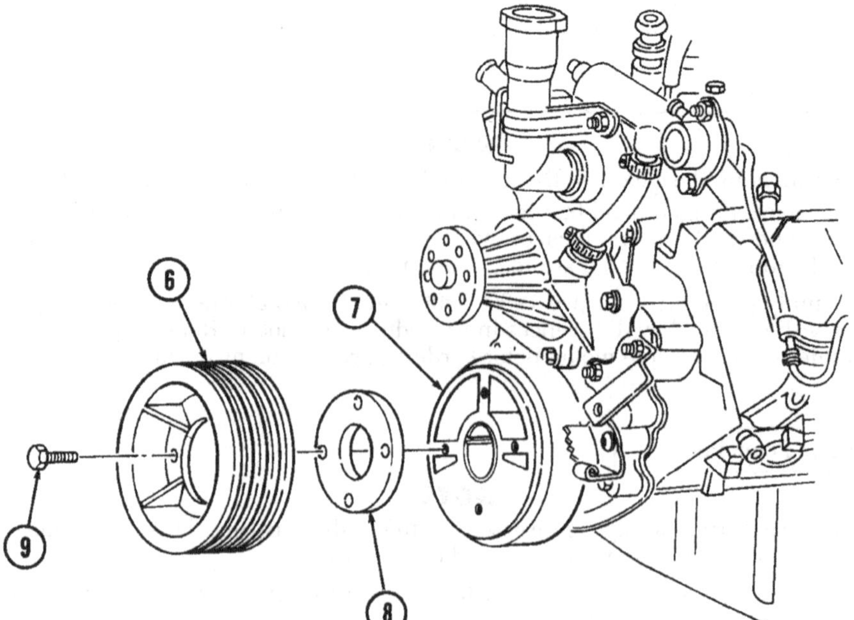

2-15. ENGINE DISASSEMBLY INTO SUBASSEMBLIES (Cont'd)

c. Torsional Damper

1. Hold flywheel stationary with pry bar.
2. Remove capscrew (1) and washer (2) from torsional damper (3) and crankshaft (5).
3. Reinstall capscrew (1) and remove torsional damper (3) with puller.
4. Remove capscrew (1) and woodruff key (4) from crankshaft (5). Discard woodruff key (4).

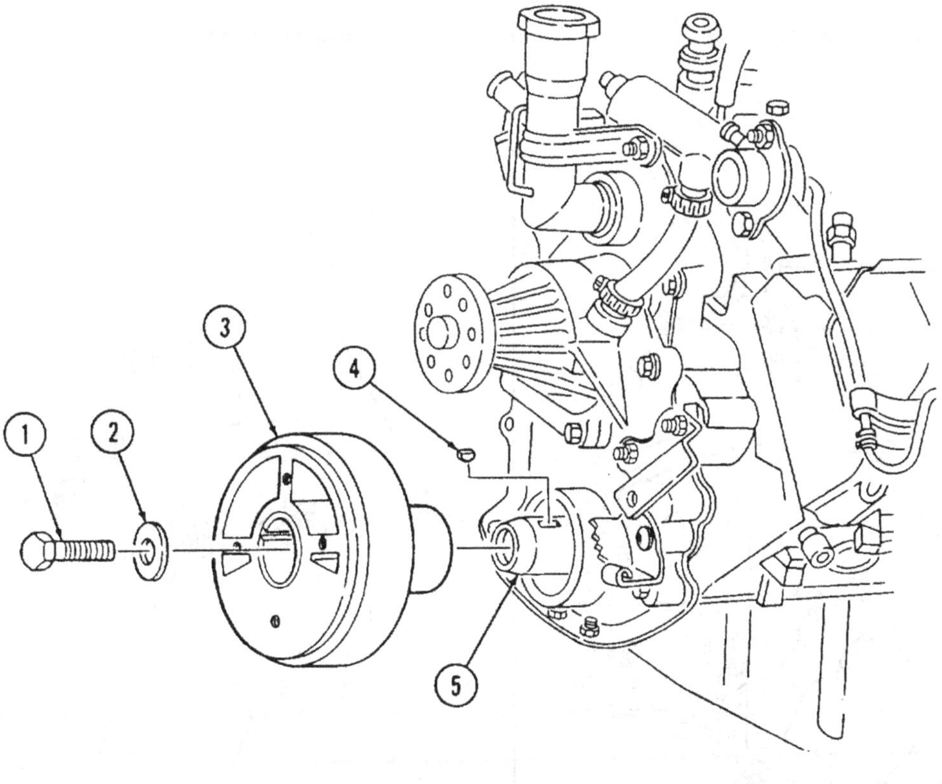

2 - 1 5ENGINE DISASSEMBLY INTO SUBASSEMBLIES (Cont'd)

NOTE

Perform tasks c.1 through c.4 for 6.5L turbo engines only.

c.1. Wastegate Actuator

1. Remove clip (11) and actuator rod (6) from wastegate bellcrank (12).
2. Remove two capscrews (8) from wastegate actuator (7) and wastegate housing (10).
3. Remove capscrew (13) and wastegate actuator (7) from cylinder head (9).

c.2. Wastegate Housing

Remove four capscrews (14) and wastegate housing (10) from turbocharger (15).

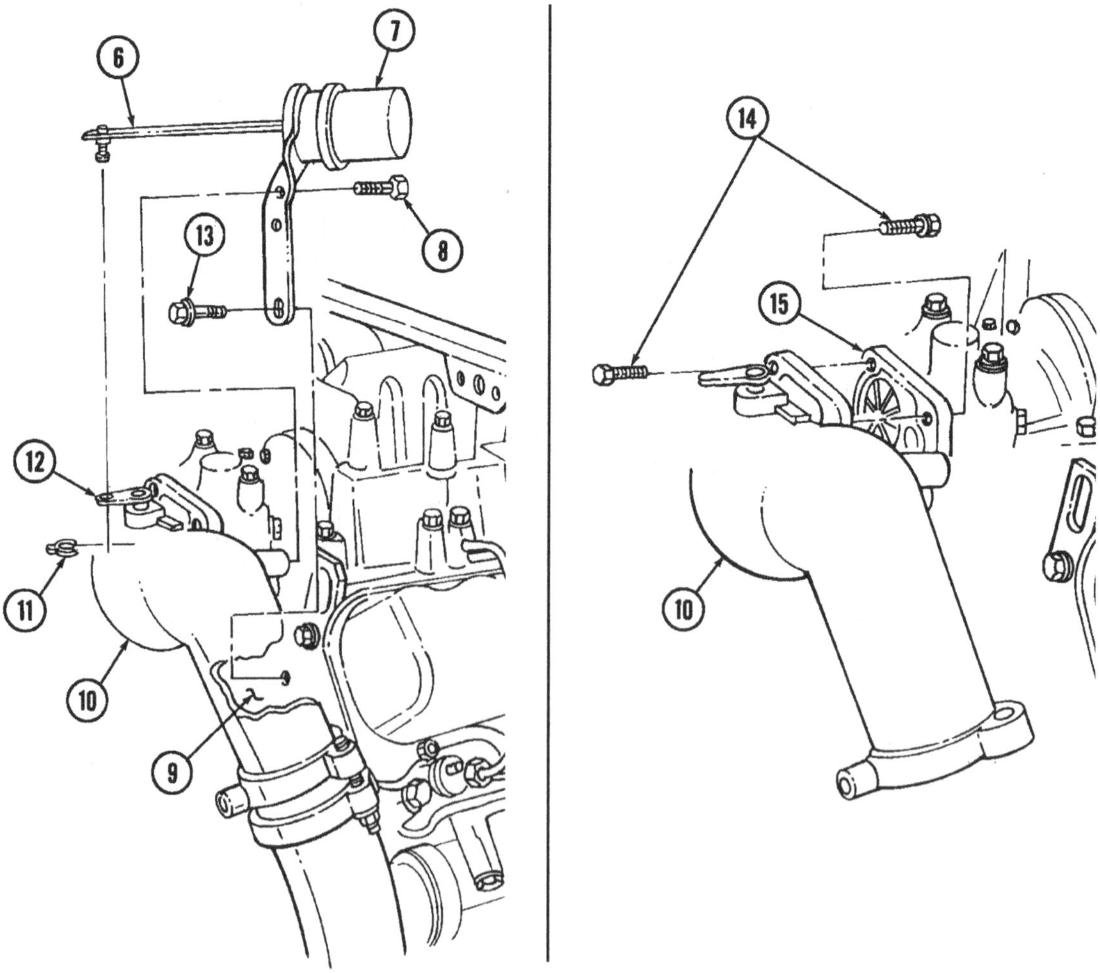

2 - 1 5ENGINE DISASSEMBLY INTO SUBASSEMBLIES (Cont'd)

c.3. Manifold-to-Turbocharger Exhaust Pipe

1. Loosen clamp (5) and remove exhaust pipe (1) from exhaust manifold (4). Slide clamp (5) onto exhaust pipe (1).
2. Loosen clamp (2) and remove exhaust pipe (1) from turbocharger (3). Slide clamp (2) onto exhaust pipe (1).
3. Remove clamps (2) and (5) from exhaust pipe (1).
4. Repeat steps 1 through 3 for opposite side of engine.

c.4. Turbocharger

NOTE

Center inlet manifold must be removed before removing turbocharger.

1. Remove two capscrews (5.2) and turbocharger (5.4) from engine block (5.5).
2. Remove turbocharger oil gasket (5.1) from engine block (5.5). Discard gasket (5.1).
3. Remove O-ring (5.3) from turbocharger (5.4) outlet. Discard O-ring (5.3).

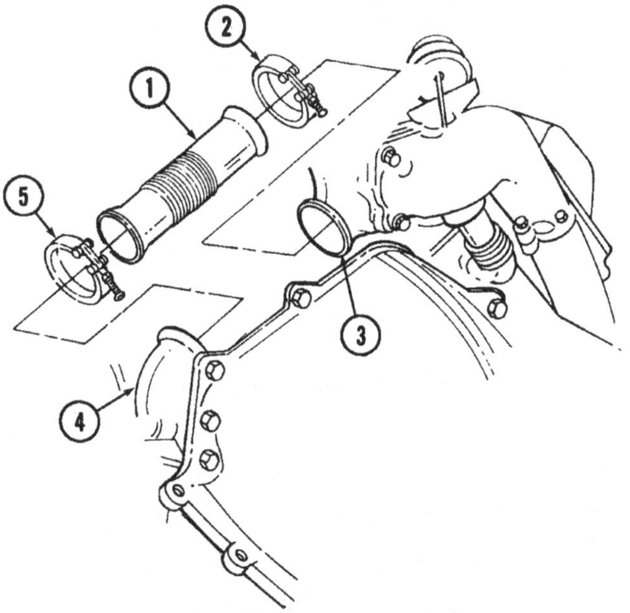

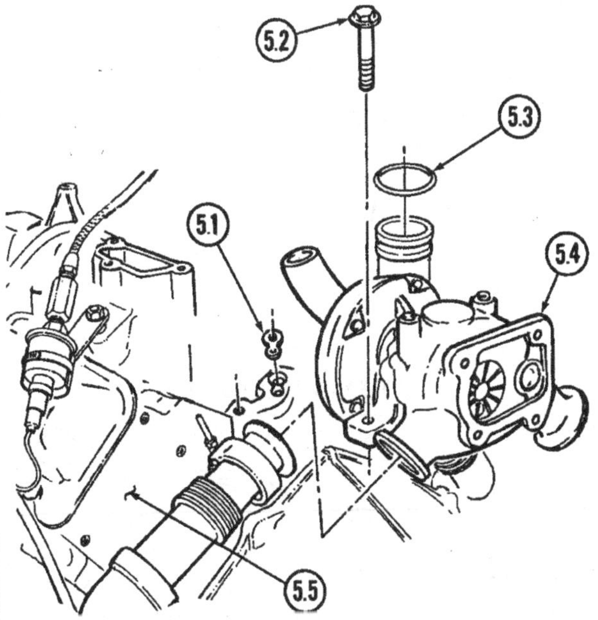

2 - 1 5ENGINE DISASSEMBLY INTO SUBASSEMBLIES (Cont'd)

d. Exhaust Manifolds

NOTE

Early production vehicles may have a socket-head screw in place of stud.

1. Remove stud (6) and washer (7) securing left exhaust manifold (8) to cylinder head (9).

NOTE

Turbo exhaust manifolds use hex-head capscrews and do not have gaskets as indicated in steps 2 and 3.

2. Using 8-mm hex-head driver, remove seven socket-head screws (12), washers (11), left exhaust manifold (8), and gasket (10) from cylinder head (9). Discard gasket (10).
3. Using 8-mm hex-head driver, remove eight socket-head screws (17), washers (16), right exhaust manifold (15), and gasket (14) from cylinder head (13). Discard gasket (14).

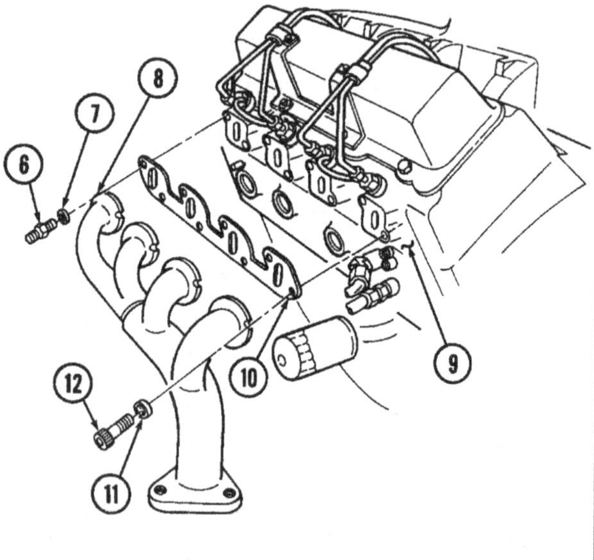

LEFT EXHAUST MANIFOLD

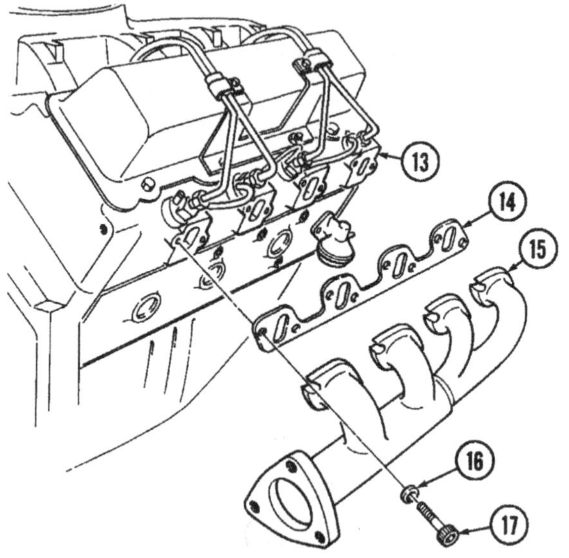

RIGHT EXHAUST MANIFOLD

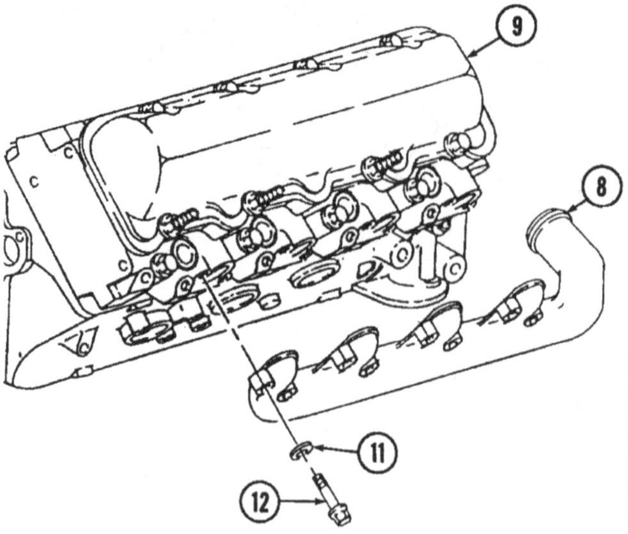

TURBO LEFT EXHAUST MANIFOLD

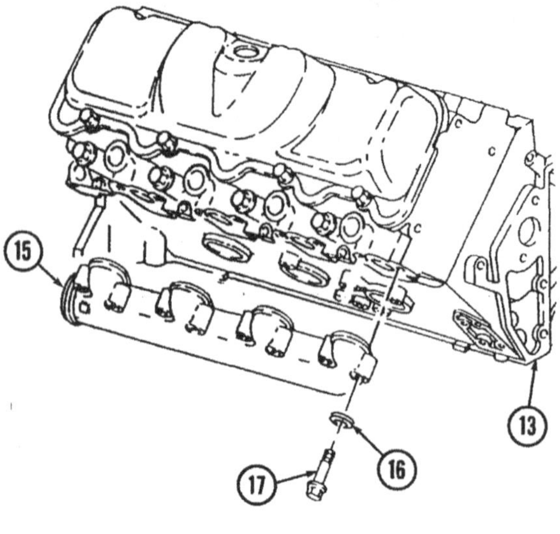

TURBO RIGHT EXHAUST MANIFOLD

2 - 1 5ENGINE DISASSEMBLY INTO SUBASSEMBLIES (Cont'd)

e. Intake Manifold

1. Remove fuel filter line clamp (1) from stud (2).

NOTE

- Note location of studs for installation.
- Perform step 2 for 6.2L engines. Perform step 3 for 6.5L engines.

2. Remove six intake manifold capscrews (5), ten studs (2), and sixteen washers (3).
3. Remove four intake manifold capscrews (5), twelve studs (2), and sixteen washers (3).
4. Remove four injection line clips (4).
5. Remove intake manifold (6) and two gaskets (7) from cylinder heads (8). Discard gaskets (7).

NOTE

- Perform steps 6 through 8 for turbo engines.
- Outer intake manifold is removed before engine is removed from vehicle.

6. Remove eight intake manifold studs (8.1) from intake manifold (8.2).

WARNING

Gaskets installed on some 6.2L engines assembled prior to 1991 may contain asbestos. Gaskets should be removed with a scraper or putty knife and then be disposed of IAW current directives. Inhalation of asbestos fibers can cause respiratory ailments.

7. Remove intake manifold (8.2) and gasket (8.3) from cylinder head (8.4). Discard gasket (8.3).
8. Repeat steps 6 and 7 for opposite side.

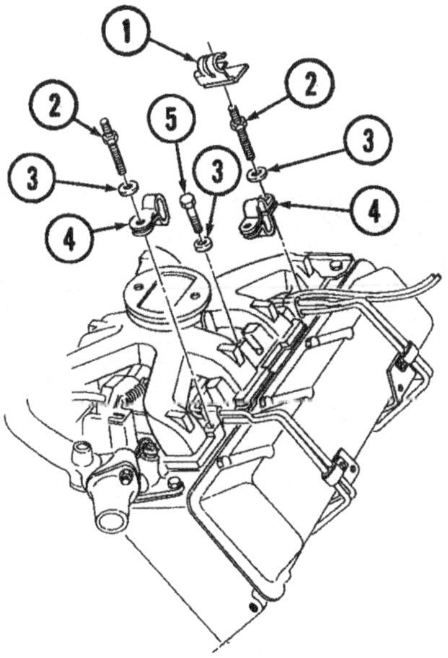

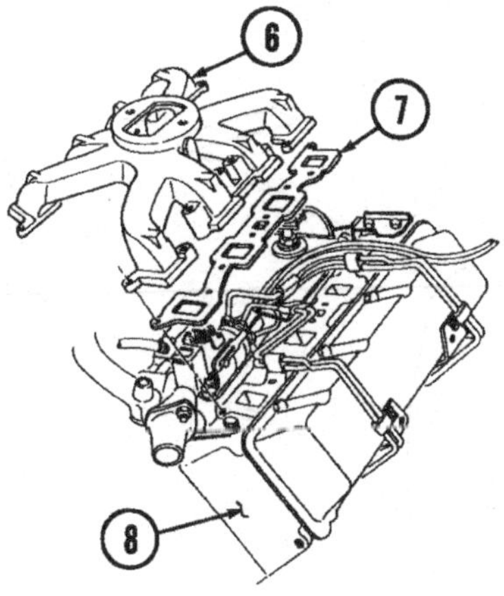

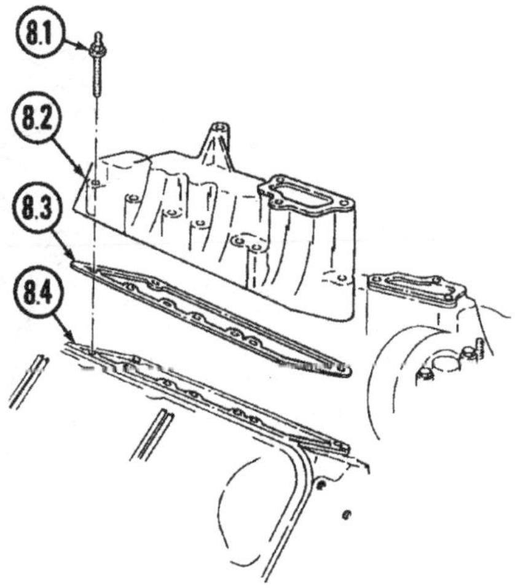

2-15 ENGINE DISASSEMBLY INTO SUBASSEMBLIES (Cont'd)

f. Water Crossover

1. Remove four capscrews (9) securing water crossover (14) to cylinder heads (8).
2. Loosen hose clamp (12) and disconnect thermostat bypass hose (11) from water crossover nipple (13).

WARNING

Gaskets installed on some 6.2L engines assembled prior to 1991 may contain asbestos. Gaskets should be removed with a scraper or putty knife and then be disposed of IAW current directives. Inhalation of asbestos fibers can cause respiratory ailments.

3. Remove water crossover (14) and two gaskets (10). Discard gaskets (10).

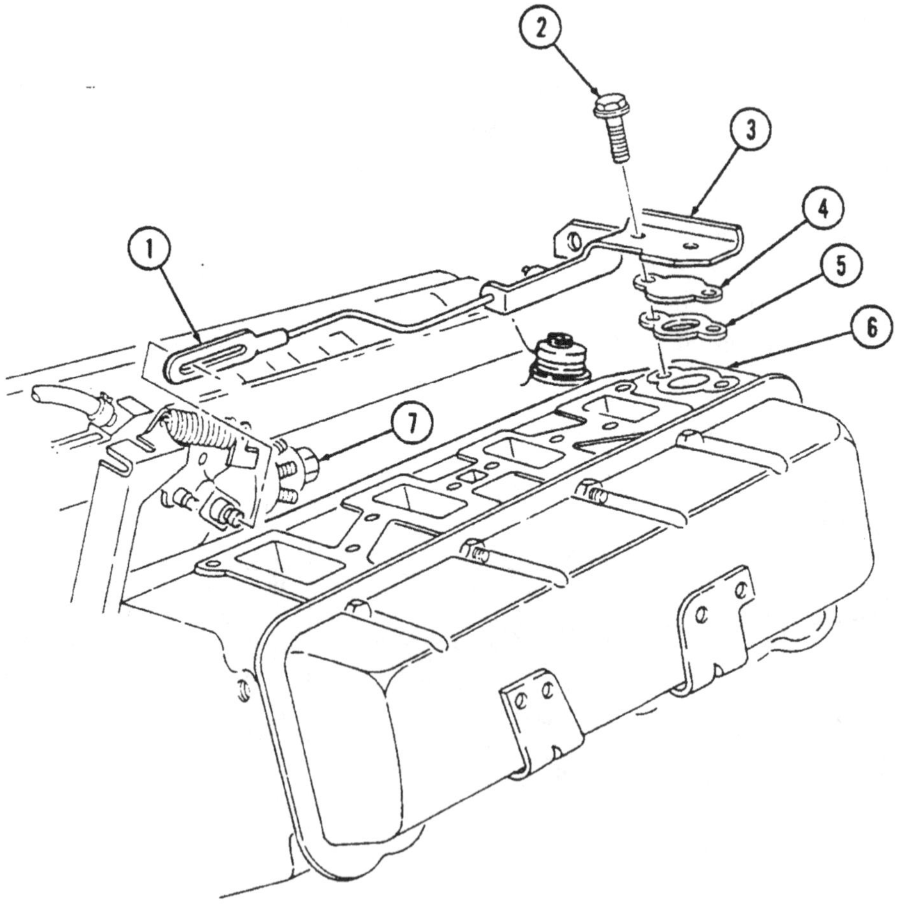

2-15. ENGINE DISASSEMBLY INTO SUBASSEMBLIES (Cont'd)

g. Fuel Injection Lines

WARNING

Diesel fuel is highly flammable. Do not perform this procedure near fire, flame, or sparks. Severe injury or death by result.

1. Loosen and disconnect eight fuel injection line nuts (4) at injection nozzles (3).
2. Remove four screw-assembled washers (5) and clamps (6) from support brackets (2).

CAUTION

Plug fuel injection pump openings to prevent contamination.

NOTE

Tag lines by cylinder number for assembly.

3. Remove clamp (8) and boot (9) from fuel injection pump (1). Disconnect and remove eight fuel injection lines (7) at fuel injection pump (1).

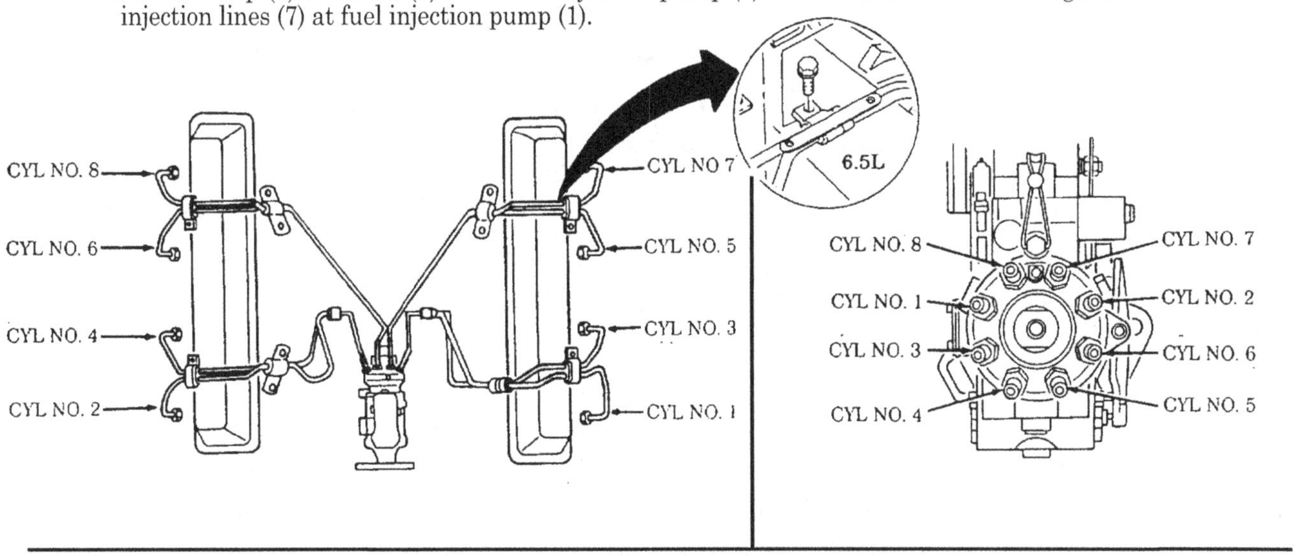

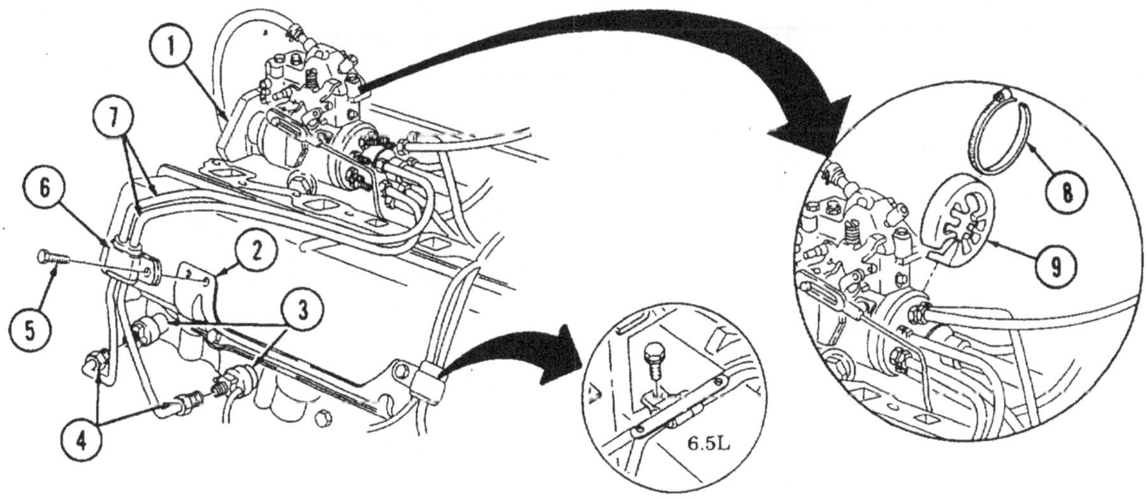

2-15. ENGINE DISASSEMBLY INTO SUBASSEMBLIES (Cont'd)

h. Fuel Supply and Return Lines

1. Loosen clamp (11) and remove fuel filter inlet hose (12) from fuel supplY line (10).
2. Disconnect fuel supply line (10) at fuel pump (27).
3. Remove stud (24), washer (25), two clamps (28) and fuel supply line (10) from fuel return line (18) and cylinder block (29).
4. Remove nut (19), washer (20), and clamp (21) from fuel return line (18) and rocker arm cover stud (22).
5. Remove capscrew (13), washer (14), and clamp (15) from fuel return line (18) and timing gear cover (26).
6. Loosen three clamps (16) and disconnect three hoses (17) at fuel injection pump (1) and front fuel injection nozzles (23).
7. Remove fuel return line (18).

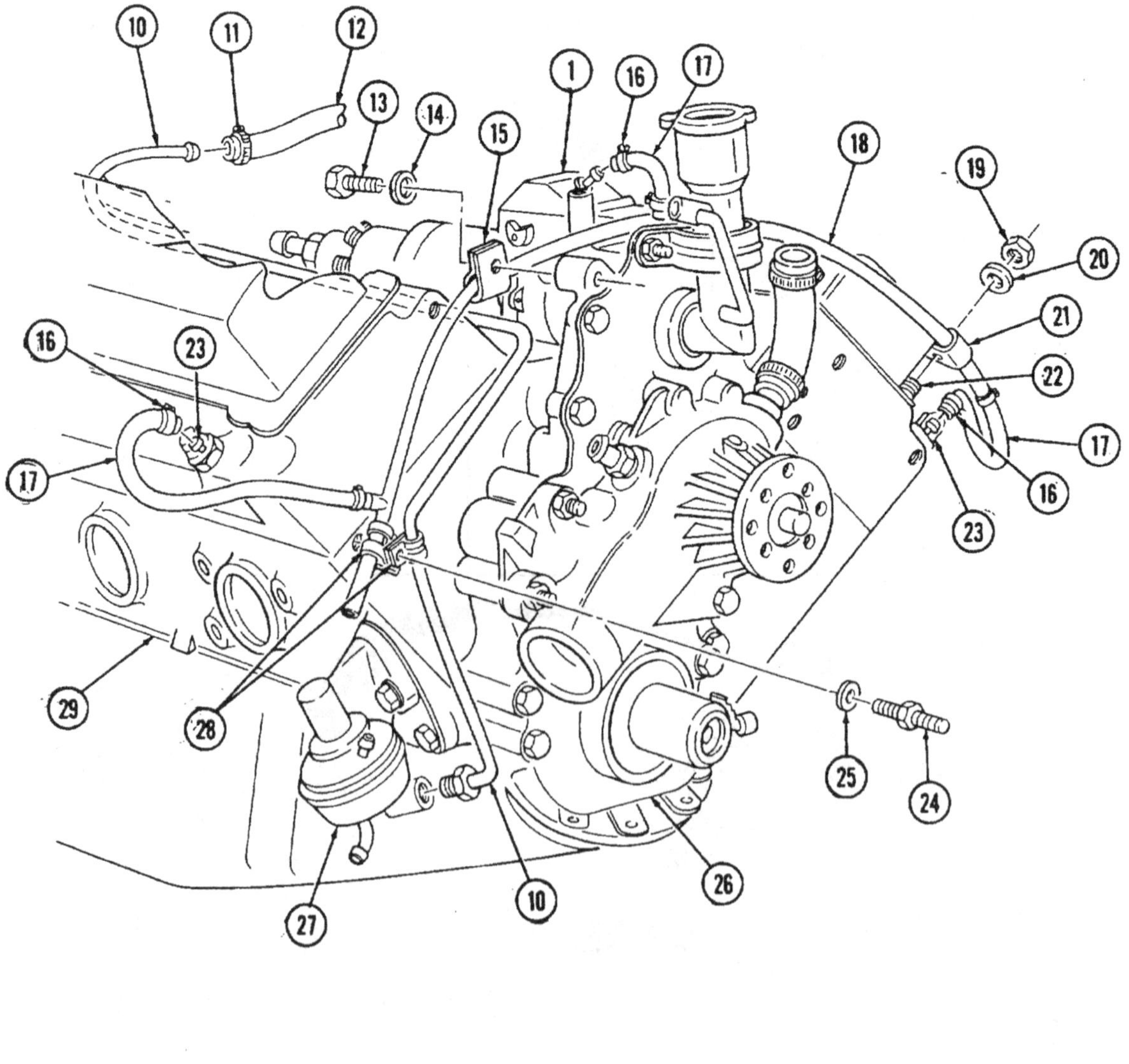

2 - 1 5ENGINE DISASSEMBLY INTO SUBASSEMBLIES (Cont'd)

i. Modulator Link

NOTE

Removal of the modulator link is applicable only to 6.2L engines.

1. Remove two capscrews (2) securing modulator link bracket (3) to cylinder head (6).

WARNING

Gaskets installed on some 6.2L engines assembled prior to 1991 may contain asbestos. Gaskets should be removed with a scraper or putty knife and then be disposed of IAW current directives. Inhalation of asbestos fibers can cause respiratory ailments.

2. Slide modulator link (1) forward and disconnect from fuel injection pump (7) and remove bracket (3), modulator link (1), water jacket cover (4), and gasket (5). Discard gasket (5).

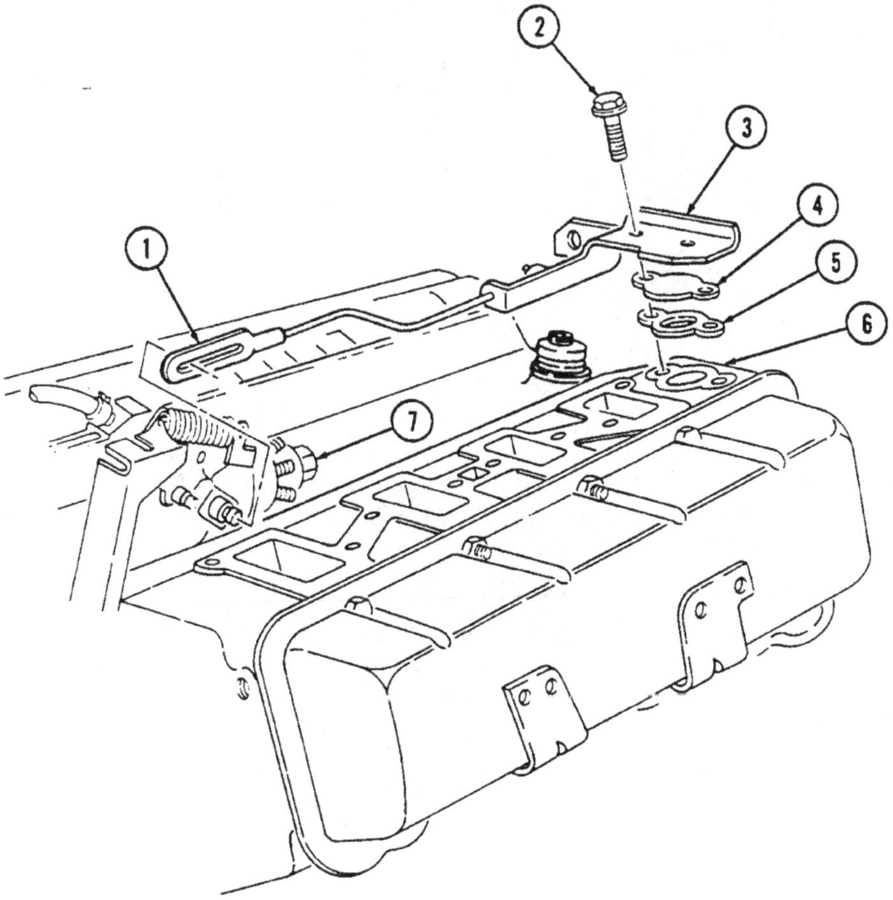

2-15. ENGINE DISASSEMBLY INTO SUBASSEMBLIES (Cont'd)

j. Rocker Arm Covers

1. Remove two nuts (8) and injection line support bracket (9) from rocker arm cover studs (14).
2. Remove a combination of three capscrews (10), five studs (14), eight washers (11), rocker arm cover (12), and gasket (13) (if applicable) from cylinder head (6).
3. Repeat steps 1 and 2 for opposite side.

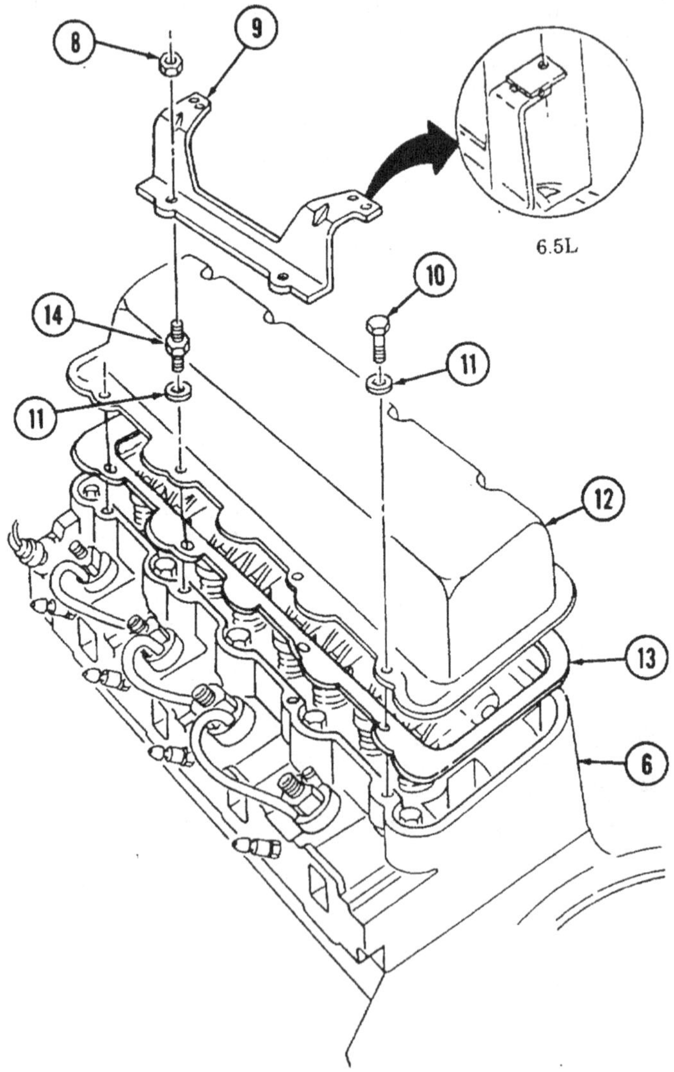

2-15. ENGINE DISASSEMBLY INTO SUBASSEMBLIES (Cont'd)

k. Rocker Arm Shafts and Pushrods

1. Remove four capscrews (3) and four retainers (2) from rocker arm and shaft assembly (4) and cylinder head (5).

NOTE

Tag rocker arm and shaft assembly for assembly.

2. Remove two rocker arm and shaft assemblies (4).

NOTE

Tops of pushrods are hardened and must be tagged for assembly.

3. Remove eight pushrods (1).
4. Repeat steps 1 through 3 for opposite side.

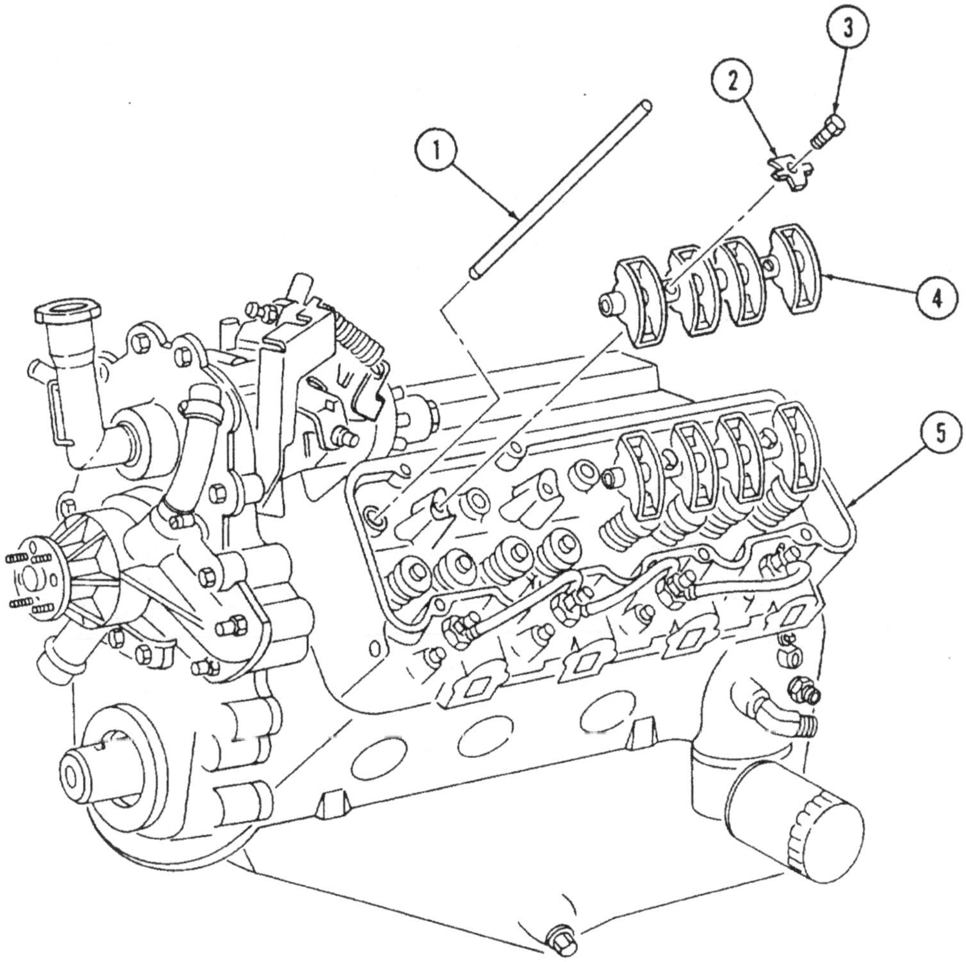

2-15. ENGINE DISASSEMBLY INTO SUBASSEMBLIES (Cont'd)

l. Fuel Injection Nozzles

1. Remove seven clamps (10), one cap (9), and three hoses (8) from fuel injection nozzles (6).
2. Using injection nozzle socket, remove four fuel injection nozzles (6) and gaskets (7) from cylinder head (5). Discard gaskets (7).
3. Repeat steps 1 and 2 for opposite side.

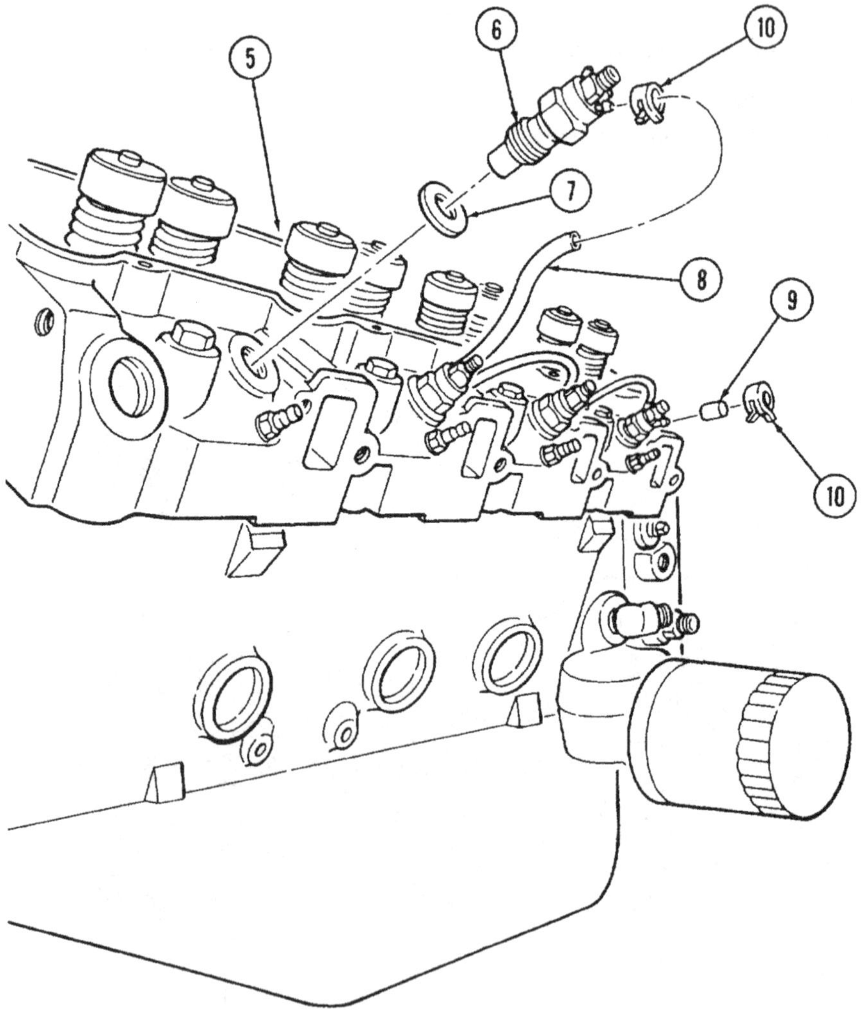

2-15. ENGINE DISASSEMBLY INTO SUBASSEMBLIES (Cont'd)

m. Glow Plugs

1. Using glow plug socket, remove four glow plugs (1) from cylinder head (2).
2. Repeat step 1 for opposite side.

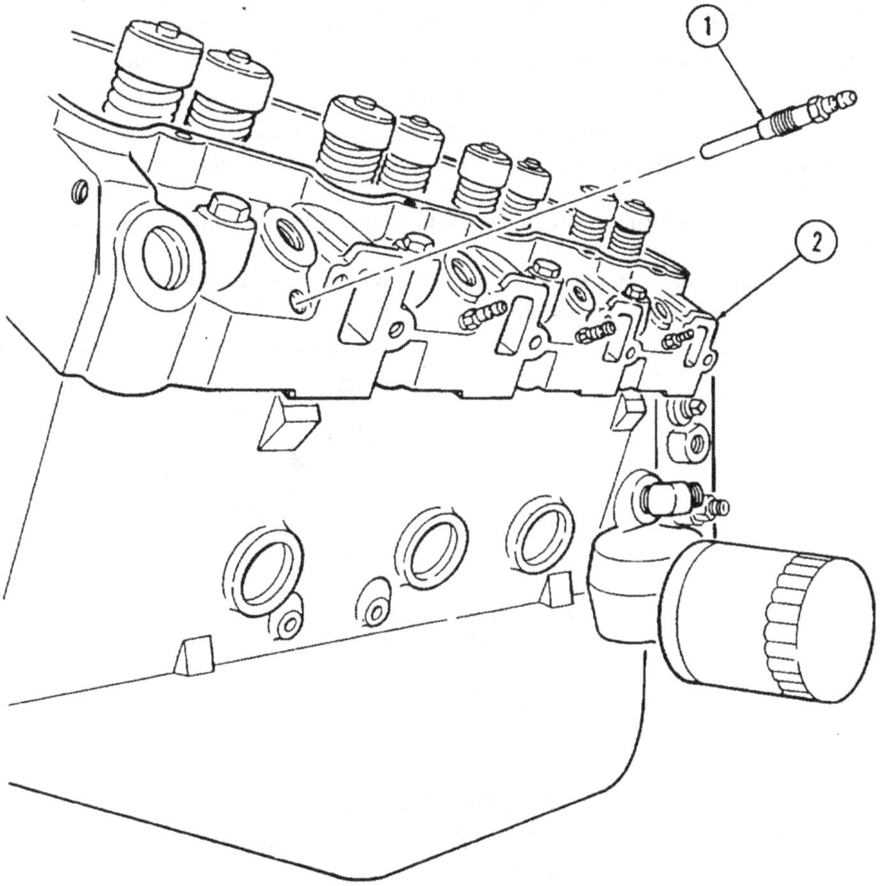

2-15. ENGINE DISASSEMBLY INTO SUBASSEMBLIES (Cont'd)

n. Cylinder Heads

WARNING

Gaskets installed on some 6.2L engines assembled prior to 1991 may contain asbestos. Gaskets should be removed with a scraper or putty knife and then be disposed of IAW current directives. Inhalation of asbestos fibers can cause respiratory ailments.

1. Remove seventeen capscrews (3) from cylinder head (2) and cylinder block (5).

NOTE

Use of cylinder head lifting device is optional.

1.1. Install cylinder head lifting device on center of cylinder head (2) with two washers (3.2) and capscrews (3.1).

1.2. Using a hoist or other lifting device, remove cylinder head (2) and cylinder head gasket (4) from cylinder block (5).

1.3. Remove two capscrews (3.1), washers (3.2), and cylinder head lifting device from cylinder head (2).

2. Repeat steps 1 through 1.3 for opposite side.

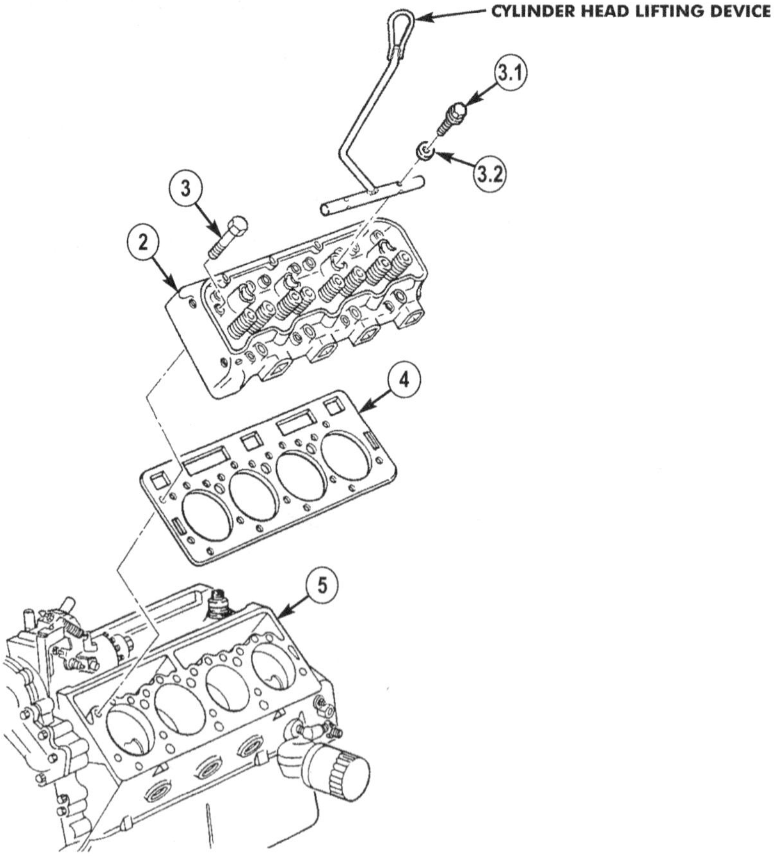

2-15. ENGINE DISASSEMBLY INTO SUBASSEMBLIES (Cont'd)

o. Valve Lifters

1. Remove two capscrews (3), guide plate clamps (2), and four guide plates (4) from cylinder block (5).

NOTE

Tag lifters for assembly.

2. Remove eight valve lifters (1) from cylinder block (5) with hydraulic valve lifter remover.
3. Repeat steps 1 and 2 for opposite side.

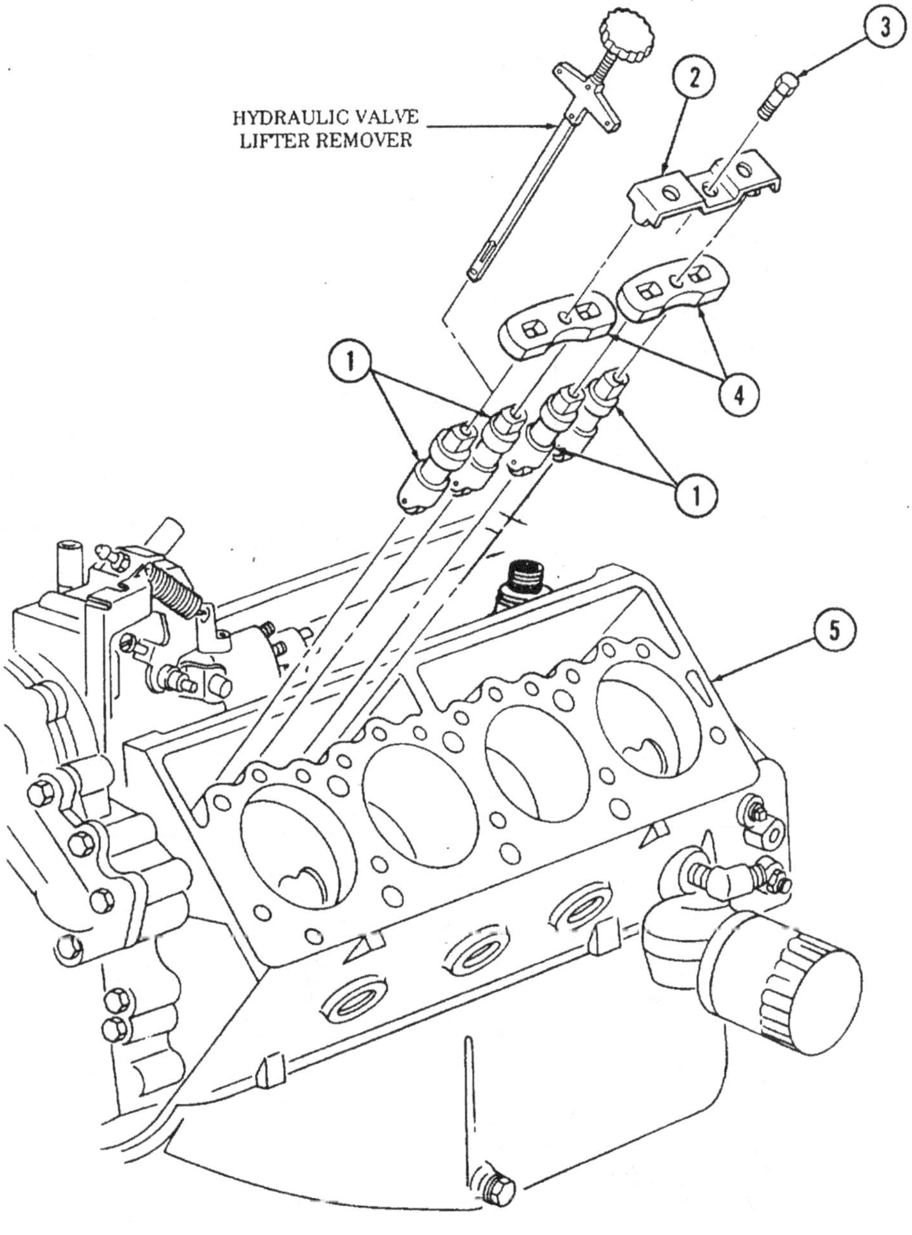

2 - 1 5ENGINE DISASSEMBLY INTO SUBASSEMBLIES (Cont'd)

p. Water Pump and Adapter Plate

1. Remove two nuts (6), washers (7), and oil fill tube (8) from adapter plate (9).
2. Remove two studs (10), (14), and (17), four capscrews (18), washers (11), two capscrews (15), capscrew (16), water pump (13), and adapter plate (9) from timing gear cover (12).

WARNING

Gaskets installed on some 6.2L engines assembled prior to 1991 may contain asbestos. Gaskets should be removed with a scraper or putty knife and then be disposed of IAW current directives. Inhalation of asbestos fibers can cause respiratory ailments.

3. Remove seven capscrews (20) from water pump (13) and separate water pump (13) and gasket (19) from adapter plate (9). Discard gasket (19).

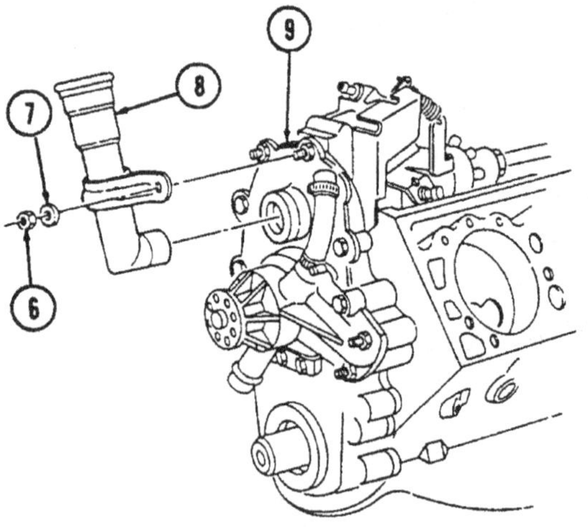

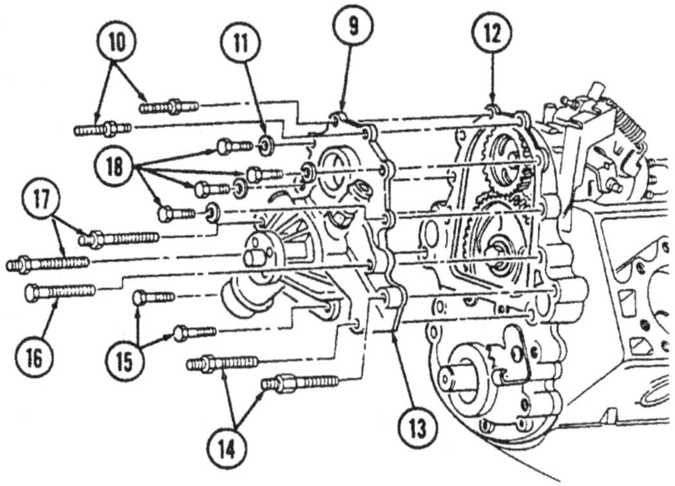

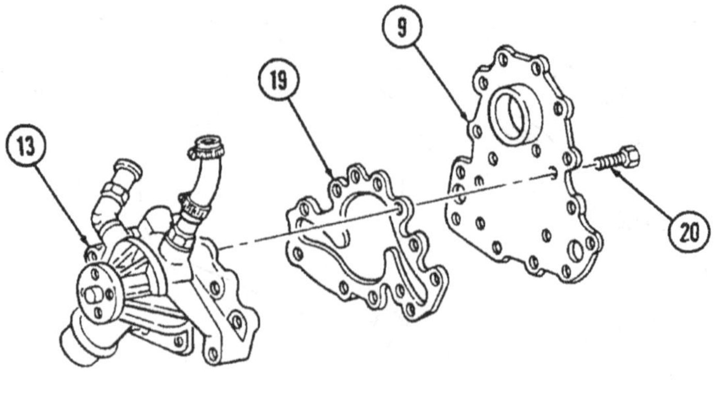

2 - 1 5ENGINE DISASSEMBLY INTO SUBASSEMBLIES (Cont'd)

q. Fuel Injection Pump

1. Remove three capscrews (1) and pump driven gear (2) from fuel injection pump (5).
2. Remove idle return spring (7) from throttle lever (6).
3. Remove three nuts (8) and washers (9) from fuel injection pump (5) and timing gear cover (3).
4. Remove two capscrews (10) and accelerator cable bracket (11) from fuel injection pump (5).

WARNING

Gaskets installed on some 6.2L engines assembled prior to 1991 may contain asbestos. Gaskets should be removed with a scraper or putty knife and then be disposed of IAW current directives. Inhalation of asbestos fibers can cause respiratory ailments.

5. Remove the fuel injection pump (5) and gasket (4) from cover (3). Discard gasket (4).

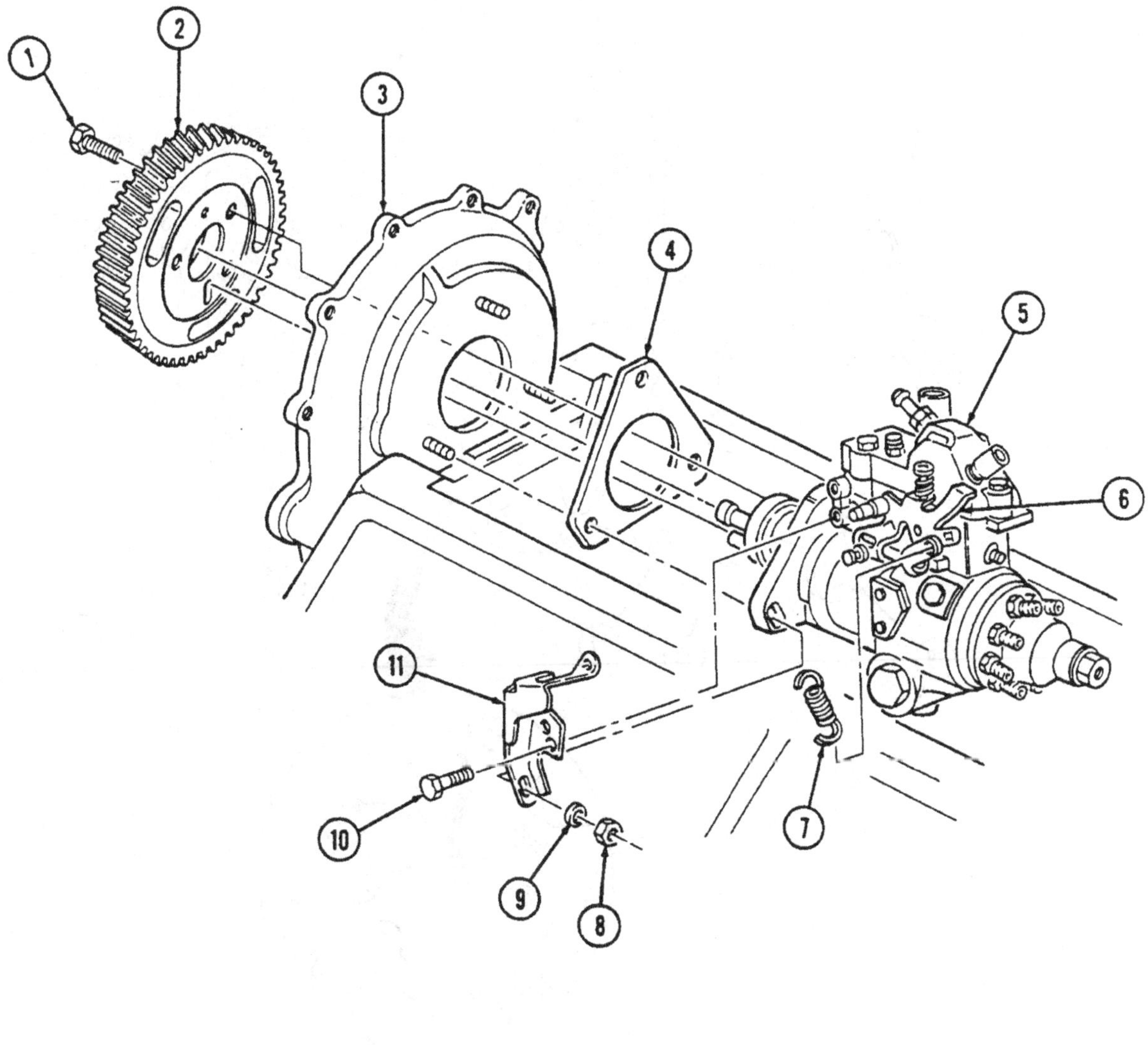

2-15. ENGINE DISASSEMBLY INTO SUBASSEMBLIES (Cont'd)

r. Timing Gear Cover

NOTE

Perform step 1 for 6.5L engines.

1. Remove capscrew (21) and rpm sensor (22) from timing gear cover (3).
2. Remove two capscrews (20), nut (12), and baffle (13) from timing gear cover (3).
3. Remove four capscrews (15) from oil pan (16) and timing gear cover (3).
4. Remove five capscrews (18), four washers (17), and timing gear cover (3) from cylinder block (14).
5. Remove front cover seal (19) from timing gear cover (3). Discard seal (19).

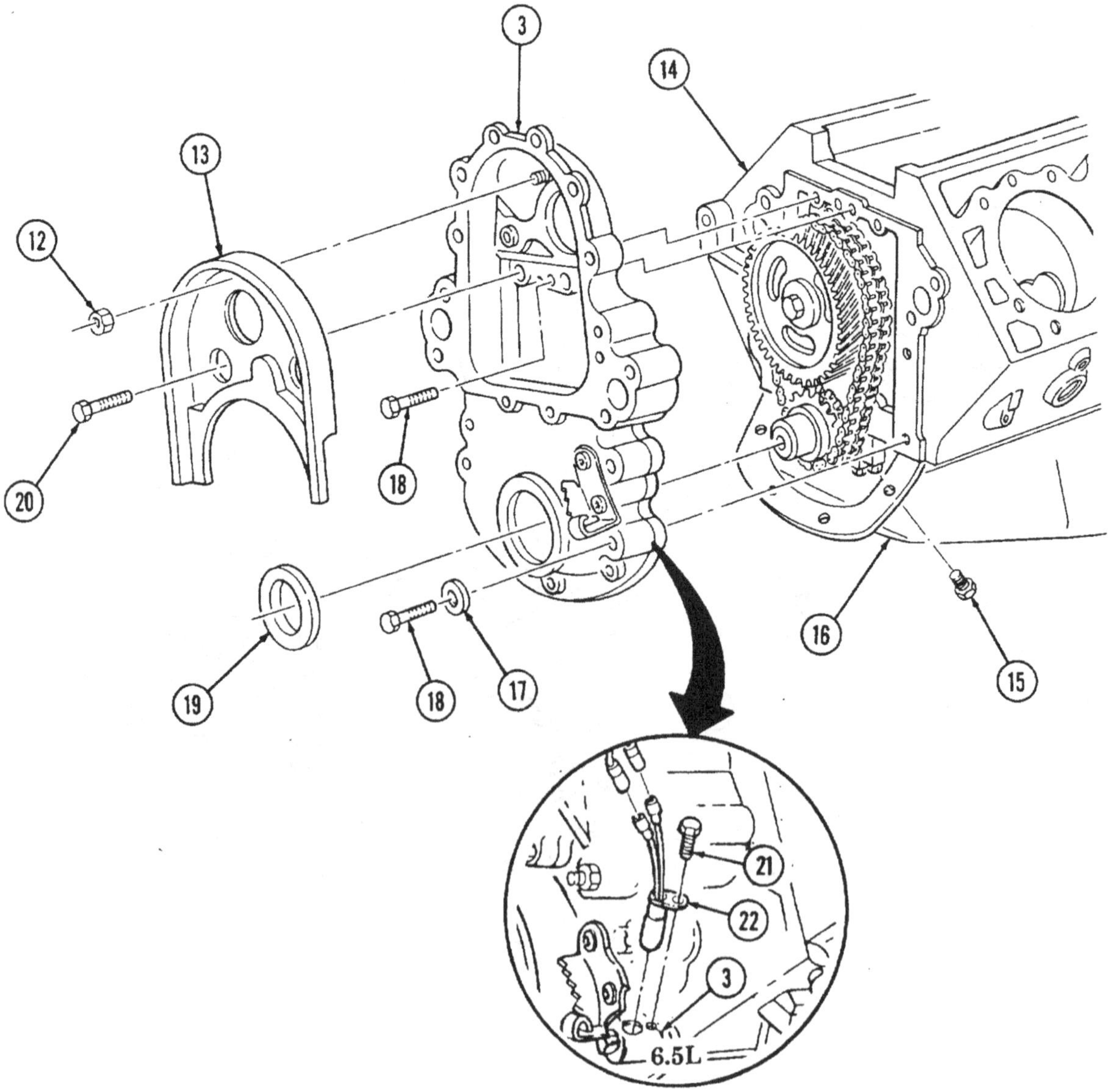

2-15. ENGINE DISASSEMBLY INTO SUBASSEMBLIES (Cont'd)

s. Timing Chain and Drive Sprockets

NOTE

When measuring timing chain deflection, slack should be removed from one side before measurement is taken on opposite side.

1. Using dial indicator, check timing chain (10) deflection midway between camshaft sprocket (4) and crankshaft sprocket (11). Total deflection must not exceed 0.810-in. (20.6 mm). If deflection exceeds specification, timing chain (10) must be replaced.
2. Using dial indicator, check camshaft end play. Camshaft end play must not be more than 0.012-in. (0.3 mm). If end play exceeds specification, camshaft sprocket (4), thrust plate (7), and spacer must be inspected for wear after removal.
3. Remove capscrew (1), washer (2), and pump drive gear (3) from camshaft (5).
4. Remove crankshaft sprocket (11), camshaft sprocket (4), and timing chain (10) as an assembly.

NOTE

Cover oil pan opening to prevent woodruff key from falling into oil pan.

5. Remove woodruff key (8) from crankshaft (9) and woodruff key (6) from camshaft (5). Discard woodruff keys (8) and (6).

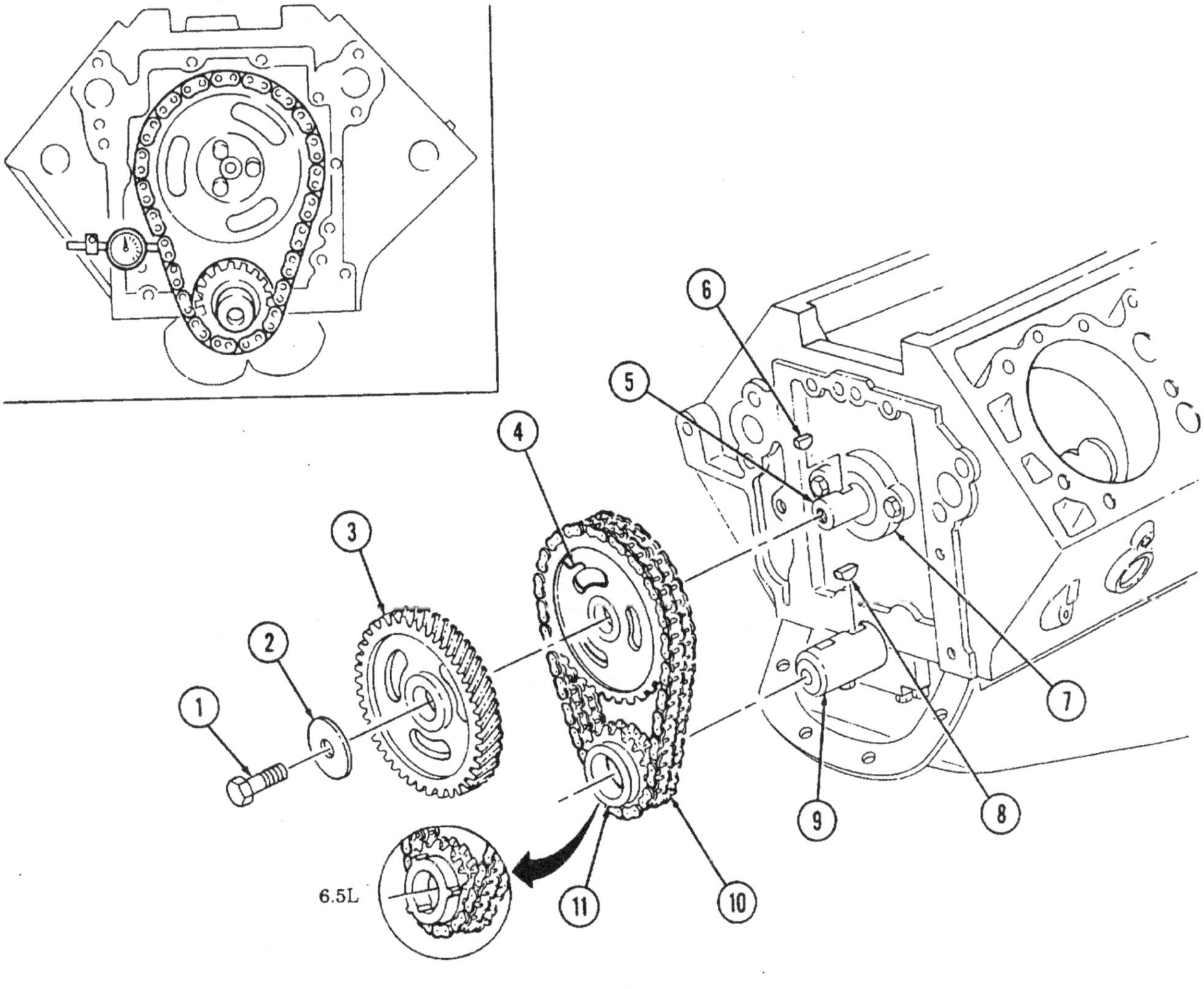

2-15. ENGINE DISASSEMBLY INTO SUBASSEMBLIES (Cont'd)

t. Oil Filter, Adaptter, and Oil Pressure Sending Unit

NOTE

Have drainage container ready to catch oil.

1. Remove oil filter (19) from adapter (15). Discard oil filter (19).
2. Remove adapter bolt (17), gasket (16), and adapter (15) from cylinder block (21). Discard gasket (16).
3. Remove two O-rings (18) from adapter bolt (17). Discard o-rings (18).
4. Remove O-ring (14) from adapter (15). Discard O-ring (14).
5. Remove two oil cooler line fittings (20) from cylinder block (21).
6. Remove oil pressure sending unit (12) from fitting (13).
7. Remove fitting (13) from cylinder block (21).

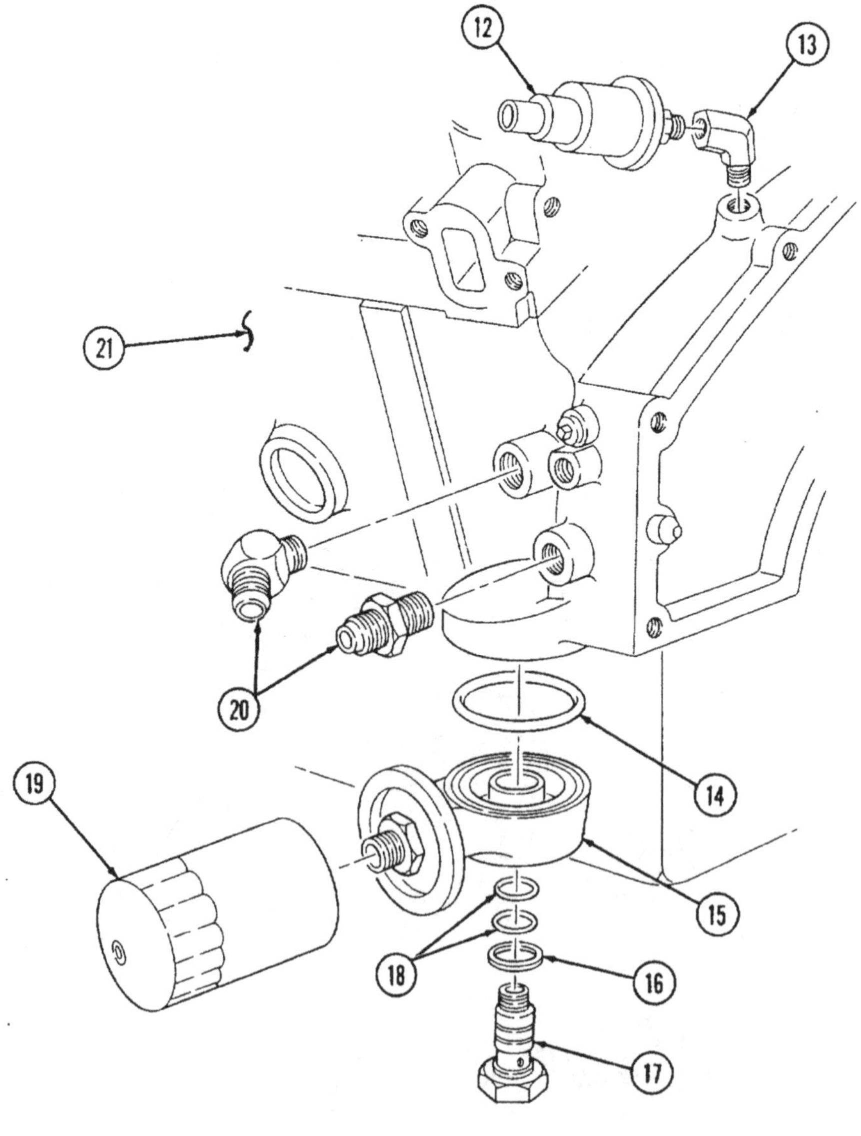

2-15. ENGINE DISASSEMBLY INTO SUBASSEMBLIES (Cont'd)

u. Oil Pan

NOTE

Have drainage container ready to catch oil.

1. Remove oil drainplug (2) and gasket (3) from oil pan (5) and drain oil. Inspect gasket (3) and discard if damaged.
2. Rotate cylinder block (7) 180°.
3. Remove eighteen capscrews (i) and two studs (4) from oil pan (5) and cylinder block (7).
4. Remove oil pan (5), gasket (8) (if installed), and oil pan rear seal (6) from cylinder block (7). Discard gasket (8) and seal (6).

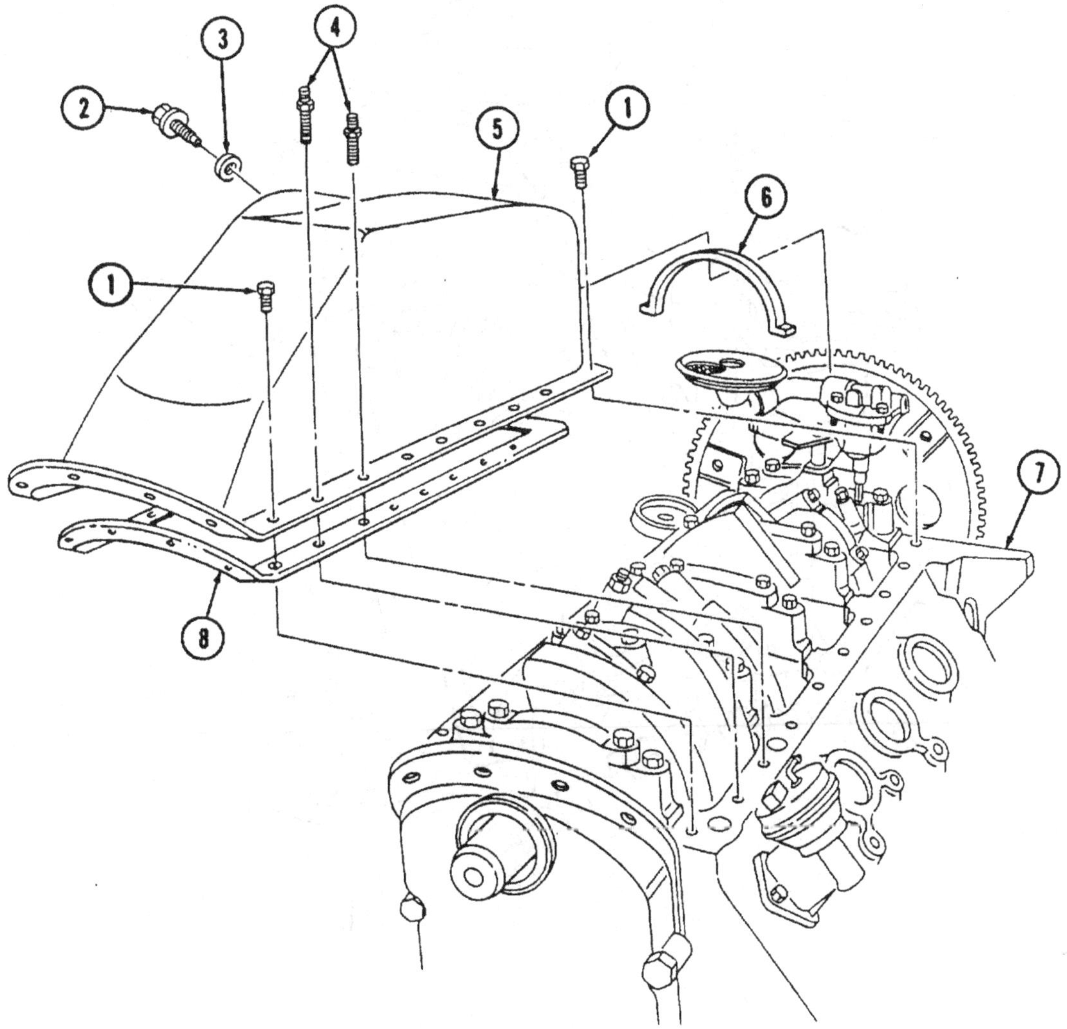

2-15. ENGINE DISASSEMBLY INTO SUBASSEMBLIES (Cont'd)

v. Oil Pump

1. Remove nut (10) securing bracket (12) to stud (9).
2. Loossencapscrew (13) from bracket (12) and oil screen (11) and slide bracket (12) off stud (9).
3. Remove stud (9) from oil pump (8) and rear main bearing cap (14).
4. Remove oil pump (8) and shaft (15) from cylinder block (7).

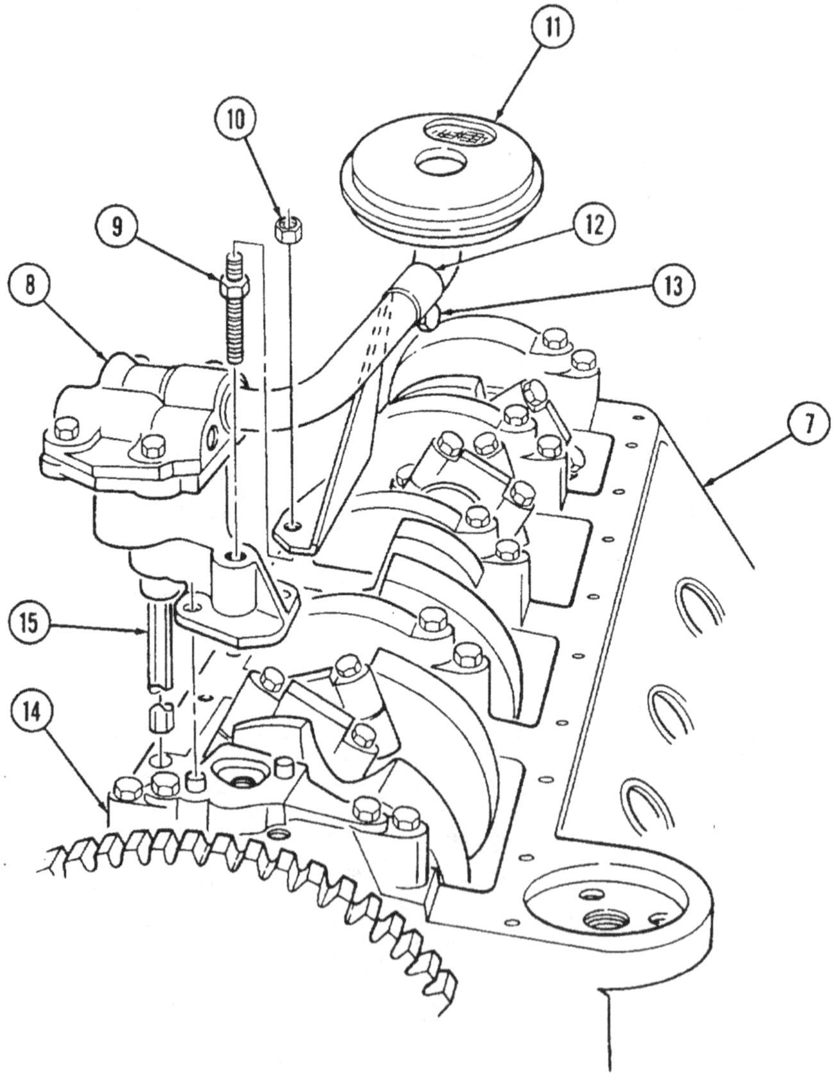

2 - 1 5ENGINE DISASSEMBLY INTO SUBASSEMBLIES (Cont'd)

w. Fuel Pump

1. Rotate cylinder block (7) 180°.

WARNING

Gaskets installed on some 6.2L engines assembled prior to 1991 may contain asbestos. Gaskets in steps 2 and 3 should be removed with a scraper or putty knife and then be disposed of IAW current directives. Inhalation of asbestos fibers can cause respiratory ailments.

2. Remove two capscrews (1), fuel pump (2), and gasket (3) from cylinder block (7). Discard gasket (3).
3. Remove two capscrews (8), adapter plate (4), gasket (5), and pushrod (6) from cylinder block (7). Discard gasket (5).

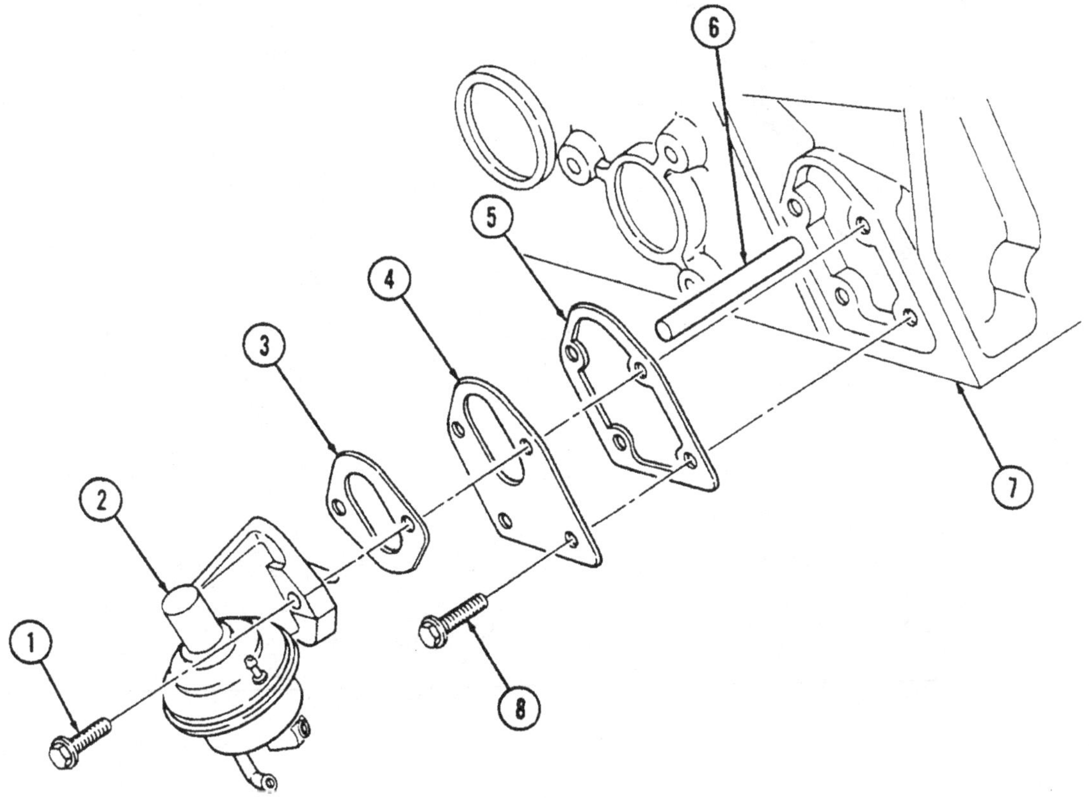

2 - 1 5ENGINE DISASSEMBLY INTO SUBASSEMBLIES (Cont'd)

x. Oil Pump Drive

1. Remove capscrew (9) and clamp (12) from cylinder block (7).
2. Remove oil pump drive (10) and gasket (11) from cylinder block (7). Discard gasket (11).

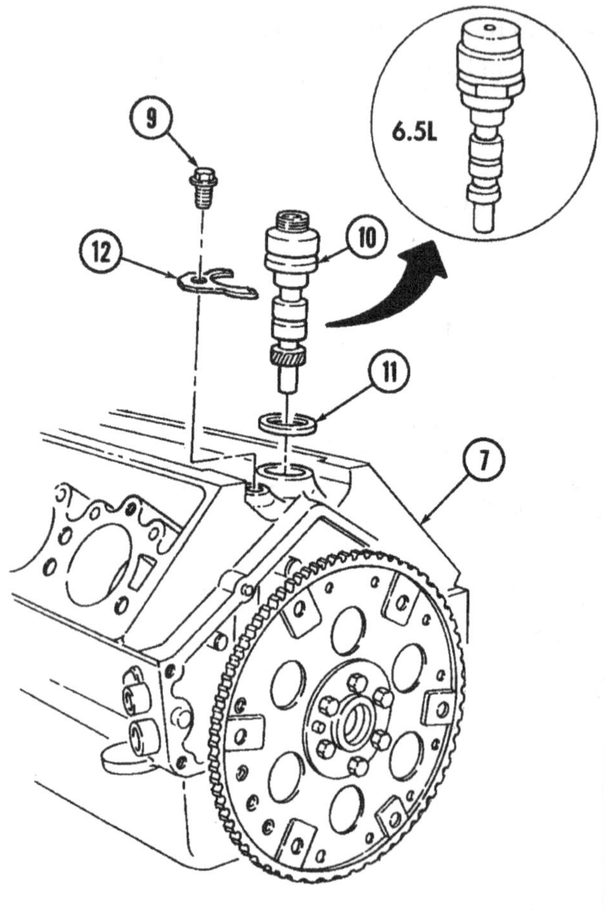

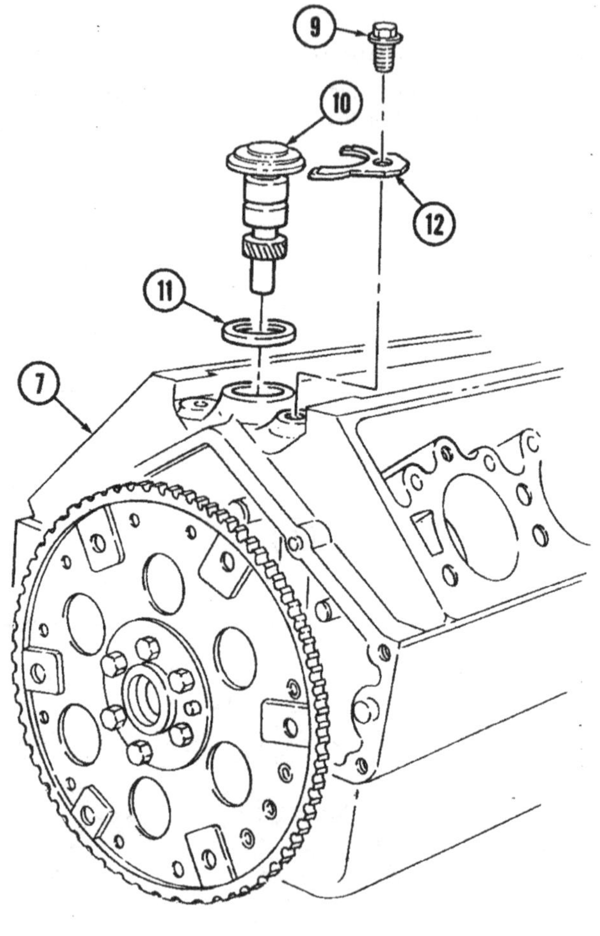

TURBO ENGINE

2-15. ENGINE DISASSEMBLY INTO SUBASSEMBLIES (Cont'd)

y. Camshaft

1. Remove two capscrews (5), camshaft thrust plate (1), and spacer (2) from cylinder block (4).

CAUTION

Support camshaft during removal to prevent damage to camshaft bearings.

2. Remove camshaft (3) from cylinder block (4).

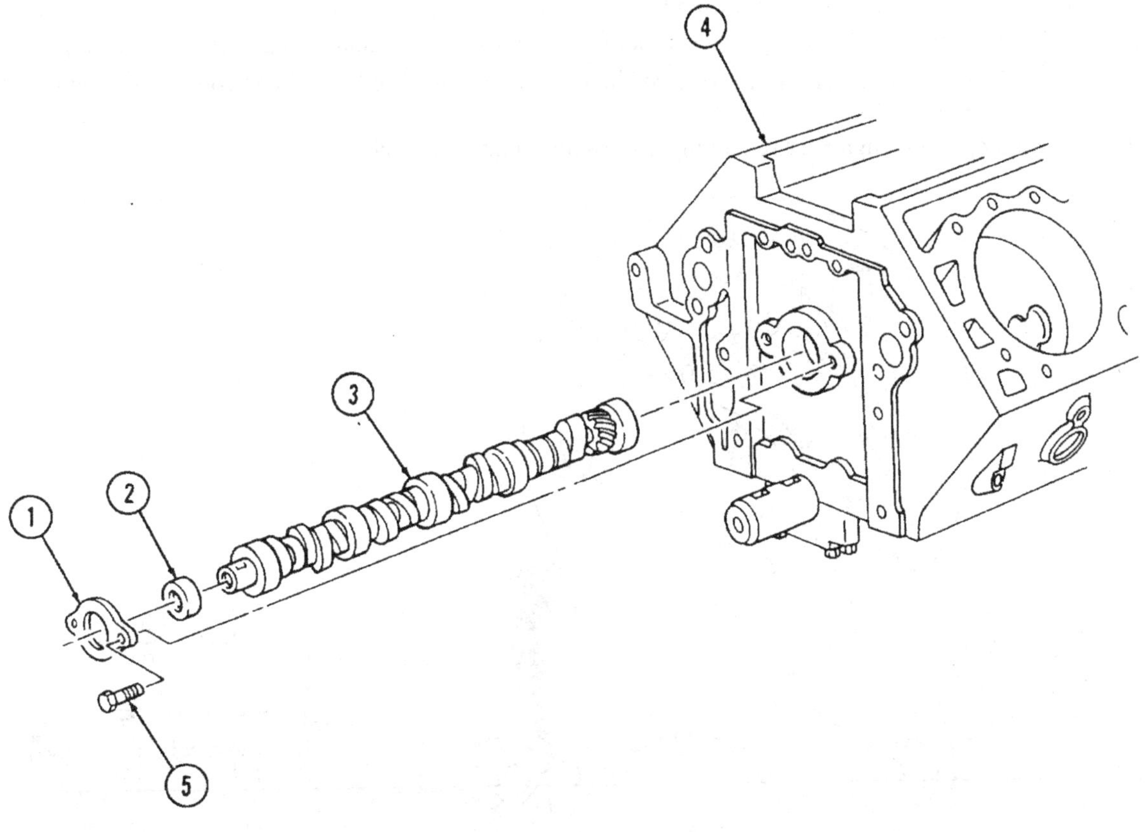

2-15. ENGINE DISASSEMBLY INTO SUBASSEMBLIES (Cont'd)

z. Pistons and Connecting Rods

1. Remove ridge from upper inside wall of each cylinder (12) with ridge reamer.
2. Rotate cylinder block (4) approximately 90°.
3. Rotate crankshaft (13) to gain access to two nuts (6).
4. Check sides of connecting rod (10) and rod cap (7) for matching identification numbers. If no numbers are present, stamp connecting rod (10) and rod cap (7) with cylinder number for assembly.
5. Remove two nuts (6) securing rod cap (7) to connecting rod (10) and remove rod cap (7) and two bearings (8). Discard bearings (8).
6. Cover connecting rod bolts (9) with 3/8-in. rubber hose.
7. Push connecting rod (10) and piston (11) out from the inside and remove from cylinder block (4).
8. Remove rubber hose from connecting rod bolts (9) and install rod cap (7) on connecting rod (10) and secure with nuts (6).
9. Repeat steps 3 through 8 for remaining pistons and connecting rods.

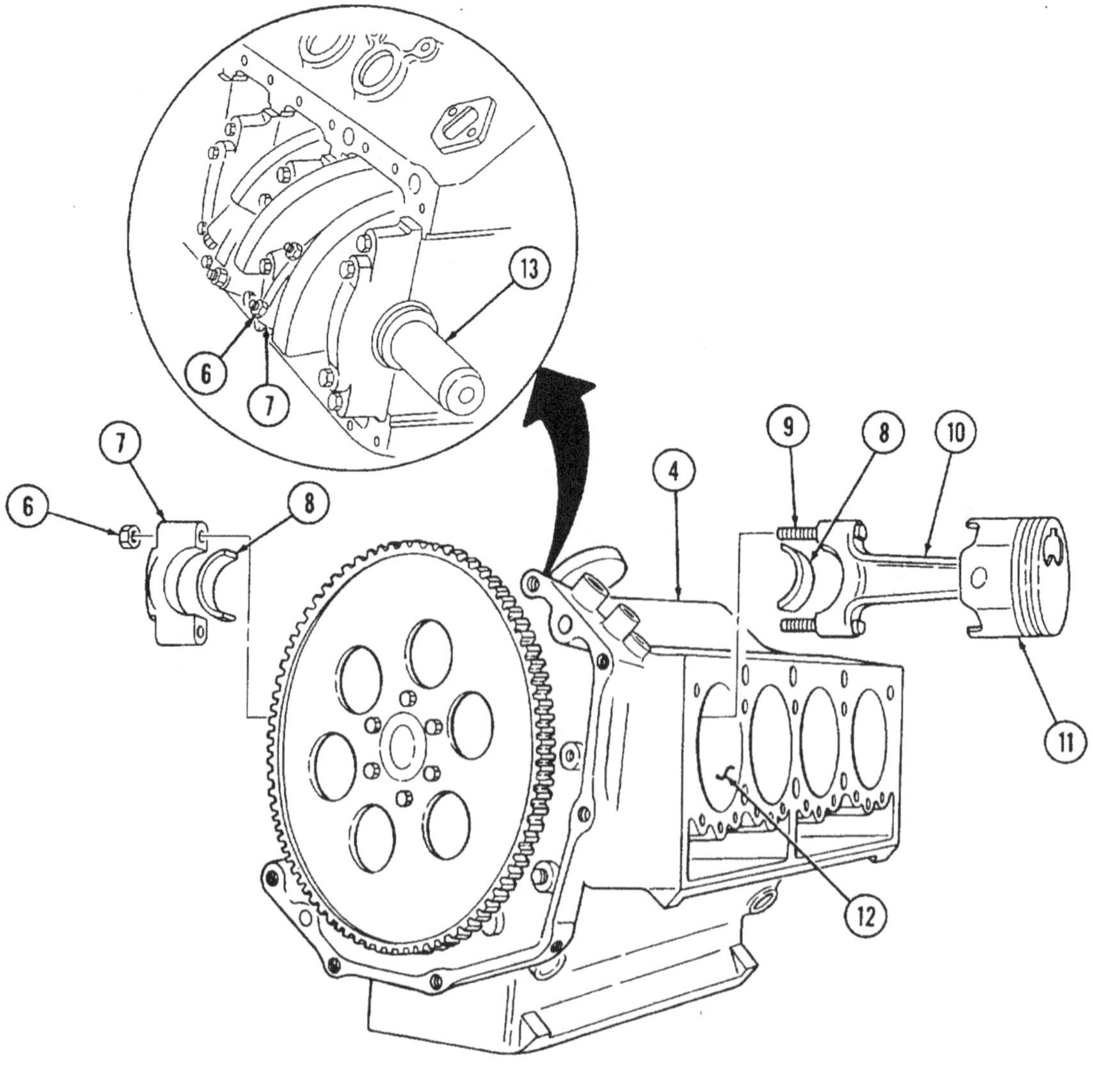

2-15. ENGINE DISASSEMBLY INTO SUBASSEMBLIES (Cont'd)

aa. Flywheel

Keep crankshaft (3) from rotating and remove six capscrews (2) and flywheel (1) from crankshaft (3).

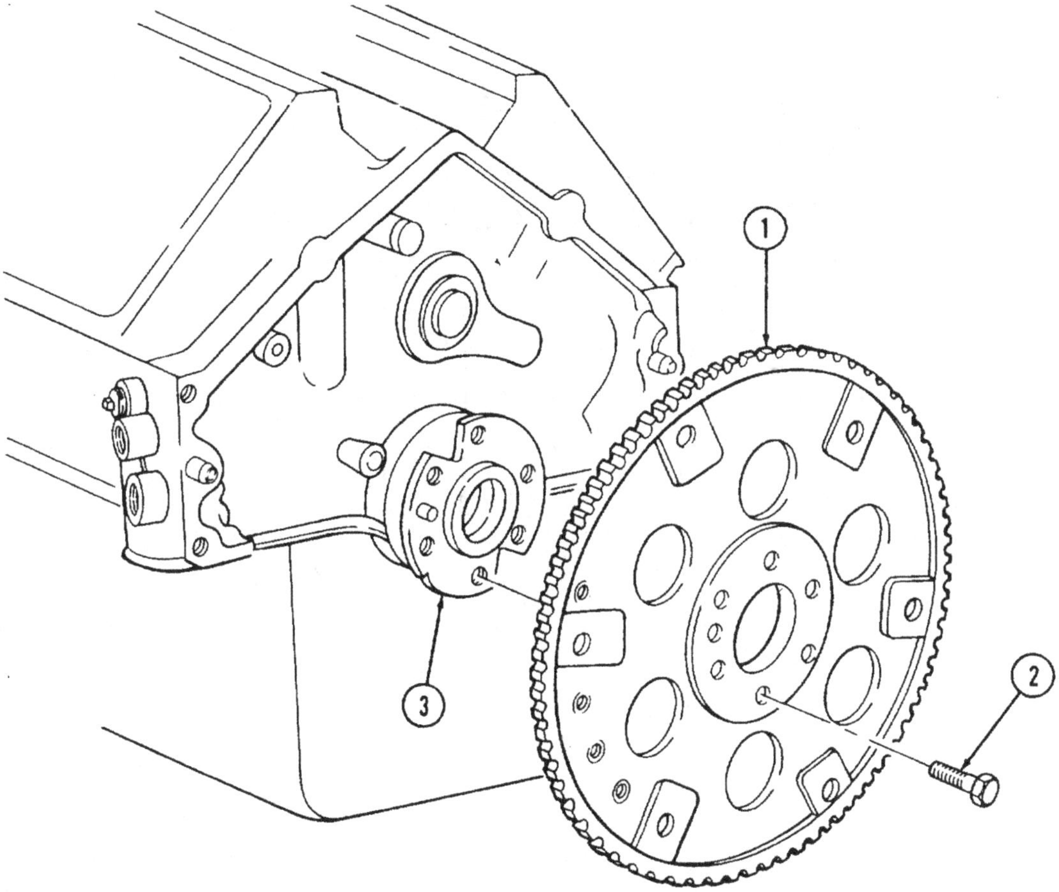

2-15. ENGINE DISASSEMBLY INTO SUBASSEMBLIES (Cont'd)

bb. Crankshaft and Main Bearings

1. Rotate cylinder block (6) so crankshaft (3) faces upward.
2. Check location numbers (5) on main bearing caps (4). If location numbers (5) are not clear, stamp main bearing caps (4) for assembly.

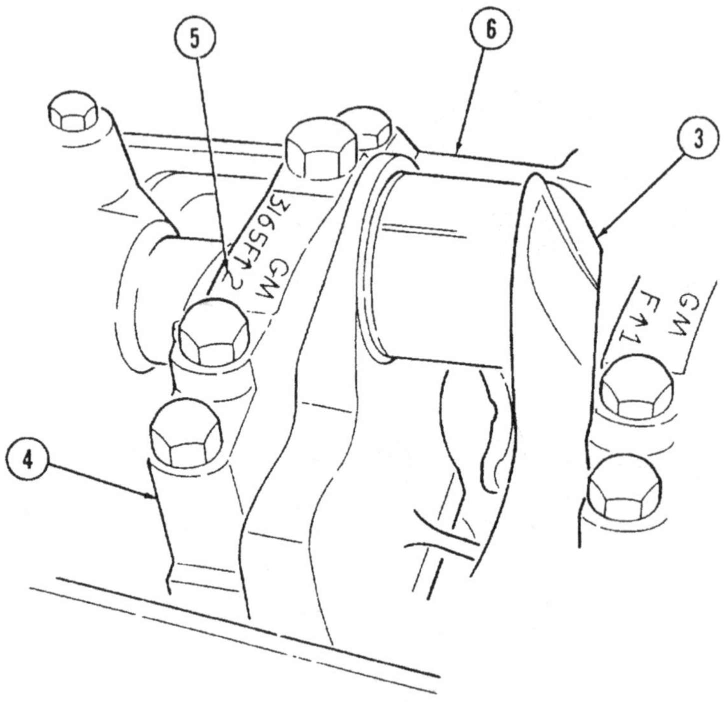

2 - 1 5ENGINE DISASSEMBLY INTO SUBASSEMBLIES (Cont'd)

NOTE

Perform step 3 for 6.5L engines only.

3. Remove one-piece crankshaft rear oil seal (9) from rear main bearing cap (1) and cylinder block (7). Discard one-piece crankshaft rear oil seal (9).
4. Remove four capscrews (3) securing main bearing cap (2) to cylinder block (7) and remove cap (2) and bearing (4). Discard bearing (4).
5. Repeat step (4) for remaining main bearing caps (2), rear main bearing cap (1), and bearings (4).

WARNING

Crankshaft must be supported during removal and installation. Failure to support crankshaft may cause injury to personnel or damage to equipment.

6. Remove crankshaft (5) and bearings (6) from cylinder block (7). Discard bearings (6).

WARNING

Gaskets installed on some 6.2L engines assembled prior to 1991 may contain asbestos. Gaskets should be removed with a scraper or putty knife and then be disposed of IAW current directives. Inhalation of asbestos fibers can cause respiratory ailments.

NOTE

Perform step 7 for 6.2L engines only.

7. Remove two-piece crankshaft rear oil seal (8) from rear main bearing cap (1) and cylinder block (7). Discard two-piece crankshaft rear oil seal (8).
8. Install bearing caps (1) and (2) on cylinder block (7) and secure with twenty capscrews (3).

2 - 1 5ENGINE DISASSEMBLY INTO SUBASSEMBLIES (Cont'd)

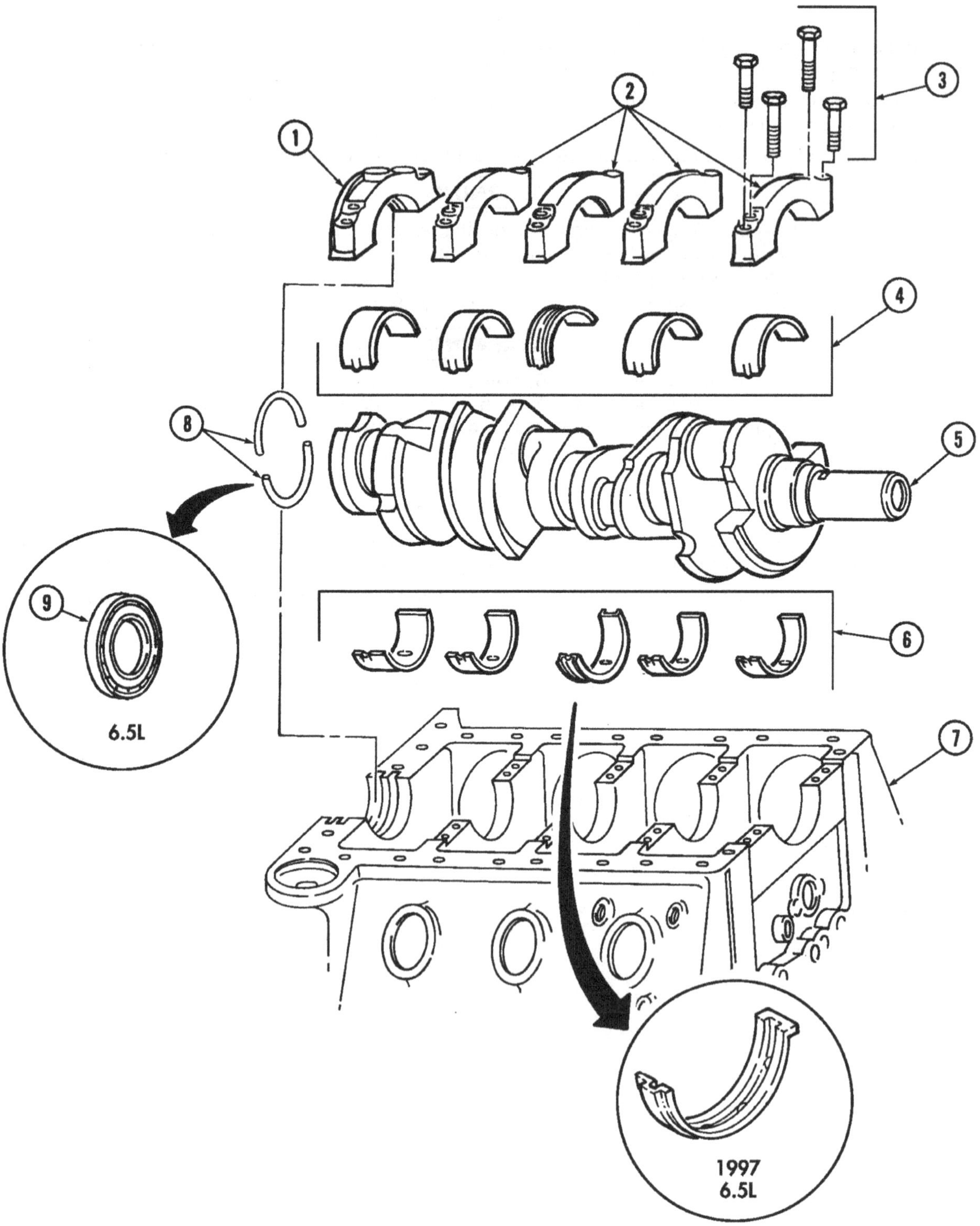

2-16. CYLINDER BLOCK REPAIR

This task covers:

a. **Preinspection**
b. **Cleaning**
c. **Inspection**
d. **Repair**

INITIAL SETUP:

Tools

General mechanic's tool kit:
automotive (Appendix E, Item 1)
Straightedge (Appendix E, Item 15)
Feeler gauge (Appendix E, Item 13)
Cylinder bore gauge (Appendix E, Item 14)
Micrometer (Appendix E, Item 16)
Cylinder boring machine
(Appendix E, Item 17)
Cylinder honing unit (Appendix E, Item 18)

Special Tools

Camshaft bearing remover and installer
(Appendix E, Item 12)

Materials/Parts

Six protective dust caps (Appendix D, Item 5)
Lubricating oil OE/HDO (Appendix B, Item 13)
Pipe sealing compound (Appendix B, Item 17)
Sealing compound, anaerobic (Appendix B, Item 18)
Sealing compound, high temperature
(Appendix B, Item 21)

Manual References

TM 9-2815-237-34P

Equipment Condition

Engine disassembled into subassemblies (para. 2-15).

General Safety Instructions

- Compressed air for cleaning purposes will not exceed 30 psi (207 kPa).
- Protective clothing must be used with steam cleaning.
- Wear eyeshields when using compressed air.

a. Preinspection

1. Inspect camshaft bearings (2) for chips, cracks, or scoring. If any bearing (2) is chipped, cracked, or scored, remove all camshaft bearings (2) (para. 2-16.d.3) and hot tank clean cylinder block (1) (para. 2-16.b.1).
2. Measure inside diameter of all camshaft bearings (2). Number one through number four camshaft bearings must be 2.167-2.170 in. (55.041-55.118 mm) and number five camshaft bearing (rear) must be 2.009-2.012-in. (51.028-51.104 mm). If any camshaft bearings (2) do not meet specifications, replace all camshaft bearings (2) (para. 2-16.d.3) and hot tank clean cylinder block (1) (para. 2-16.b.1).
3. If all bearings (2) are to be reused, steam clean cylinder block (1) (para. 2-16.b.2).

b. Cleaning

1. Hot tank cleaning.
   a. Remove six protective dust caps (3) and all plugs (4) from cylinder block (1).
   b. Clean cylinder block (1) by submerging in tank of heated cleaning solution.
   c. Run rods with brushes through all oil passages and repeat step 1.b.

2-16. CYLINDER BLOCK REPAIR (Cont'd)

WARNING

Compressed air used for cleaning purposes will not exceed 30 psi (207 kPa). Use only with effective chip guarding and personal protective equipment (goggles/shield, gloves, etc.).

d. Blow all dirt and cleaning solution from capscrew holes and oil passages with compressed air.

e. Refer to para. 2-9 for additional cleaning instructions.

f. Coat all cylinder bores, lifter bores, and machined surfaces with OE/HDO.

g. Apply pipe sealing compound to six protective dust caps (3) and all plugs (4) and install in cylinder block (1).

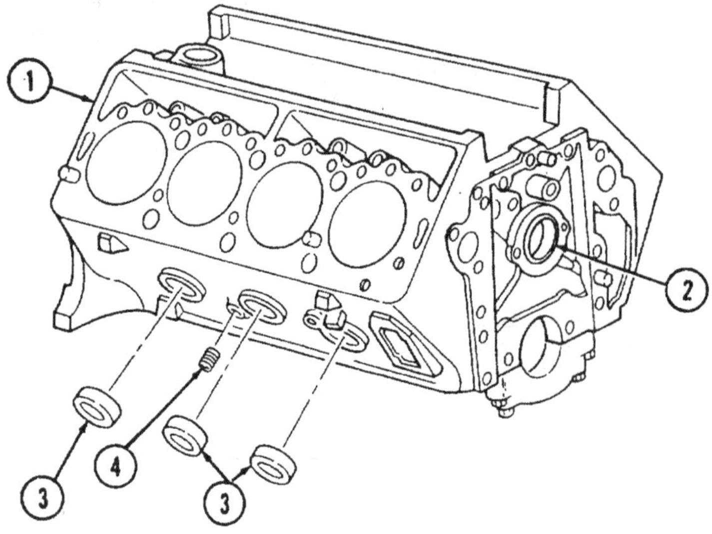

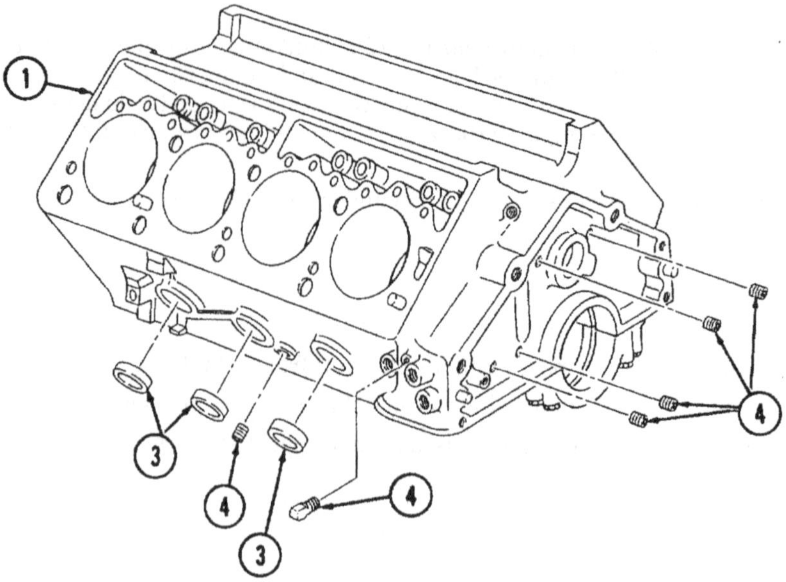

2 - 1 6CYLINDER BLOCK REPAIR (Cont'd)

2. Steam cleaning.
   a. Remove six protective dust caps (9) and all plugs (10) from cylinder block (1).

WARNING

When steam cleaning, protective clothing must be used. Failure to use protective clothing may result in serious injury.

   b. Thoroughly steam clean cylinder block (1).
   c. Run rods with brushes through all oil passages and repeat step b.

WARNING

Compressed air used for cleaning purposes will not exceed 30 psi (207 kPa). Use only with effective chip guarding and personnel protective equipment (goggles/shield, gloves, etc.).

   d. Blow all dirt and cleaning solution from capscrew holes and oil passages with compressed air.
   e. Refer to para. 2-9 for additional cleaning instructions.
   f. Coat all cylinder bores, lifter bores, and machined surfaces with OE/HDO.
   g. Apply pipe sealing compound to six protective dust caps (9) and all plugs (10) and install in cylinder block (1).

c. Inspection

NOTE

- For general inspection instructions, refer to para. 2-10.
- If cylinder block inspection indicates that cylinder block is suitable for continued use, except for out-of-round, taper, or minor nicks or scratches, it can be conditioned by honing or boring. The cylinder block can then be fitted with a Hi-Limit or oversized piston (refer to para. 2-45, table 2-4 or 2-5). The 6.5L engines do not have Hi-Limit pistons.

1. Inspect all surfaces and general overall condition of the cylinder block (1).
2. Visually check the cylinder block (1) for pitting in cylinder walls (4) and lifter bores (2). Replace engine if pitted or distorted.
3. Inspect the cylinder block (1) for cracks in cylinder walls (4), water jacket (5), lifter bores (2), and main bearing webs (8). Replace engine if cracked.
4. Inspect main bearing caps (6) for cracks, scoring, or damage. Replace engine if any are cracked, scored, or damaged.
5. Inspect main bearing cap bolts (7) for damage. Replace if damaged.
6. Inspect dowel pins (3) for damage. Replace if damaged.
7. Using a straightedge and feeler gauge, check flatness of cylinder head sealing surfaces on cylinder block (1). Flatness on cylinder block (1) must not vary more than 0.002 in. (0.051 mm) in a distance of 6 in. (15.2 mm) or more than 0.006 in. (0.152 mm) overall. Repair cylinder or replace engine if flatness does not meet above specifications.

NOTE

Perform step 7.1 for 1997 6.5L engines only.

7.1. Inspect piston oil nozzles (11) and bores (12) for damage. Replace piston oil nozzles (11) if damaged.

2 - 1 6CYLINDER BLOCK REPAIR (Cont'd)

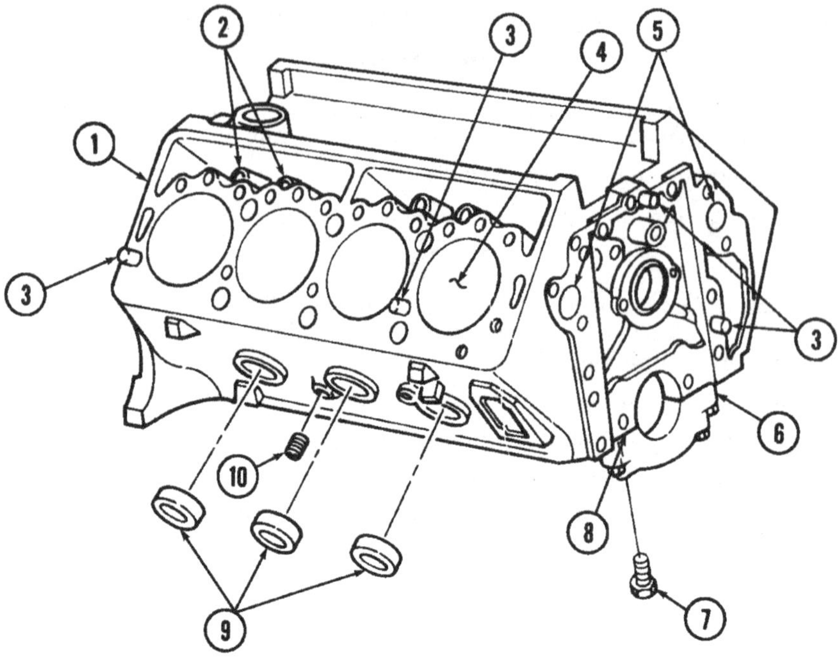

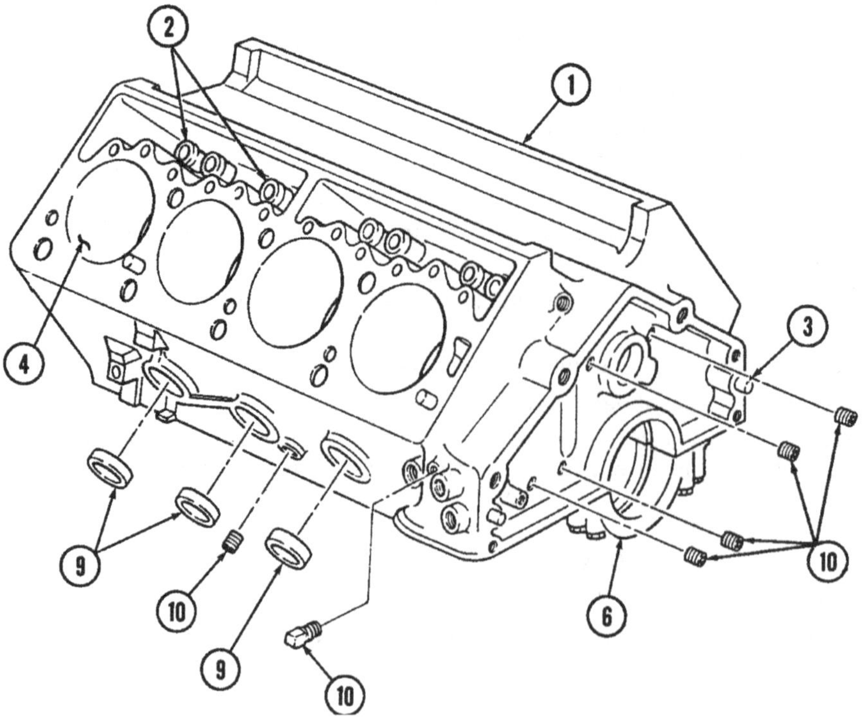

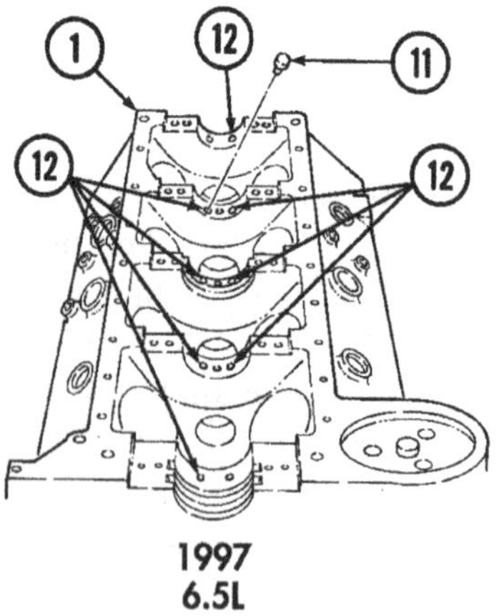

1997
6.5L

2 - 1 6CYLINDER BLOCK REPAIR (Cont'd)

NOTE

If cylinder bores are found to have less than 0.005 in. (0.127 mm) wear, taper, or out-of-round, they may be honed oversize (refer to para. 2-16.d.1) and fitted with "Hi-Limit" standard size pistons.

8. Using cylinder bore gauge, measure all cylinder bores (2) for taper. Measurements must be taken at point "A" (4) at top and bottom of cylinder bore (2). If any cylinder exceeds specifications (table 2-2 or 2-3) by 0.005 in. (0.127 mm) or more, repair cylinder or replace engine.
9. Using cylinder bore gauge, measure all cylinder bores (2) for out-of-round. Measurement must be taken at point "A" (4) and point "B" (3) at top and bottom of cylinder bore (2). If any cylinder exceeds specifications (table 2-2 or 2-3) by 0.005 in. (0.127 mm) or more, repair cylinder or replace engine.
10. Using cylinder bore gauge, measure all cylinder bores (2) for bore diameter. Measurement must be taken at point "A" (4) 2-1/2 in. (64 mm) from top of cylinder bore (2). If any cylinder exceeds specifications (table 2-2 or 2-3) by 0.005 in. (0.127 mm) or more, repair cylinder or replace engine.

d. Repair

NOTE

If cylinder block inspection indicated that the cylinder block was suitable for continued use except for out-of-round, taper, or minor nicks or scratches, it can be conditioned by honing or boring.

1. Boring.
   a. Before using any type of boring bar, the top of the cylinder block should be filed off to remove any dirt or burrs. This is very important. If not checked, the boring bar may be tilted, which would result in the rebored cylinder wall not being at right angles to the crankshaft.
   b. Measure piston diameter of the piston to be used at the skirt across the center line of the piston pin.
   c. Using a boring bar, bore the cylinder to the same diameter as the piston. Hone the cylinder to get proper piston-to-bore clearance.
   d. Repeat steps b. and c. for remaining cylinders.
2. Honing.
   a. Using cylinder bore gauge, measure cylinder bore (2) for cylinder bore diameter. Measurement must be taken at point "A" (4) 2-1/2 in. (64 mm) from top of cylinder bore (2).
   b. Measure "piston diameter" of piston (5) removed from corresponding cylinder using a micrometer at piston skirt (7) across center line of piston pin (6).
   c. Subtract "piston diameter" from "cylinder bore diameter" to determine "piston-to-bore clearance".
   d. Repeat steps a. through c. for remaining cylinder bores and pistons.

NOTE

- Cylinders can be conditioned with a hone or boring bar and fitted with a standard "Hi-Limit" piston or oversized piston, if necessary.
- If a cylinder is pitted or worn such that it cannot be honed and fitted with a "Hi-Limit" piston or oversized piston, a cylinder sleeve may be used to repair the engine block.
- 6.5L engines do not have "Hi-Limit" pistons.

   e. If all cylinder bores (2) have proper "piston-to-bore clearance" of 0.004-0.005 in. (0.102-0.127 mm) or if proper "piston-to-bore clearance" may be obtained using a standard "Hi-Limit" piston or oversized piston, go to step f. If not, engine may be repaired using a cylinder sleeve (para. 2-16.d.2).

2 - 1 6CYLINDER BLOCK REPAIR (Cont'd)

NOTE

Occasionally during the honing operation, cylinder bore should be thoroughly cleaned, and the piston-to-bore clearance checked for correct fit.

f. Using cylinder hone, refinish cylinder bore (2). Hone should be moved up and down at sufficient speed to obtain very fine uniform surface finish marks, in a cross-hatch pattern of approximately 45° to 65° included angle. The finish marks should be clean, not sharp, and free from imbedded particles and torn or folded material.

g. If a "Hi-Limit" piston or oversized piston (5) was fitted to a cylinder bore (2), permanently mark the piston (5) for the cylinder bore (2) to which it has been fitted.

h. Repeat steps f. and g. for remaining cylinder bores.

i. Thoroughly clean the cylinder block (1) with hot water and detergent. Scrub cylinder bores (2) with a stiff brush and rinse thoroughly with hot water. The cylinder bores (2) should be coated with OE/HDO and wiped with a clean, dry cloth.

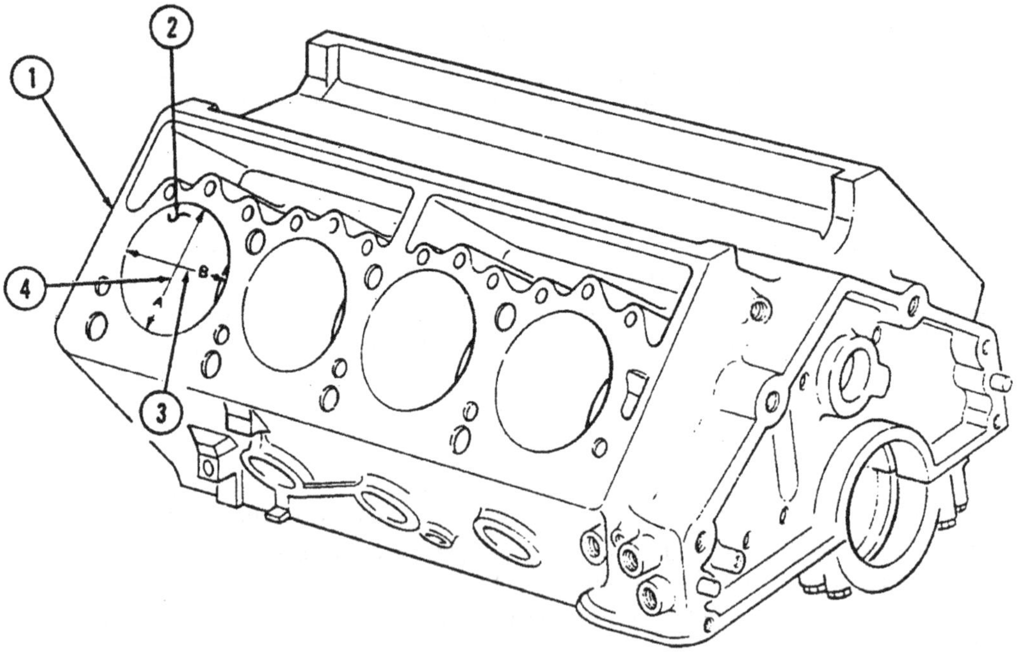

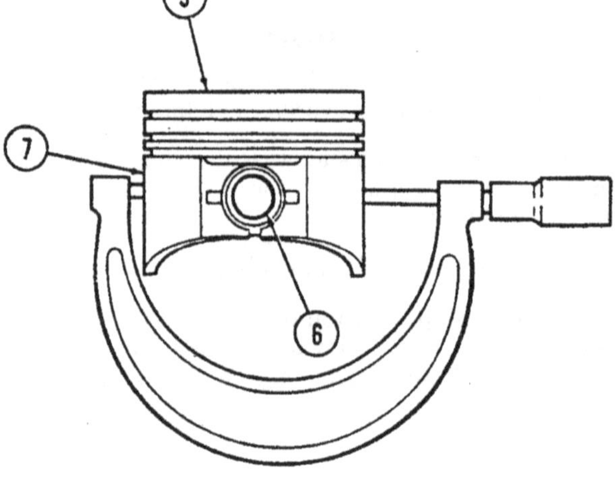

2 - 1 6CYLINDER BLOCK REPAIR (Cont'd)

2. Installation of cylinder sleeves.

CAUTION

Do not use cylinder sleeves to repair a cracked engine block.

NOTE

Cylinders that still show damage or wear after being honed to a maximum oversize of 0.030 in. (0.762 mm) will require sleeves.

a. Identify cylinders that require sleeves.

b. Using micrometer, measure outside diameter of cylinder sleeve (1).

NOTE

- Before using any type of boring bar, the top of the cylinder block should be filed off to remove any dirt or burrs. This is very important. If not checked, the boring bar may be tilted, which would result in the rebored cylinder wall not being at right angle to crankshaft.
- The cylinder must be bored to within 0.125 in. (3.18 mm) of the rod relief in the bottom of the cylinder. Ensure the depth of the bore does not exceed the length of the sleeve. Ensure to cut the bottom of the bore square. This will provide a stop, or ledge, for the sleeve to bottom out on.

c. Using cylinder boring machine, bore cylinder (2) to diameter of sleeve (1) less 0.002 to 0.003 in. (0.051 to 0.076 mm) to create an interference fit between sleeve (1) and cylinder (2).

NOTE

Store the sleeve in a freezer at a temperature of 28 to 30°F (-2.2 to -1.1°C) for several hours before installing it into the engine block.

d. Apply bead of high temperature sealing compound to top and bottom of cylinder (2) bore. Press frozen sleeve (1) into cylinder (2) until it bottoms out on ledge at bottom of cylinder (2) bore.

e. Remove any excess sleeve material protruding above the head deck surface (4). Finish the sleeve (1) flush with deck of block (3), taking care not to damage deck surface (4).

f. Using cylinder boring machine, bore the sleeved cylinder (1) to appropriate piston size.

g. Finish cylinder surface (refer to para. 2-16.d.1).

2 - 1 6CYLINDER BLOCK REPAIR (Cont'd)

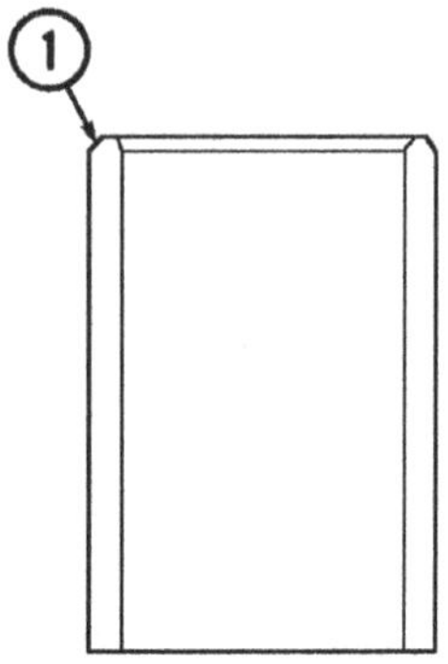

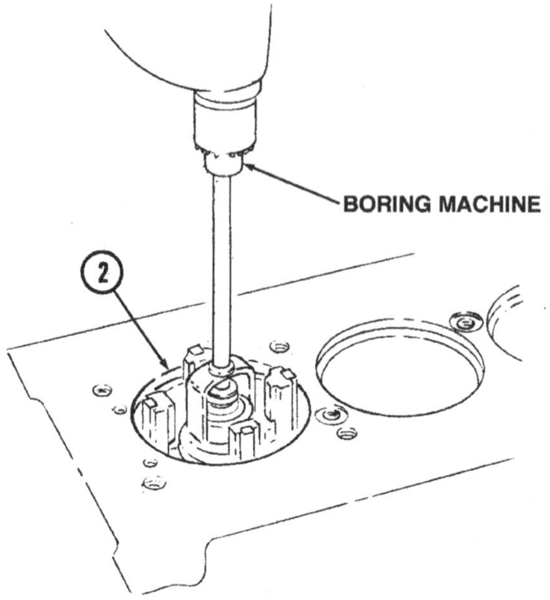

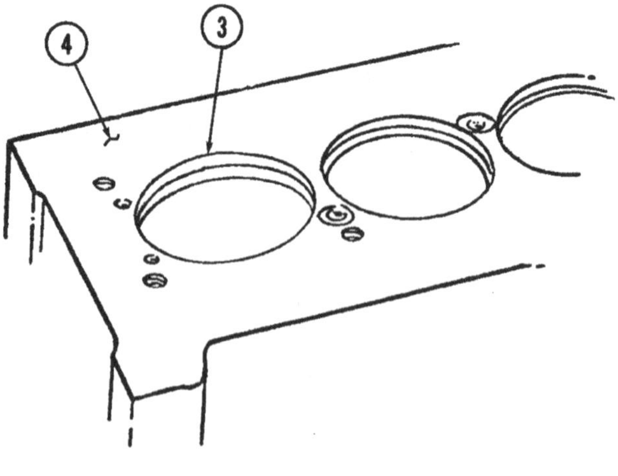

2-16. CYLINDER BLOCK REPAIR (Cont'd)

3. Camshaft bearing replacement.

WARNING

Cylinder block must be supported during removal and installation from engine stand. Failure to support cylinder block may cause injury to personnel or damage to equipment.

NOTE

The following steps are to be performed only if camshaft bearings are to be replaced.

a. Remove cylinder block (2) from engine stand.

b. Remove twenty capscrews (3) and five main bearing caps (4) from cylinder block (2).

c. Drive camshaft rear plug (1) from cylinder block (2). Discard plug (1).

d. Install adapter J 6098-11 in camshaft bearing number two (6).

e. Install pilot in number one camshaft bearing (5). Insert threaded rod through pilot and thread into adapter.

f. Hold threaded rod stationary and tighten long hex nut to remove camshaft bearing (6) from cylinder block (2). When camshaft bearing (6) has been completely removed from cylinder block (2), remove threaded rod, pilot, and adapter. Discard bearing (6).

g. Repeat steps d. through f. for number three camshaft bearing (7).

h. Repeat steps d. through f., working from rear of cylinder block (2), for number four camshaft bearing (8).

i. Using driver handle and adapter J6098-11, remove number one camshaft bearing (5) from cylinder block (2). Discard bearing (5).

j. Using driver handle and adapter J 6098-12, remove number five camshaft bearing (9) from cylinder block (2). Discard bearing (9).

2-16. CYLINDER BLOCK REPAIR (Cont'd)

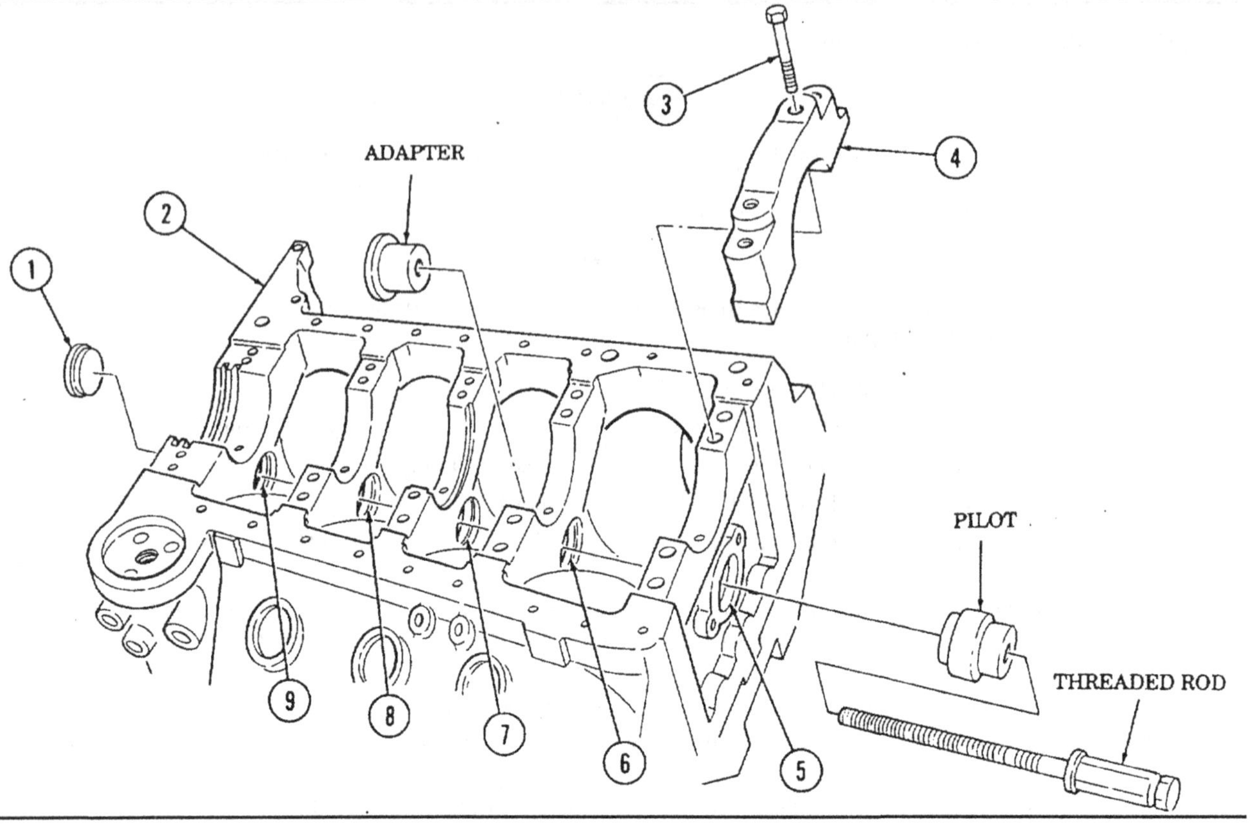

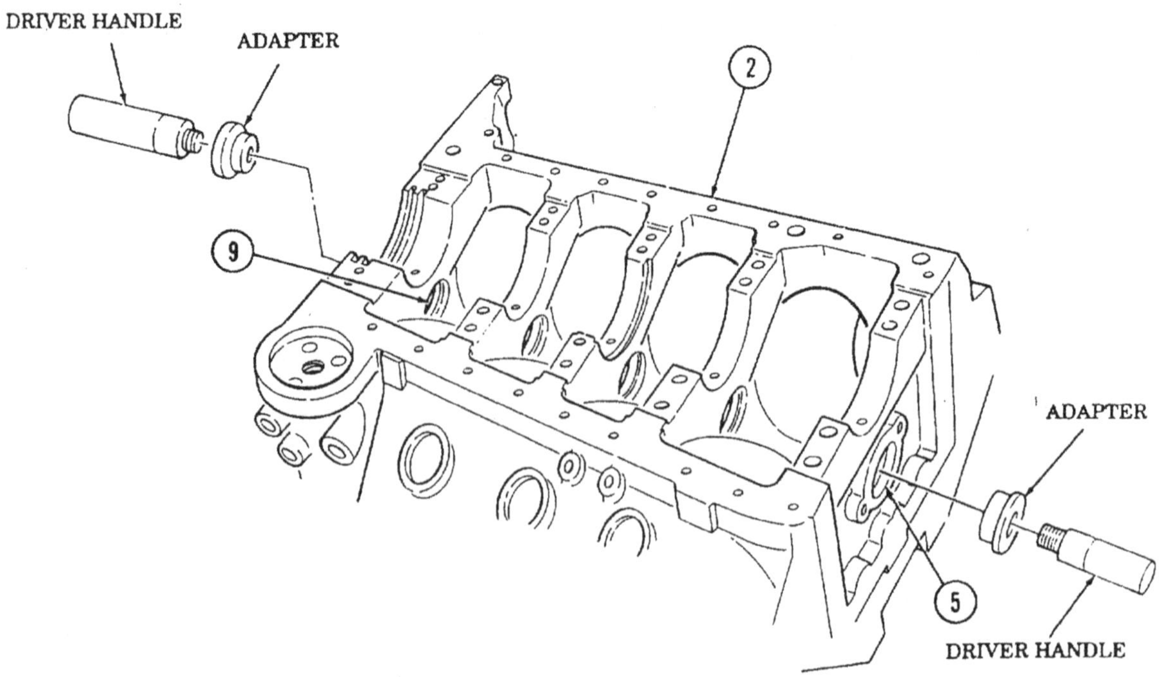

2-16. CYLINDER BLOCK REPAIR (Cont'd)

CAUTION

- All five bearings must have an oil hole at the approximate ten o'clock position, when viewed from the front with the block in an upside down position, or engine damage will result.
- The seam in the bearing must always be located in the lower half of the block face, when viewed with the block in an upside down position, or engine damage will result.
- The front bearing has an additional oil hole which will be located between the six and seven o'clock positions. This bearing also has a notch which must be positioned toward the front of the block, or engine damage will result.
- All oil holes in camshaft bearings must be aligned with grooves in camshaft bearing bores after installation, or engine damage will result.

k. Using driver handle and adapter J 6098-11, align number one camshaft bearing (4) so oil hole (5) is facing down and oil hole (6) is at or near the ten o'clock position and install into cylinder block (1). Remove installation tools and check to be sure oil holes (5) and (6) align with groove (3) in bearing bore (2).

l. Using driver handle and adapter J 6098-12, align number five camshaft bearing (9) so oil hole (10) is at or near the ten o'clock position and install into cylinder block (1). Remove installation tools and check to be sure oil hole (10) is aligned with groove (7) in bearing bore (8).

m. Install number two camshaft bearing (11) on adapter J 6098-11 and position in cylinder block (1) so oil hole (12) is at or near the ten o'clock position. Install pilot in number one camshaft bearing (4) and install threaded rod through pilot and thread into adapter.

n. Hold threaded rod stationary and tighten long hex nut to install camshaft bearing (11) in cylinder block (1), remove installation tools, and check to be sure oil hole (12) is aligned with groove (16) in bearing bore (13).

o. Repeat steps m. and n. for number three camshaft bearing (17).

p. Working from rear of cylinder block (1), repeat steps m. and n. for number four camshaft bearing (18).

q. Apply anaerobic sealing compound to outside diameter of rear plug (19) and install in cylinder block (1) until flush or 1/32-in. (0.79 mm) deep.

r. Install five main bearing caps (15) on cylinder block (1) and secure with twenty capscrews (14).

s. Install cylinder block (1) on engine stand.

2-16. CYLINDER BLOCK REPAIR (Cont'd)

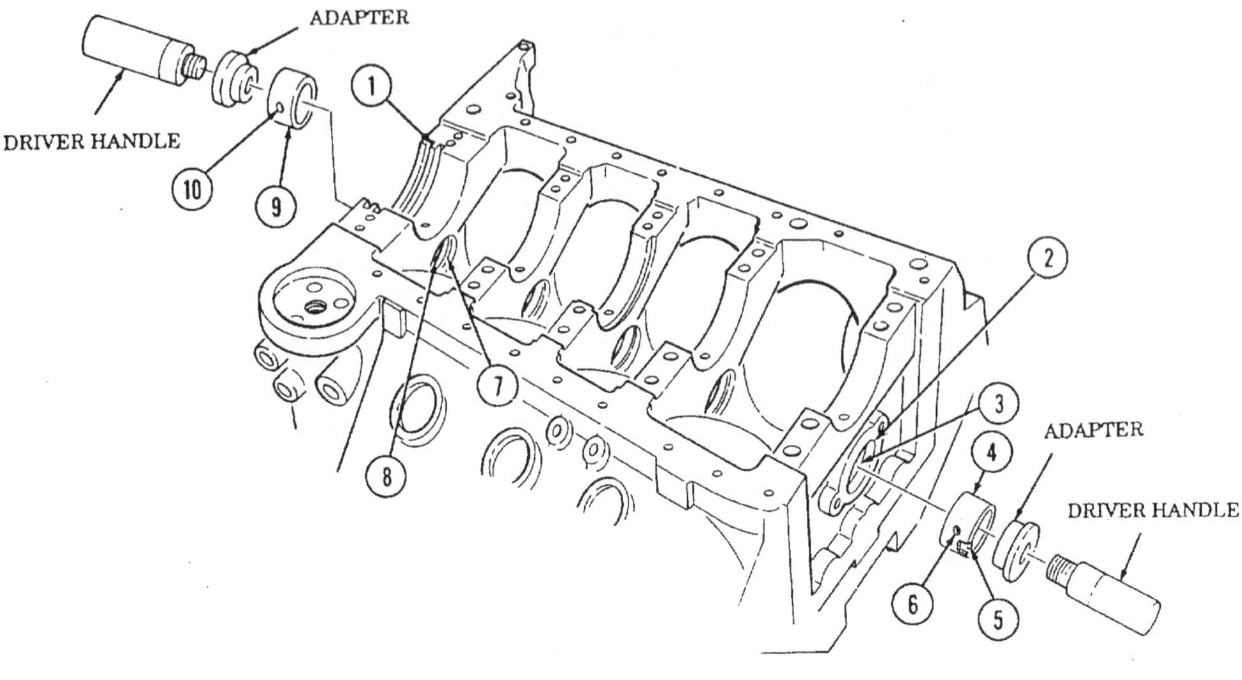

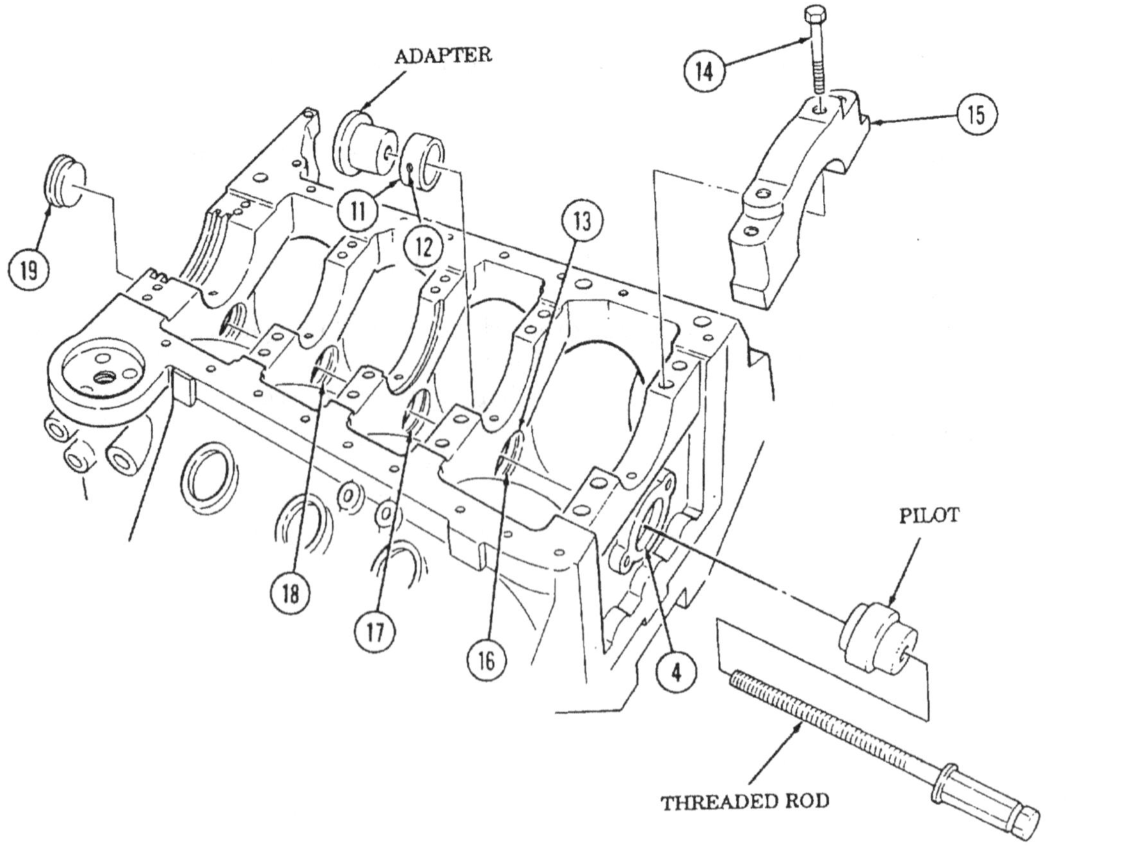

2-17. CRANKSHAFT REPAIR

This task covers:

a. Cleaning
b. Inspection
c. Repair

INITIAL SETUP

Tools
General mechanic's tool kit:
automotive (Appendix E, Item 1)
Micrometer (Appendix E, Item 16)

Materials/Parts
Abrasive crocus cloth (Appendix B, Item 5)

Manual References
TM 9-2815-237-34P

Equipment Condition
Engine disassembled into subassemblies (para. 2-15).

a. Cleaning

1. Clean all drilled oil passages (5) with a rod and cloth.
2. Refer to para. 2-9 for additional cleaning information.

b. Inspection

NOTE

For general inspection instructions, refer to para. 2-10.

1. Inspect crankshaft (1) for scratches, nicks, and cracks. Replace if scratched, nicked, or cracked.
2. Check main bearing journals (4) and connecting rod journals (6) for taper, out-of-round, and diameter using an outside micrometer. Replace crankshaft (1) if any bearing journal is tapered, out-of-round, or out of specifications (table 2-2 or 2-3).
3. Inspect rear seal area (3) for roughness, scoring, grooving, flaking, pitting, and galling. Replace crankshaft (1) if rough, scored, grooved, flaked, pitted, or galled.
4. Inspect threaded holes (2) and (7) in crankshaft (1) for damage. Replace crankshaft (1) if threads are damaged.

c. Repair

Buff all minor nicks and scratches using abrasive crocus cloth.

2-17. CRANKSHAFT REPAIR (Cont'd)

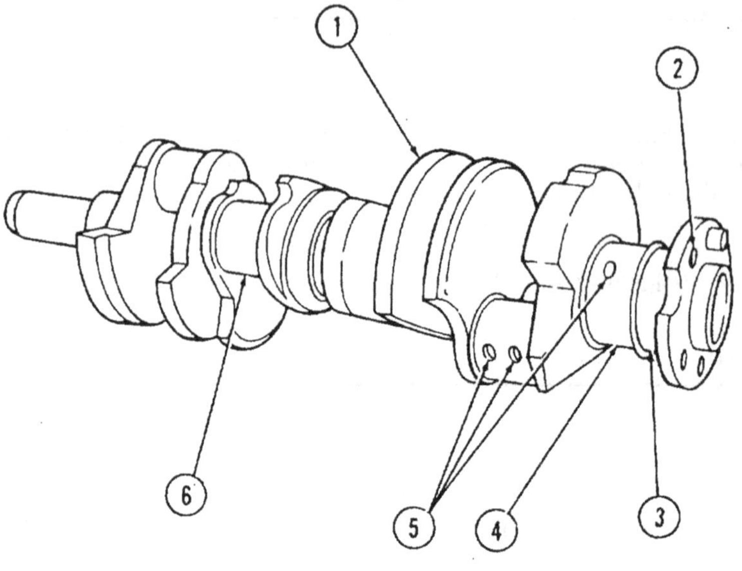

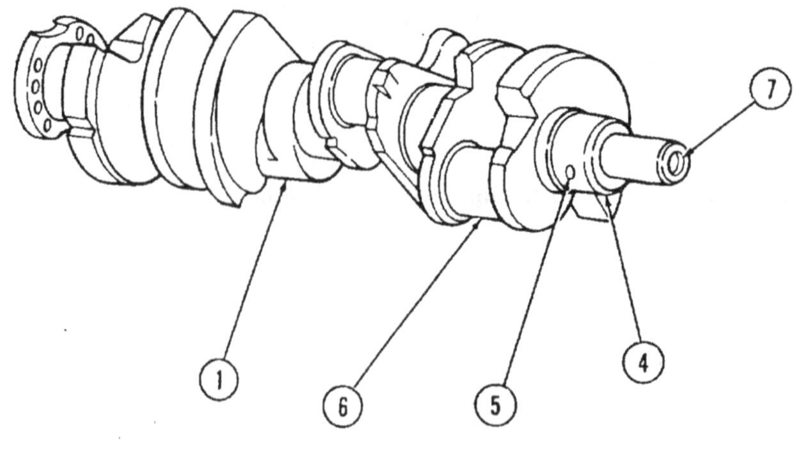

2 - 1 8CONNECTING ROD AND PISTON R&H R

This task covers:

a. **Disassembly**
b. **Cleaning**
c. **Inspection**
d. **Assembly**

INITIAL SETUP:

Tools
General mechanic's tool kit:
automotive (Appendix E, Item 1)
Feeler gauge (Appendix E, Item 13)
Micrometer (Appendix E, Item 16)
Piston ring expander (Appendix E, Item 20)

Materials/Parts
Sixteen snaprings (Appendix D, Item 49)
Eight piston ring sets
(Appendix D, Item 37 or 48)
Lubricating oil OE/HDO (Appendix B, Item 13)

Manual References
TM 9-2815-237-34P

Equipment Condition
Engine disassembled into subassemblies (para. 2-15).

a. Disassembly

1. Remove top compression ring (1), second compression ring (2), and oil ring (3) from piston (4). Discard rings (1), (2), and (3).
2. Remove two snaprings (5), piston pin (15), and connecting rod (13) from piston (4). Discard snaprings (5).
3. Remove two nuts (11) and rod cap (12) from connecting rod (13).
4. Repeat steps 1 through 3 for remaining pistons and connecting rods.

b. Cleaning

Clean all components in accordance with para. 2-9.

c. Inspection

NOTE

- For general inspection instructions, refer to para. 2-10.
- To identify correct piston size for standard, Hi-Limit, and oversize pistons, check code stamped on face of piston, and determine correct piston size. Refer to para. 2-45, table 2-4, for piston-cylinder specifications.
- 6.5L engines do not have Hi-Limit pistons.

1. Inspect piston (4) for cracks at ring lands (6), skirt (7), and pin bosses (16). Replace piston (4) if cracked.
2. Inspect piston (4) for scoring, galling, and scuffing. Replace if scored, galled, or scuffed.
3. Measure piston (4) at center (8) and bottom (9) of piston skirt (7) using micrometer. If measurement at bottom (9) is smaller than measurement at center (8), piston skirt (7) is collapsed and piston (4) must be replaced.
4. Inspect connecting rod (13) and rod cap (12) for distortion or damage. Replace connecting rod (13) if either are damaged or distorted.

2 - 1 8CONNECTING ROD AND PISTON REPAIR (Cont'd)

5. Inspect connecting rod bolts (10) and nuts (11) for damage. Replace if damaged.
6. Inspect bronze bushing (14) in small end of connecting rod (13) for damage. Replace connecting rod (13) if damaged.
7. Inspect piston pin (15) for galling, scoring, and damage. Replace piston (4) if scored, galled, or damaged.
8. Measure pin boss (16) inside diameter and piston pin (15) outside diameter to determine if clearance is within specifications (table 2-2 or 2-3). If measurement exceeds specifications, replace piston (4).
9. Measure piston pin (15) outside diameter and bushing (14) inside diameter to determine if fit to rod is within specifications (table 2-2 or 2-3). If measurement exceeds specifications, replace connecting rod (13).
10. Check ring groove clearance. If measurement exceeds specifications (table 2-2 or 2-3), replace piston (4).
11. Repeat steps 1 through 10 for remaining pistons and rods.

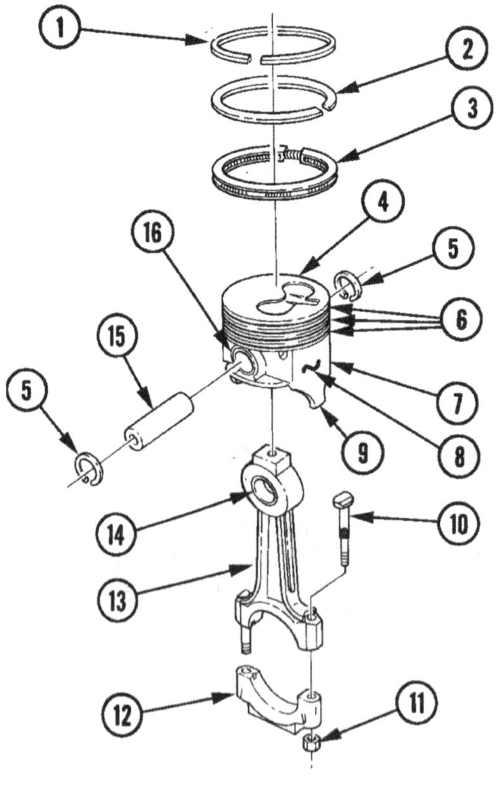

2 - 1 8CONNECTING ROD AND PISTON REPAIR (Cont'd)

d. Assembly

NOTE

- For general assembly instructions, refer to para. 2-12.
- Top compression ring is thicker than second compression ring.
- If proper end gap cannot be obtained, refer to para. 2-45, table 2-4 or 2-5 for piston-to-cylinder applications. Refer to para. 2-16.d. for cylinder block repair.

1. Place top compression ring (1) in cylinder bore (13) in which it will be used.
2. Slide piston ring (1) 1 in. (25.4 mm) into cylinder bore (13). Be sure piston ring (1) is square with cylinder bore (13).
3. Check gap in piston ring (1) using feeler gauge.
4. Top compression ring (1) end gap must be 0.012-0.022 in. (0.30-0.55 mm). If not, replace ring (1) and check gap.
5. Repeat steps 1 through 3 using second compression ring (2). End gap must be 0.030-0.039 in. (0.75-1.00 mm). If not, replace ring (2) and check gap.

NOTE

Spring must be removed from oil ring when checking end gap.

6. Repeat steps 1 through 3 using oil ring (4). End gap must be 0.010-0.020 in. (0.25-0.51 mm). If not, replace ring (4) and check gap. Reinstall spring (3).
7. Repeat steps 1 through 6 for remaining piston rings.
8. Install connecting rod (7) in piston (5) so that identification number (8) and depression (12) on top of piston (5) are on the same side.
9. Coat piston pin (11) with OE/HDO and install in piston (5) and connecting rod (7) and secure with two snaprings (6).
10. Using piston ring expander, install oil ring (4) on piston (5).
11. Using piston ring expander, install second compression ring (2) on piston (5). Be sure marked side of piston ring (2) faces up.
12. Using piston ring expander, install top compression ring (1) on piston (5). Be sure marked side of piston ring (1) faces up.
13. Install rod cap (10) on connecting rod (7) and secure with two nuts (9).
14. Repeat steps 8 through 13 for remaining pistons and connecting rods.

2-18. CONNECTING ROD AND PISTON REPAIR (Cont'd)

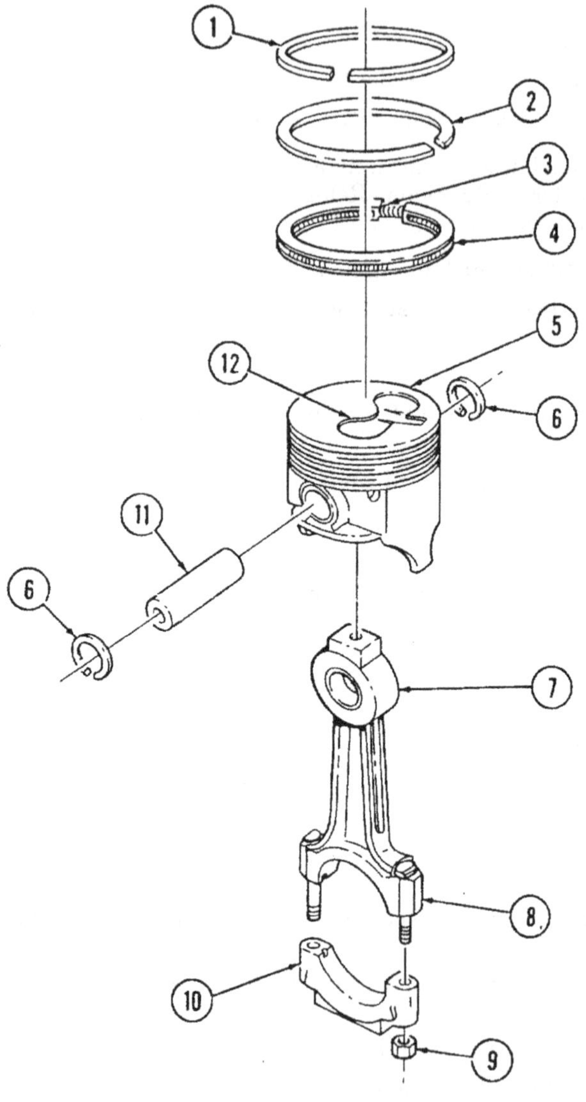

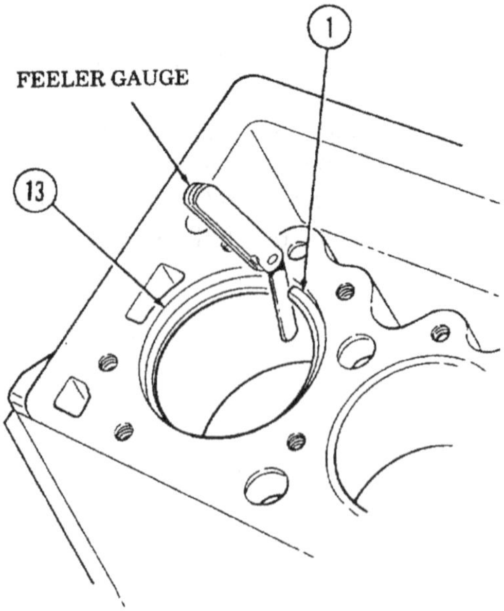

2-19. CAMSHAFT, TIMING CHAIN, AND DRIVE GEARS REPAIR

This task covers:

a. Cleaning **b. Inspection**

INITIAL SETUP

Tools
General mechanic's tool kit:
automotive (Appendix E, Item 1)
Micrometer (Appendix E, Item 16)
"V" blocks (Appendix E, Item 21)
Dial indicator (Appendix E, Item 10)

Manual References
TM 9-2815-237-34P

Equipment Condition
Engine disassembled into subassemblies (para. 2-15).

a. Cleaning

Clean all components in accordance with para. 2-9.

b. Inspection

NOTE

- For general inspection instructions, refer to para. 2-10.
- All lifters must be replaced when replacing camshaft.

1. Inspect camshaft bearing journals (7) and lobes (8) for damage. Replace camshaft (6) if damaged.
2. Measure diameter of bearing journals (7) using outside micrometer. Replace camshaft (6) if any bearing journal (7) is out of specification (table 2-2 or 2-3).
3. Support camshaft (6) on "V" blocks.
4. Using dial indicator and "V" blocks, measure cam lobe (8) lift. Replace camshaft (6) if any cam lobe (8) is out of specification (table 2-2 or 2-3).
5. Inspect crankshaft sprocket (9), camshaft sprocket (3), pump drive gear (10), and pump driven gear (1) for damage. Replace any gear or sprocket if damaged.
6. Inspect timing chain (2) for cracks, breaks, or binding. Replace if cracked, broken, or binding.
7. Inspect camshaft thrust plate (4) and spacer (5) for damage. Replace if damaged.

2-19. CAMSHAFT, TIMING CHAIN, AND DRIVE GEARS REPAIR (Cont'd)

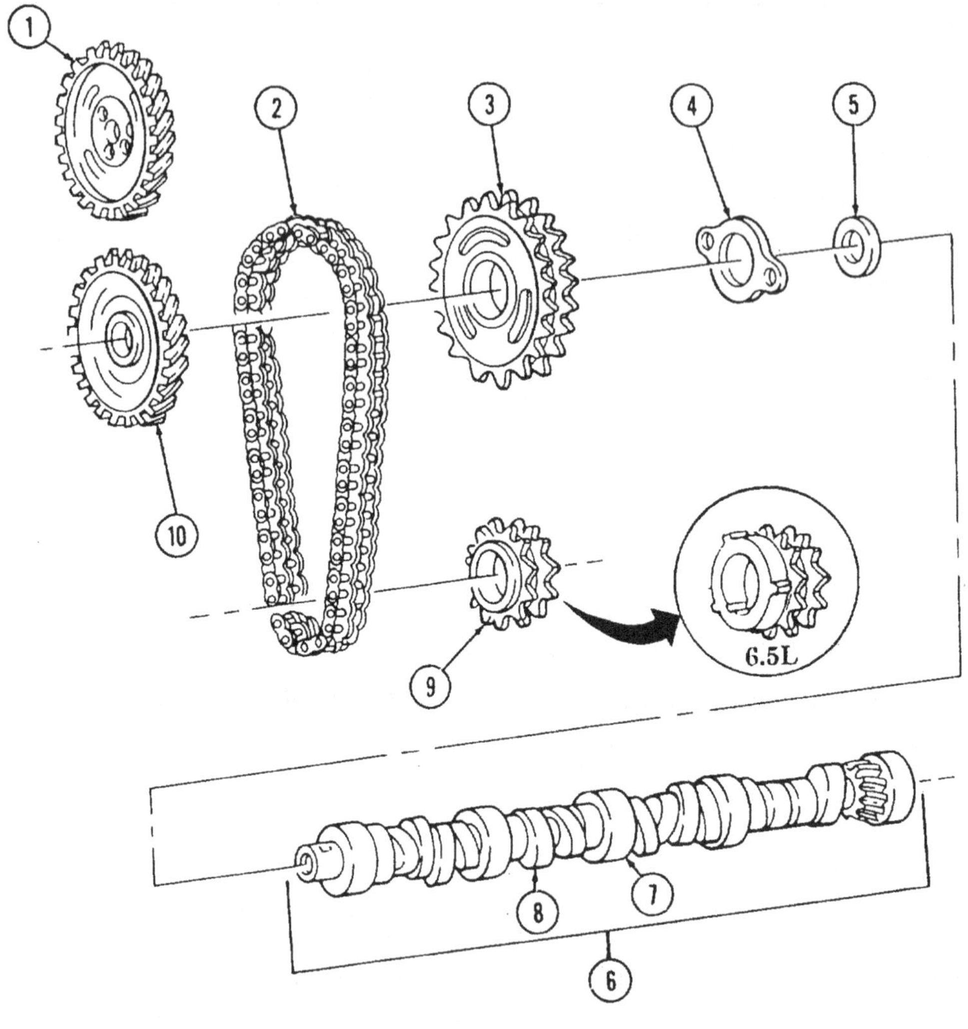

2-20. TIMING GEAR COVER REPAIR

This task covers:

a. Cleaning **b. Inspection**

INITIAL SETUP

Tools
General mechanic's tool kit:
automotive (Appendix E, Item 1)

Manual References
TM 9-2815-237-34P

Equipment Condition
Engine disassembled into subassemblies (para. 2-15).

a. Cleaning

Clean all components in accordance with para. 2-9.

b. Inspection

NOTE

For general inspection instructions, refer to para. 2-10.

1. Inspect sealing surfaces (7) and (6) on timing gear cover (5) and sealing surfaces (3) and (4) on water pump adapter plate (1) for deep scratches or dents. Replace either if deeply scratched or dented.
2. Inspect timing gear cover (5) and water pump adapter plate (1) for damage. Replace either if damaged.
3. Inspect oil filler tube grommet (2) for damage. Replace if damaged.
4. Inspect oil filler tube (9) for damage. Replace if damaged.
5. Inspect oil filler tube cap (8) for damage. Replace if damaged.

2-20. TIMING GEAR COVER REPAIR (Cont'd)

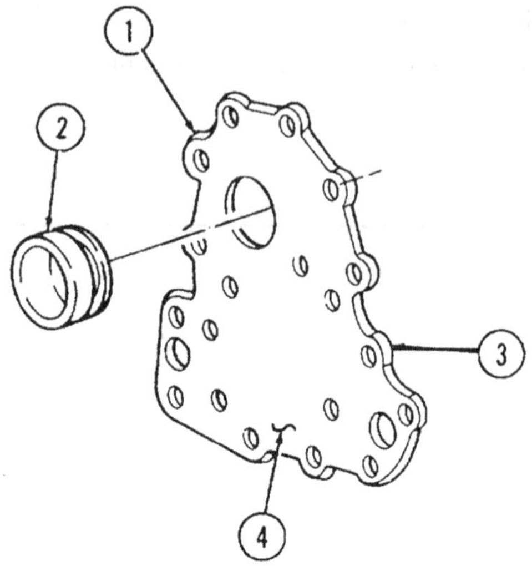

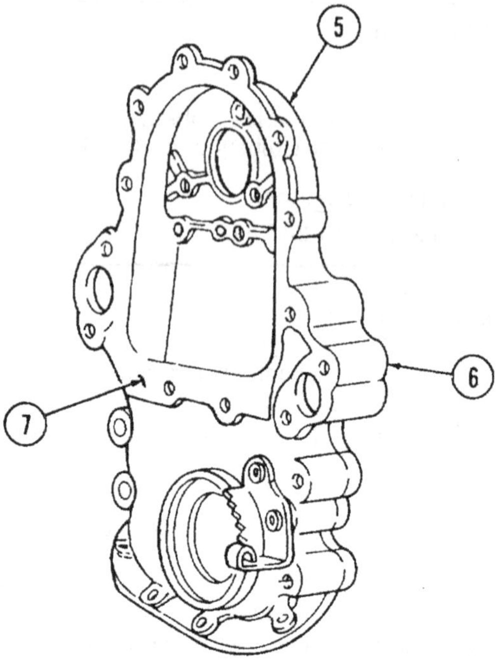

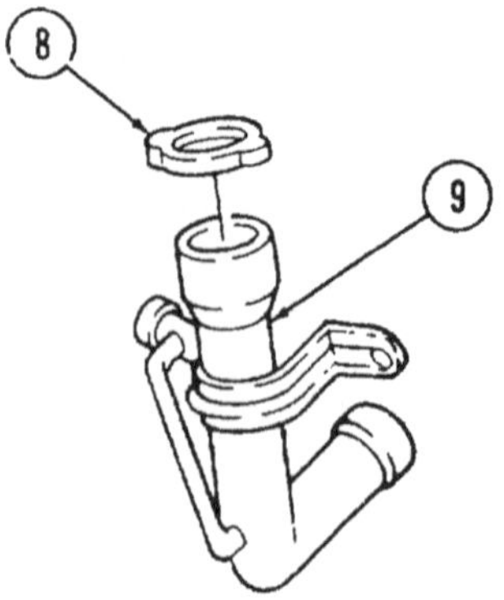

2-21. CYLINDER HEAD AND VALVE REPAIR

This task covers:

a. **Disassembly**
b. **Cleaning**
c. **Inspection**
d. **Refacing Valves**
e. **Grinding Valve Seats**
f. **Assembly**

INITIAL SETUP

Tools

General mechanic's tool kit:
automotive (Appendix E, Item 1)
Valve spring compressor (Appendix E, Item 22)
Bench grinder (Appendix E, Item 23)
Rotary wheel wire brush (Appendix E, Item 24)
Electric drill (Appendix E, Item 25)
Rotary end wire brush (Appendix E, Item 26)
Valve spring tester (Appendix E, Item 27)
Straightedge (Appendix E, Item 15)
Feeler gauge (Appendix E, Item 13)
Dial indicator (Appendix E, Item 10)
Valve face grinding machine (Appendix E, Item 28)
Valve seat grinding kit (Appendix E, Item 29)

Materials/Parts

Sixteen valve seals (Appendix D, Item 47)
Water jacket cover gasket (Appendix D, Item 20)
Lubricating oil OE/HDO (Appendix B, Item 13)
Pipe sealing compound (Appendix B, Item 17)

Manual References

TM 9-2815-237-34P

Equipment Condition

Engine disassembled into subassemblies (para. 2-15).

NOTE

The cylinder head and valve repair procedures for 6.5L engines, 1990 and above 6.2L engines and previous model engines are basically the same. Any differences in procedures are noted. Refer to engine decal model number on left rocker arm cover before ordering replacement parts.

a. Disassembly

1. Using valve spring compressor, compress intake valve spring (3) and remove valve keys (7). Release valve spring compressor slowly.
2. Remove valve cap (6), shield (4), valve spring (3), valve spring shim (2), and valve seal (5) from intake valve (16). Discard valve seal (5).

NOTE

Valves must be installed in the same location they were removed from. As each valve is removed, place on numbered board or stand.

3. Remove intake valve (16) from cylinder head (1).
4. Repeat steps 1 through 3 for remaining intake valves.
5. Using valve spring compressor, compress exhaust valve spring (13) and remove valve keys (8). Release valve spring compressor slowly.

NOTE

- Valve caps on exhaust valves are rotaters.
- Perform step 6 for 6.5L and 1990 and above 6.2L engines only. 6.5L engine has another seal as indicated in step 6.

6. Remove valve cap (9), shield (11), valve stem seal (12), valve spring (13), valve spring shim (14), seal (17), and valve seal (10) from exhaust valve (15). Discard valve seal (10).
7. Remove valve cap (9), shield (11), valve spring (13), valve spring shim (14), and valve seal (10) from exhaust valve (15). Discard valve seal (10).

2-21. CYLINDER HEAD AND VALVE REPAIR (Cont'd)

NOTE

Valves must be installed in the same location they were removed from. As each valve is removed, place on numbered board or stand.

8. Remove exhaust valve (15) from cylinder head (1).
9. Repeat steps 5 through 8 for remaining exhaust valves.

NOTE

Precombustion chambers must be installed in the same location they were removed from. Mark precombustion chambers for assembly.

10. Remove four precombustion chambers (18) from cylinder head (1).
11. Repeat steps 1 through 10 for opposite cylinder head.
12. Remove engine temperature sending unit (19) from left cylinder head (1)
13. Remove cold start advance switch (20) from right cylinder head (1).
14. Remove two capscrews (21), water jacket cover (22), and water jacket cover gasket (23) from right cylinder head (1). Discard gasket (23).

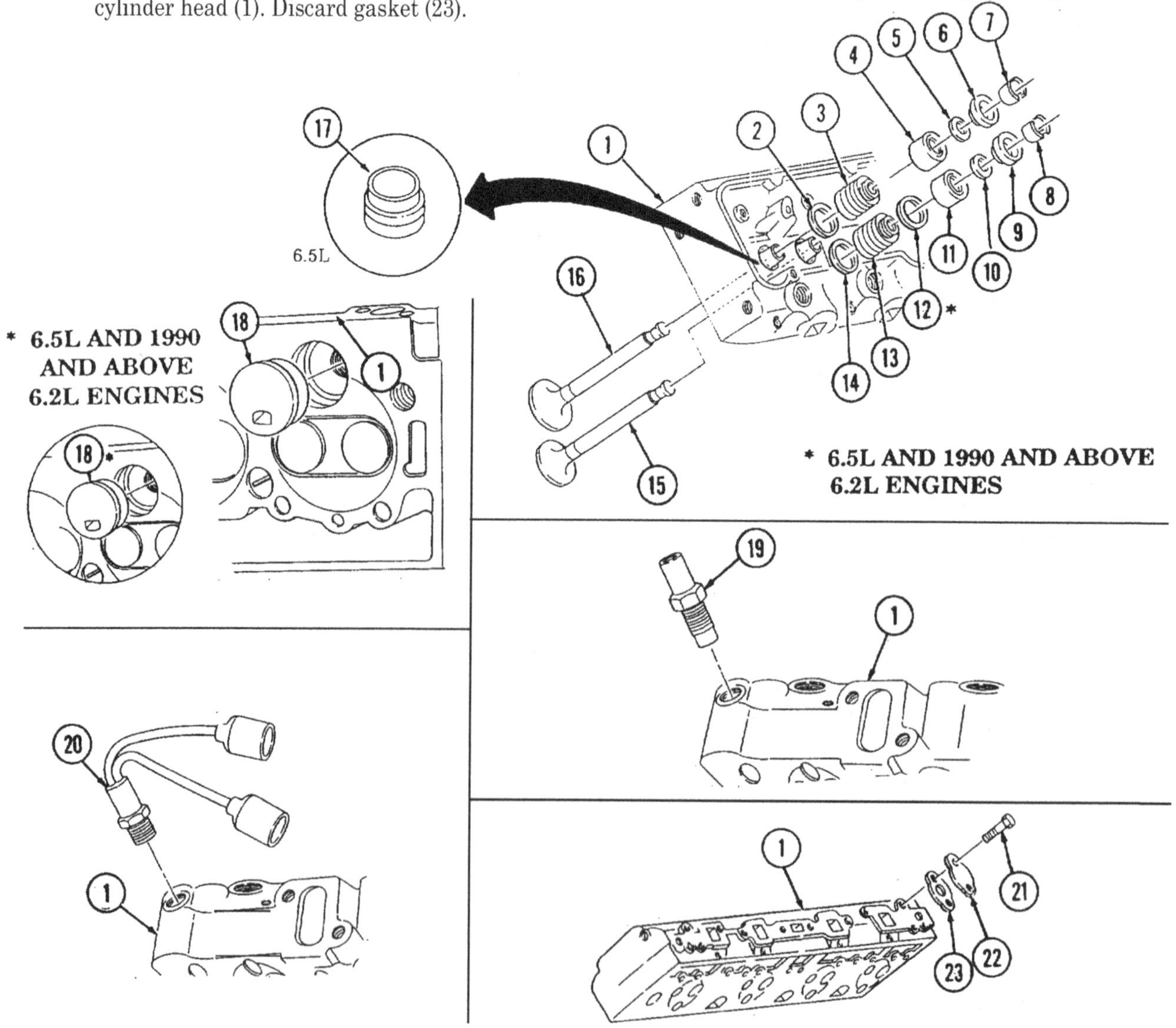

2-21. CYLINDER HEAD AND VALVE REPAIR (Cont'd)

b. Cleaning

1. Using bench grinder and rotary wheel wire brush, clean all valve stems (2) and valve heads (3).
2. Using drill motor and rotary end wire brush, clean all combustion chambers (12) and intake and exhaust ports (13).
3. Refer to para. 2-9 for additional cleaning information.
4. Repeat steps 1 through 3 for opposite cylinder head.

c. Inspection

NOTE

For general inspection instructions, refer to para. 2-10.

1. Inspect valves (1) for cracks, scoring, galling, and burned heads (3). Replace valves (1) if cracked, scored, galled, or if heads (3) are burned.
2. Inspect valve springs (8) and dampers (7) for damage. Replace valve spring (8) if either are damaged.
3. Inspect valve caps (5), shields (6), valve spring shims (9), and valve keys (4) for damage. Replace if damaged.
4. Remove dampers (7) from valve springs (8). Using valve spring tester, compress valve spring (8) to 1-13/16 in. (46 mm). If pressure is less than 70 lb (311 N), replace valve spring (8). Compress valve spring (8) to 1-25/64 in. (35 mm). If pressure is less than 220 lb (978 N), replace valve spring (8). Test remaining springs in same manner. Install dampers (7) into springs (8).
5. Inspect cylinder head (11) for cracks in valve ports (13) or combustion chambers (12) and external cracks to the coolant chamber (14). Replace cylinder head (11) if cracked.

NOTE

Precombustion chamber is available in standard and oversize. If replacing a precombustion chamber, check for marking "OS" on precombustion chamber. If marking is present, oversize precombustion chamber must be used.

6. Inspect precombustion chambers (10) for cracks or damage. Replace if cracked or damaged.
7. Inspect cylinder head (11) mating surfaces for scratches or dents. Replace cylinder head (11) if scratched or dented deeper than 0.003 in. (0.076 mm).
8. Using a straight edge and feeler gauge, check flatness of cylinder head to block mating surface. Flatness of cylinder head (11) must not vary more than 0.002-in. (0.051 mm) in distance of 6 in. (15.2 cm), or more than 0.006 in. (0.152 mm) overall. Replace cylinder head (11) if flatness does not meet specifications.

2-21. CYLINDER HEAD AND VALVE REPAIR (Cont'd)

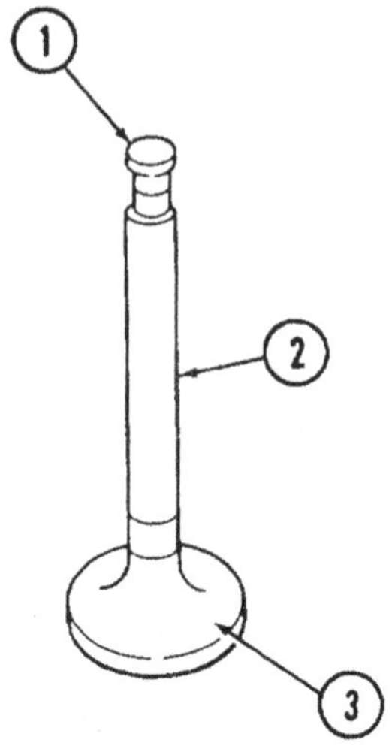

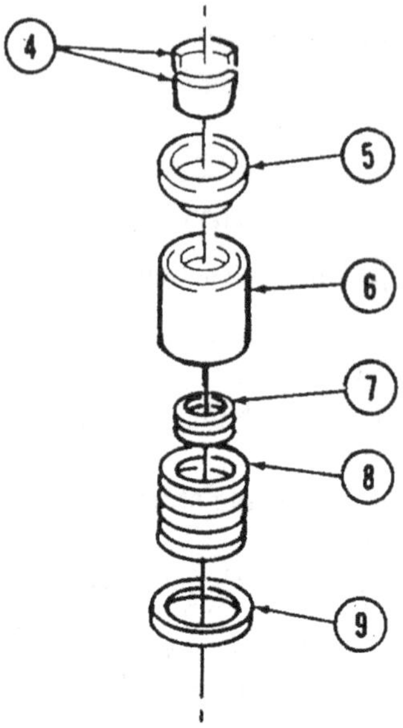

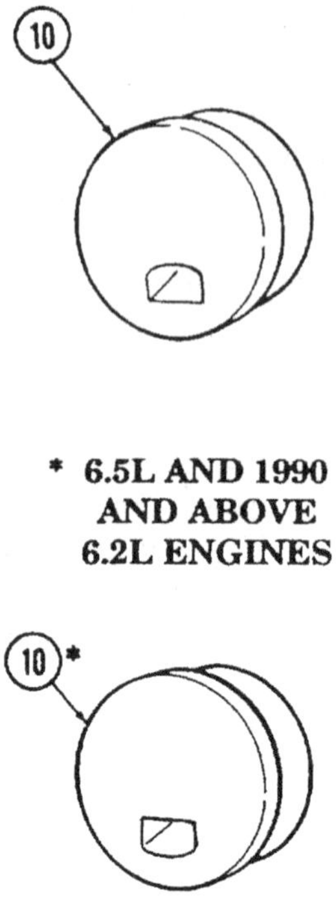

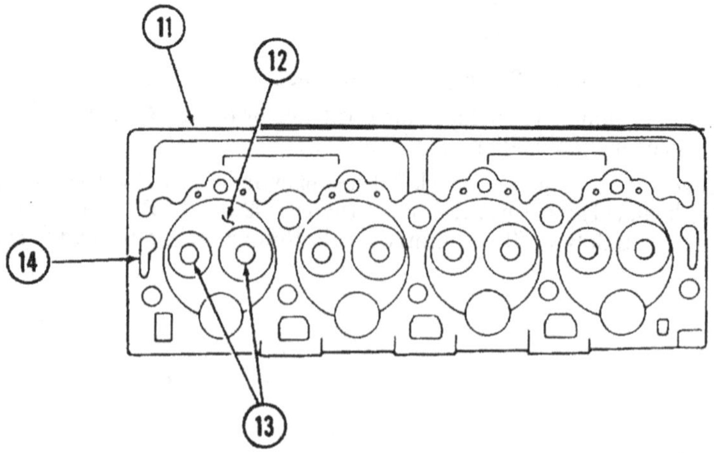

2-21. CYLINDER HEAD AND VALVE REPAIR (Cont'd)

9. Install valve (2) in valve guide (4) from which it was removed.

CAUTION

Clearance must be 0.001-0.003 in. (0.026-0.076 mm). Excessive valve stem clearance will cause excessive oil consumption and may cause valve breakage.

NOTE

- Intake and exhaust valves are available with 0.0035 in. (0.089 mm) and 0.0137 in. (0.349 mm) oversize stems.
- If valve stem clearance is more than 0.0027 in. (0.068 mm), valve guide must be reamed to next size and oversized valve installed. If unable to produce clearance with oversize valve, replace cylinder head.

10. Using dial indicator, measure valve stem clearance. Dial indicator must contact valve stem (1) just above valve guide (4) and at a right angle to the valve stem (1). Drop valve (2) approximately 1/16 in. (1.6 mm) off valve seat (3) and move side-to-side when taking measurement.
11. Repeat steps 9 and 10 for remaining valves (2) and valve guides (4).
12. Inspect cold start advance switch (6) and engine temperature sending unit (5) for damage. Replace if damaged.

d. Refacing Valves

1. Using valve face grinding machine, grind valve face (7) on all valves (2) to 45".
2. Maintain a minimum of 1/32 in. (0.80 mm) margin (8) on all valves (2). If margin falls below specifications, replace valve.

e. Grinding Valve Seats

1. Using valve seat grinding kit, grind all valve seats (3) to 46°.

NOTE

If valve seat width exceeds specifications, valve seat width must be narrowed.

2. Measure width of intake valve seats (3). Valve seats (3) must not be wider than 1/16 in. (1.53 mm).
3. Measure width of exhaust valve seats (3). Valve seats (3) must not be wider than 3/32 in. (2.36 mm).

2-21. CYLINDER HEAD AND VALVE REPAIR (Cont'd)

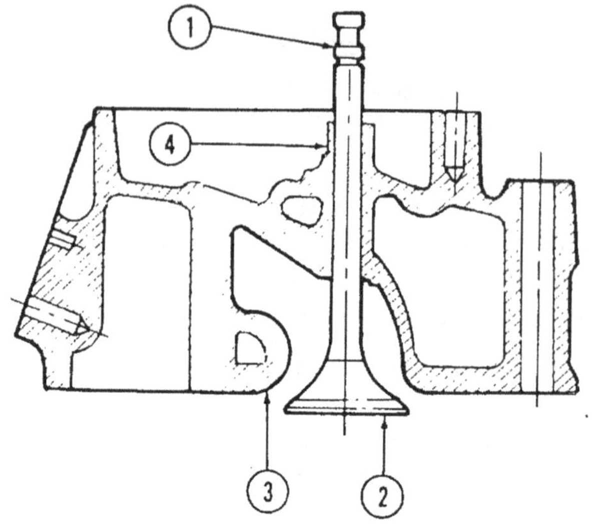

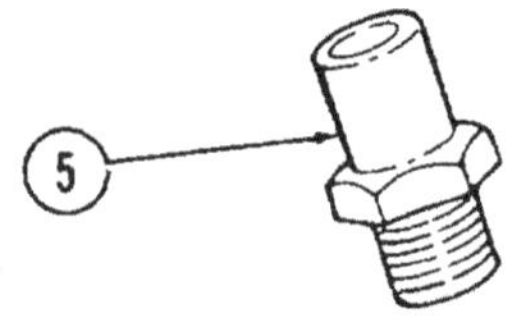

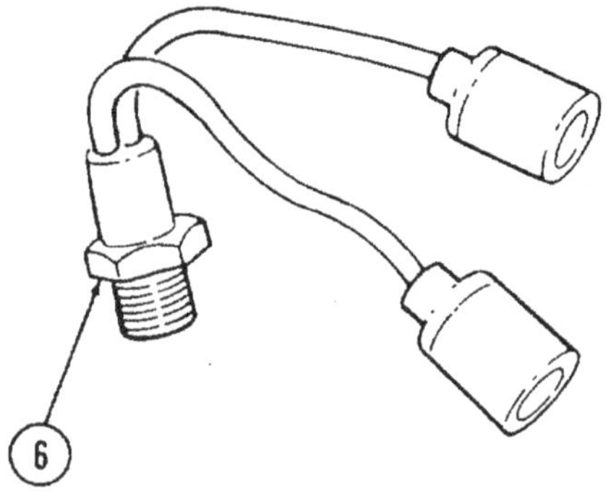

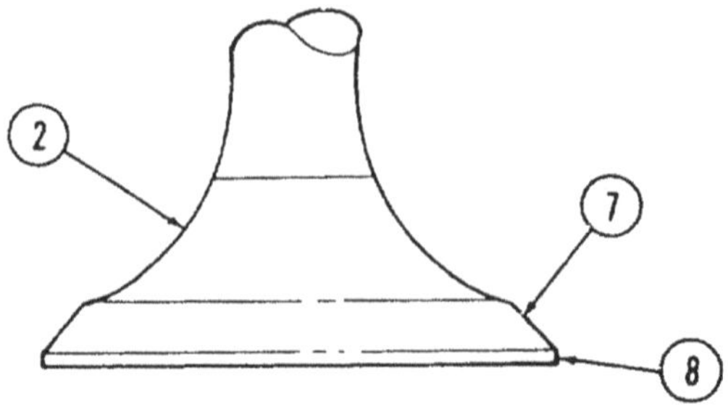

2-21. CYLINDER HEAD AND VALVE REPAIR (Cont'd)

f. Assembly

NOTE

For general assembly instructions, refer to para. 2-12.

1. Apply OE/HDO to exhaust valve (18) and install valve (18) in valve guide (15) from which it was removed.

NOTE

- Valve caps on exhaust valves are rotaters.
- Perform step 2 for 6.5L engines and 1990 and above 6.2L engines. Perform step 3 for previous model 6.2L engines.

2. Install seal (21), valve spring shim (14), valve spring (13), valve stem seal (12), shield (11), and valve cap (9) on exhaust valve (18).
3. Install valve spring shim (14), valve spring (13), shield (11), and valve cap (9) on exhaust valve (12).
4. Using spring compressor, compress valve spring (13) and install valve seal (10) in groove (17) in exhaust valve (18) and install valve keys (8). Slowly release spring compressor.
5. Apply OE/HDO to intake valve (19) and install valve (19) in valve guide (16) from which it was removed.
6. Install valve spring shim (2), valve spring (3), shield (4), and valve cap (6) to intake valve (19).
7. Using valve spring compressor, compress valve spring (3) and install valve seal (5) in groove (20) in intake valve (19) and install valve keys (7). Slowly release spring compressor.
8. Check installed height of valve springs (3) measuring from top of valve spring shim (2) to top of shield (4).
9. Spring height must be 1-13/16 in. (46 mm). If spring height is greater, remove valve spring (3) and add spring shim(s) (2) to bring spring height within specifications.
10. Repeat steps 1 through 9 for remaining valves and opposite cylinder head.

2-21. CYLINDER HEAD AND VALVE REPAIR (Cont'd)

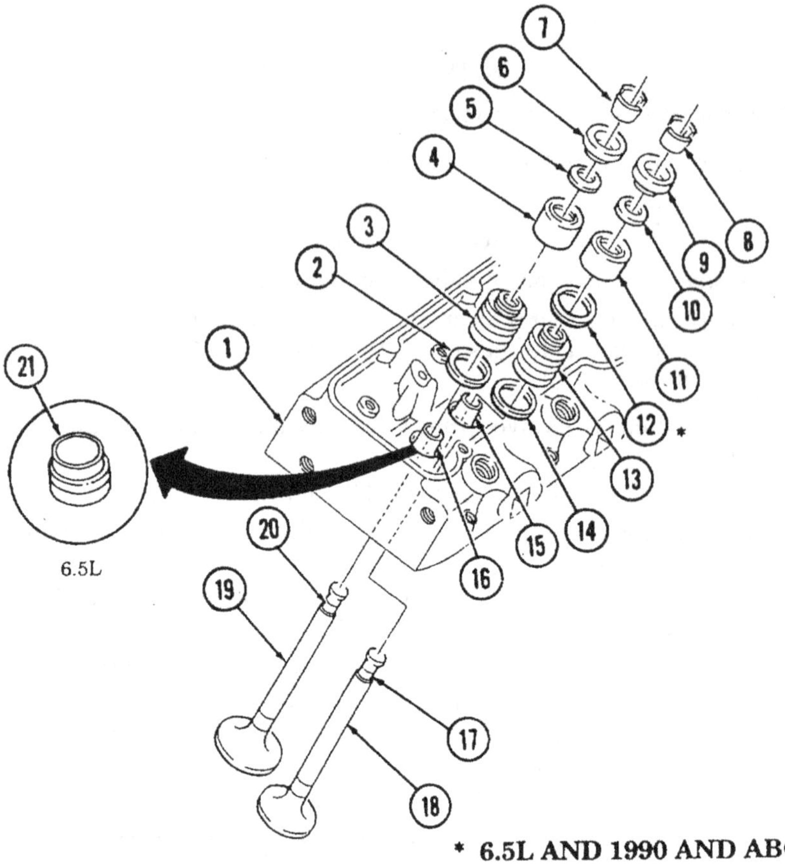

* 6.5L AND 1990 AND ABOVE 6.2L ENGINES

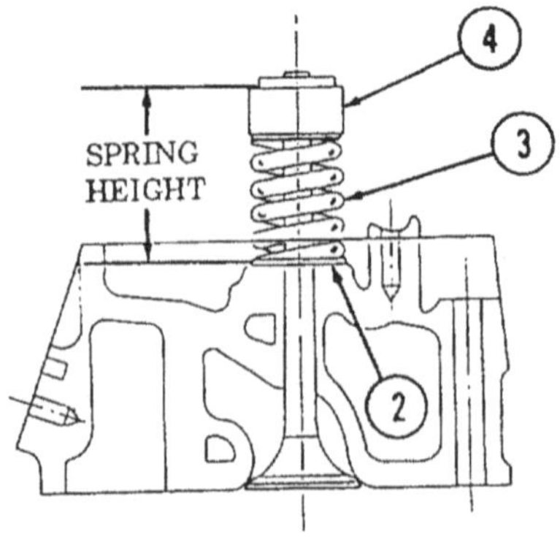

2-21. CYLINDER HEAD AND VALVE REPAIR (Cont'd)

10. Apply pipe sealing compound to cold start advance switch (1) and install in right cylinder head (2).
11. Install water jacket cover gasket (5) and water jacket cover (4) on right cylinder head (2) and secure with two capscrews (3). Tighten capscrews (3) to 25-37 lb-ft (3450 Nžm).
12. Apply pipe sealing compound to engine temperature sending unit (6) and install in left cylinder head (2).
13. Install four precombustion chambers (7) in cylinder head (2) flush to +0.002 in. (0.050 mm).
14. Repeat step 13 for opposite cylinder head.

2-21. CYLINDER HEAD AND VALVE REPAIR (Cont'd)

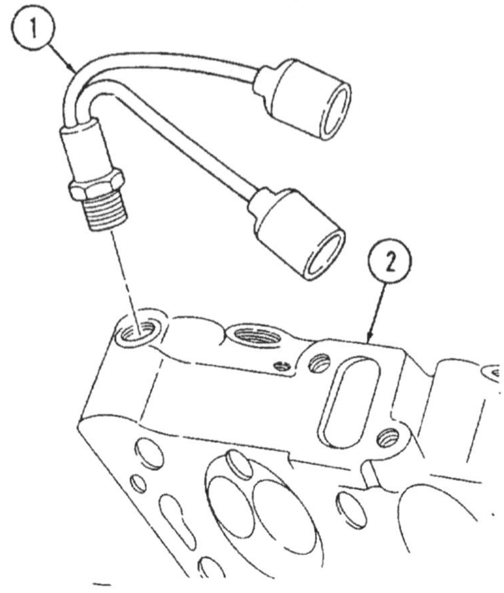

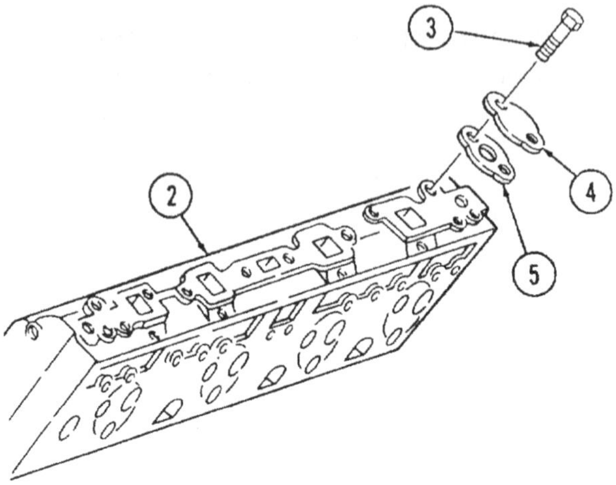

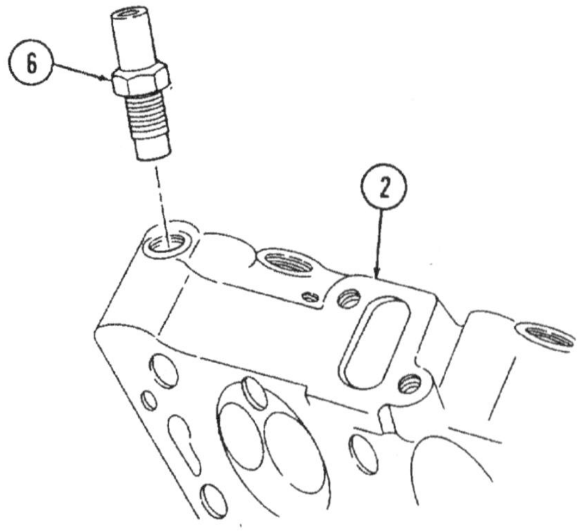

* 6.5L AND 1990 AND ABOVE 6.2L ENGINES

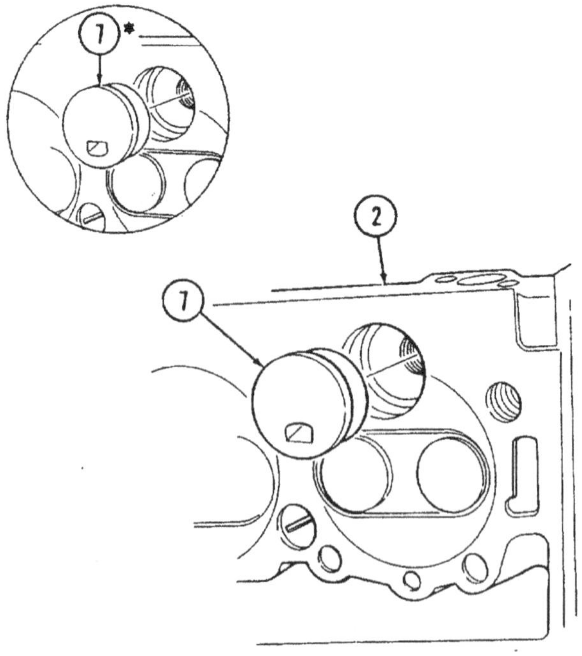

2-22. VALVE TRAIN REPAIR

This task covers:

a. Disassembly
b. Cleaning
c. Inspection
d. Assembly

INITIAL SETUP

Tools
General mechanic's tool kit:
autmotive (Appendix E, Item 1)

Materials/Parts
Four retainers (Appendix D, Item 36)

Manual References
TM 9-2815-237-34P

Equipment Condition
Engine disassembled into subassemblies (para. 2-15).

a. Disassembly

NOTE

Tag rocker arms for assembly.

1. Remove four retainers (2) and rocker arms (3) from rocker shaft (1). Discard retainers (2).
2. Repeat step 1 for remaining rocker arm and shaft assemblies.

b. Cleaning

Clean all components in accordance with para. 2-9.

c. Inspection

NOTE

For general inspection instructions, refer to para. 2-10.

1. Inspect rocker shaft (1) for damage. Replace if damaged.
2. Inspect rocker arms (3) for damage. Replace if damaged.
3. Inspect pushrods (4) for blocked oil passages and damage. Replace if damaged or oil passages are blocked.
4. Inspect lifters (5) for damage. Replace if damaged.
5. Inspect roller (6) on lifter (5) for scoring and binding. Replace lifter (5) if scored or binding.

d. Assembly

NOTE

For general assembly instructions, refer to para. 2-12.

1. Install four rocker arms (3) on rocker shaft (1) with four retainers (2).
2. Repeat step 1 for remaining rocker arm and shaft assemblies.

2-22. VALVE TRAIN REPAIR (Cont'd)

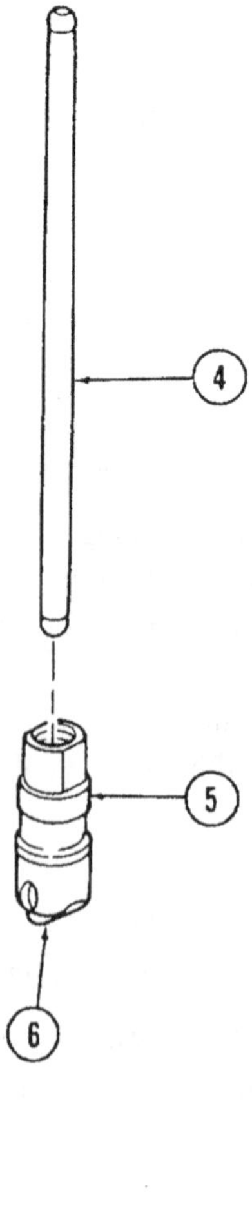

2-23. TORSIONAL DAMPER REPAIR

This task covers:

a. Cleaning **b. Inspection**

INITIAL SETUP:

Tools

General mechanic's tool kit:
automotive (Appendix E, Item 1)

Manual References

TM 9-2815-237-34 P

Equipment Condition

Engine disassembled into subassemblies (para. 2-15).

a. Cleaning

Clean torsional damper (1) in accordance with para. 2-9.

b. Inspection

NOTE

For general inspection instructions, refer to para. 2-10.

1. Inspect capscrew holes (4) for thread damage. Replace torsional damper (1) if threads are damaged.
2. Inspect front cover seal mating surface (2) for damage. Replace torsional damper (1) if damaged.
3. Inspect rubber insulator (3) between inner ring and torsional damper housing for cracks or damage. Replace torsional damper (1) if cracked or damaged.

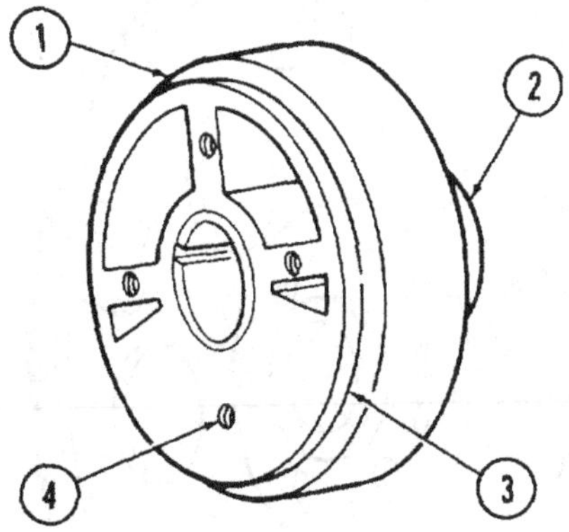

2-24. FLYWHEEL REPAIR

This task covers:

a. Cleaning **b. Inspection**

INITIAL SETUP

Tools
General mechanic's tool kit: automotive (Appendix E, Item 1)

Manual References
TM 9-2815-237-34P

Equipment Condition
Engine disassembled into subassemblies (para. 2-15).

a. Cleaning

Clean flywheel (1) in accordance with para. 2-9.

b. Inspection

NOTE

For general inspection instructions, refer to para. 2-10.

1. Inspect flywheel (1) for damaged teeth (2). Replace flywheel (1) if teeth (2) are damaged.
2. Inspect flywheel (1) for damage. Replace if damaged.

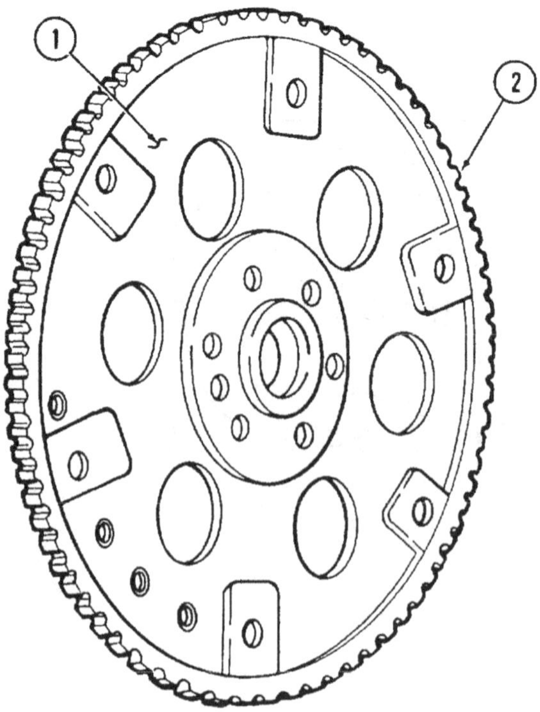

2-25. ROCKER ARM COVER REPAIR

This task covers:

a. Cleaning **b. Inspection**

INITIAL SETUP:

Tools
General mechanic's tool kit: automotive (Appendix E, Item 1)

Manual References
TM 9-2815-237-34P

Equipment Condition
Engine disassembled into subassemblies (para. 2-15).

a. Cleaning

Clean rocker arm cover (1) in accordance with para, 2-9.

b. Inspection

NOTE

For general inspection instructions, refer to para. 2-10.

Inspect rocker arm cover (1) for cracks and sealing surface distortion. Replace if cracked, or sealing surface is distorted.

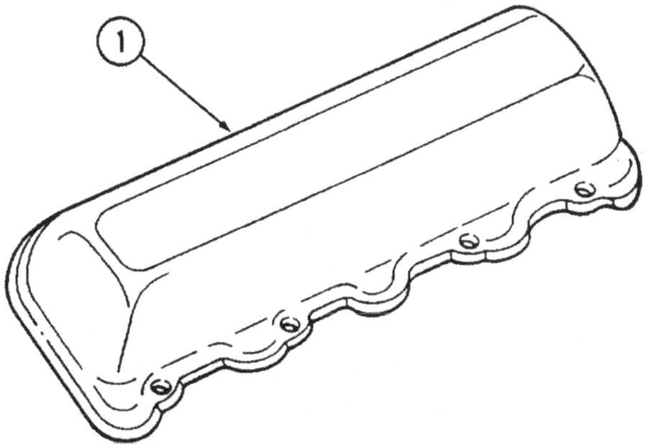

2-26. OIL PAN REPAIR

This task covers:

a. Cleaning **b. Inspection**

INITIAL SETUP

Tools

General mechanic's tool kit:
automotive (Appendix E, Item 1)

Manual References

TM 9-2815-237-34P

Equipment Condition

Engine disassembled into subassemblies (para. 2-15).

a. Cleaning

Clean oil pan (3) in accordance with para. 2-9.

b. Inspection

NOTE

For general inspection instructions, refer to para. 2-10.

1. Inspect oil pan (3) for cracks, damage, and sealing surface distortion. Replace if cracked, damaged, or sealing surface is distorted.
2. Inspect drainplug (1) for damage. Replace if damaged.
3. Inspect gasket (2) for damage. Replace if damaged.

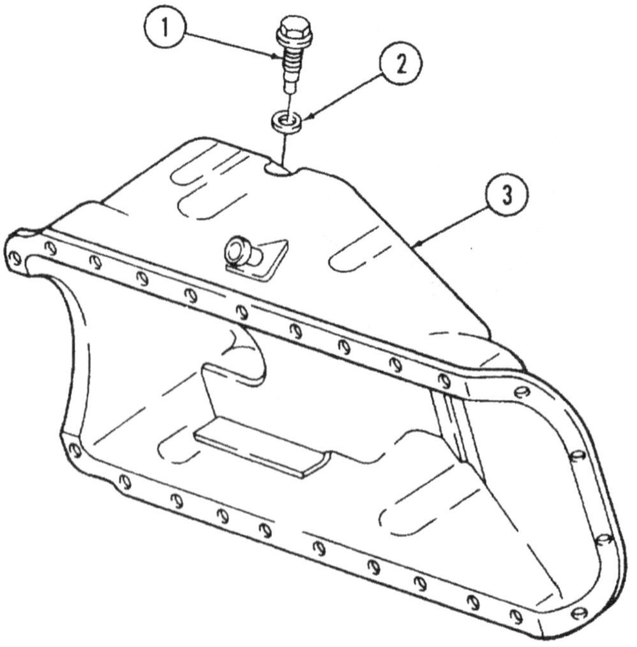

2-27. OIL PUMP REPAIR

This task covers:

a. Disassembly
b. Cleaning
c. Inspection
d. Assembly

INITIAL SETUP:

Tools
General mechanic's tool kit:
automotive (Appendix E, Item 1)
Hammer, soft head (Appendix E, Item 30)

Materials/Parts
Lubricating oil, OE/HDO (Appendix B, Item 13)

Manual References
TM 9-2815-237-34P

Equipment Condition
Engine disassembled into subassemblies (para. 2-15).

a. Disassembly

1. Remove shaft extension (10) and coupling (9) from pump body (8).
2. Remove four capscrews (3) and pump cover (5) from pump body (8).
3. Mark drive gear (7) and idler gear (6) for assembly and remove.
4. Remove retaining pin (4), regulator spring (l), and pressure regulator valve (2).

b. Cleaning

Clean all components in accordance with para. 2-9.

c. Inspection

NOTE

- For general inspection instructions, refer to para. 2-10.
- Pump body, pump cover, and pump gears are not serviced separately. If any of these parts require replacement, replace entire oil pump.

1. Inspect pump body (8) and pump cover (5) for damage. Replace oil pump if damaged.
2. Inspect drive gear (7) and idler gear (6) for damage. Replace oil pump if damaged.
3. Inspect regulator valve (2) for scoring or nicks. Replace if scored or nicked.
4. Inspect regulator spring (1) for damage. Replace if damaged.
5. Inspect shaft extension (10) and coupling (9) for damage. Replace if damaged.

2-27. OIL PUMP REPAIR (Cont'd)

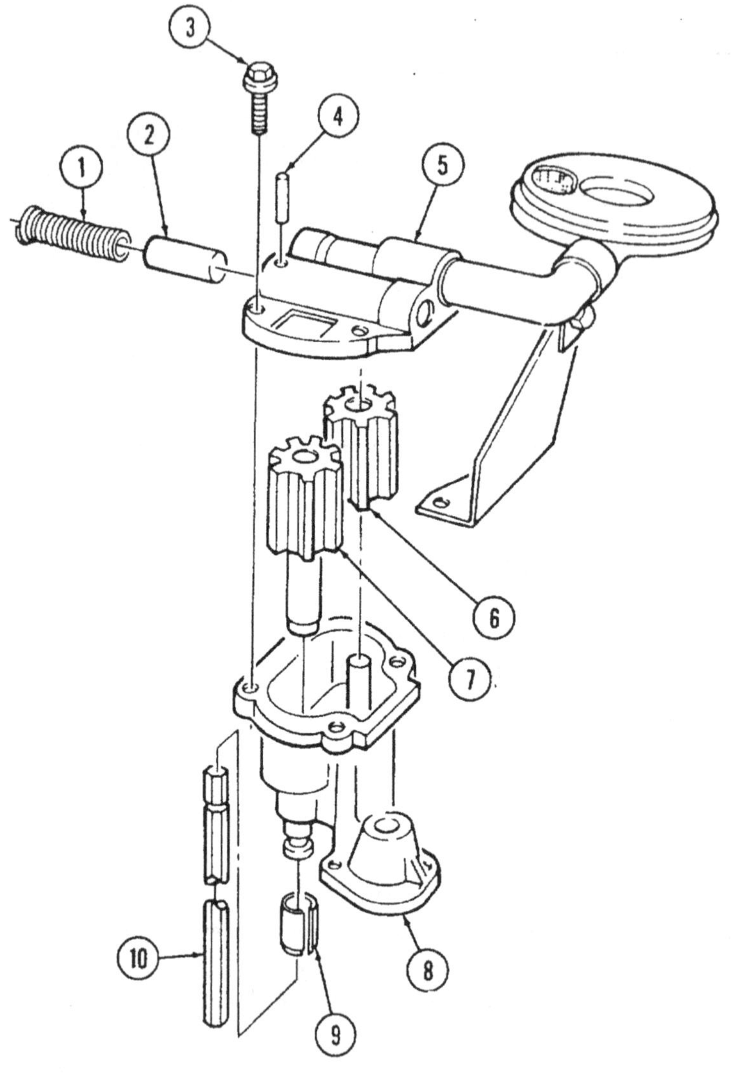

2-27. OIL PUMP REPAIR (Cont'd)

6. Inspect oil screen (7) for damage or blockage. If blocked or damaged, perform steps 8 through 12; if not, go to next step.
7. Inspect pickup tube bracket (9) for damage. If damaged, perform steps 8 and 12; if not, go to step 13.
8. Remove capscrew (8) and pickup tube bracket (9) from oil screen (7).
9. Place pump cover (5) in soft-jawed vise.
10. Using soft-head hammer, remove oil screen (7) from pump cover (5).
11. Using soft-head hammer, install oil screen (7) in pump cover (5) until bead (6) seats on pump cover (5).
12. Install pickup tube bracket (9) on oil screen (7) and install capscrew (8). Finger tighten capscrew (8).
13. Inspect regulator valve plug (10) for damage. Replace if damaged.

d. Assembly

NOTE

For general assembly instructions, refer to para. 2-12.

1. Install regulator valve (2) and regulator spring (1) in pump cover (5) with retaining pin (4).
2. Install drive gear (15) and idler gear (11) into pump body (14) and aline reference marks.
3, Apply small amount of OE/HDO into pump body (14).
4. Install pump cover (5) on pump body (14) with four capscrews (3). Tighten capscrews (3) to 120 lb-in. (14 Nžm).
5. Install coupling (12) and shaft extension (13) on end of drive gear (15).

2-27. OIL PUMP REPAIR (Cont'd)

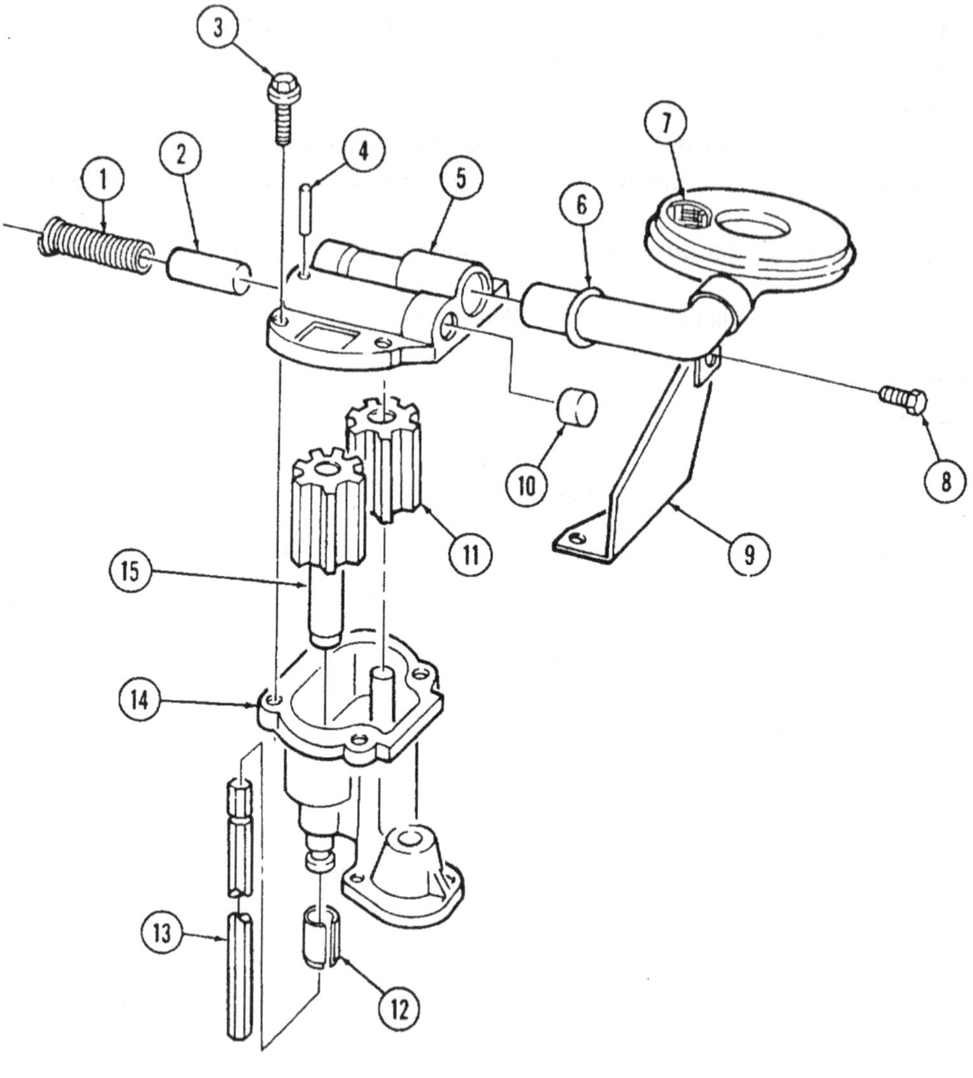

2-28. OIL FILTER ADAPTER AND OIL PRESSURE SENDING UNIT REPAIR

This task covers:

a. Cleaning **b. Inspection**

INITIAL SETUP

Tools
General mechanic's tool kit automotive (Appendix E, Item 1)

Manual References
TM 9-2815-237-34P

Equipment Condition
Engine disassembled into subassemblies (para. 2-15).

a. Cleaning

Clean all components in accordance with para. 2-9.

b. Inspection

NOTE

For general inspection instructions, refer to para. 2-10.

1. Inspect adapter bolt (2) for damaged threads. Replace if threads are damaged.
2. Inspect adapter reducer (3) for damaged threads. Replace if threads are damaged.
3. Inspect adapter (1) for damage. Replace adapter (1) if damaged and remove adapter reducer (3) for reuse.
4. Inspect fitting (5) for damage. Replace if damaged.
5. Inspect oil pressure sending unit (4) for damage. Replace if damaged.

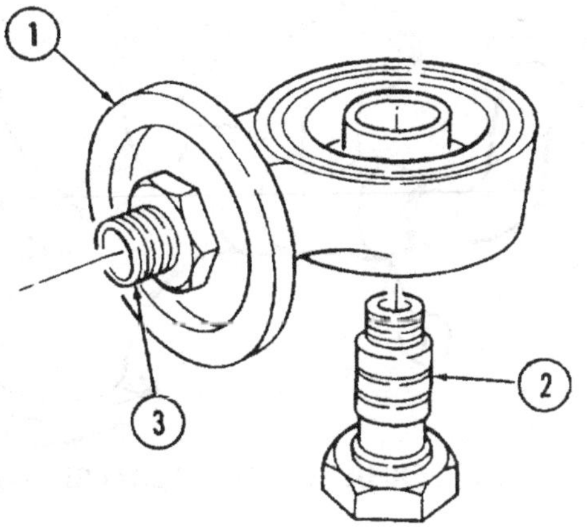

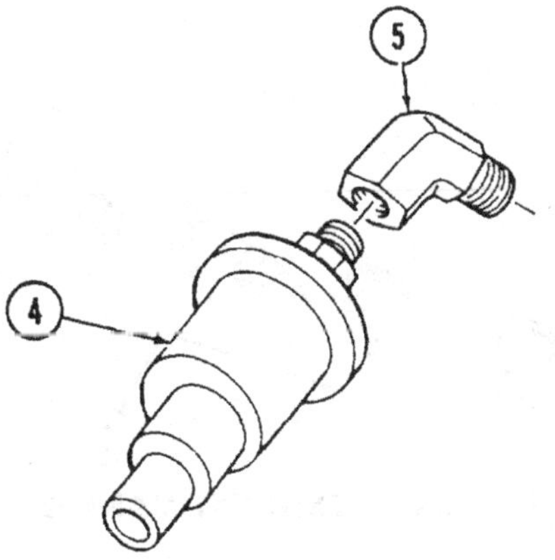

2-29 EXHAUST MANIFOLD REPAIR

This task covers:

a. **Cleaning**   b. **Inspection**

INITIAL SETUP:

Tools
General mechanic's tool kit:
automotive (Appendix E, Item 1)

Manual References
TM 9-2815-237-34P

Equipment Condition
Engine disassembled into subassemblies (para. 2-15).

NOTE

There is no repair to exhaust manifold and it must be replaced if damaged.

a. Cleaning

Clean exhaust manifolds (1) in accordance with para. 2-9.

b. Inspection

NOTE

For general inspection instructions, refer to para. 2-10.

Inspect exhaust manifold (1) for damage. Replace if damaged.

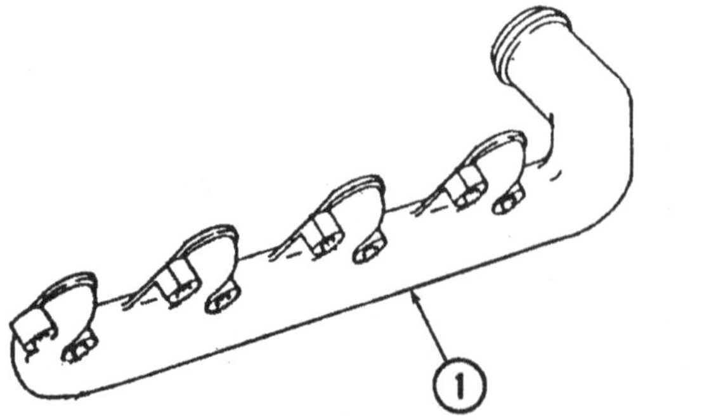

LEFT TURBO EXHAUST MANIFOLD SHOWN

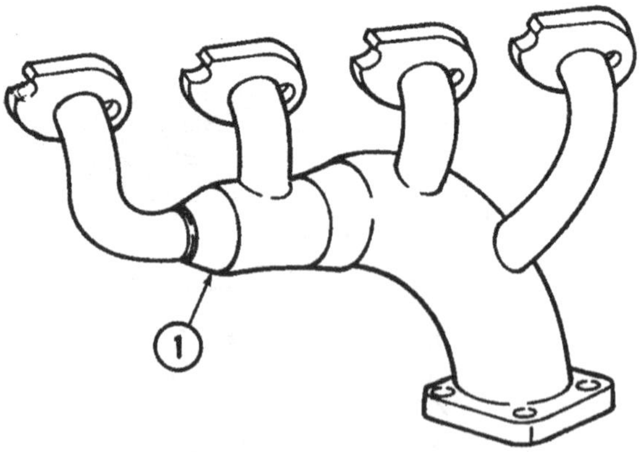

LEFT EXHAUST MANIFOLD SHOWN

2-30. INTAKE MANIFOLD REPAIR

This task covers:

a. Cleaning | b. Inspection

INITIAL SETUP:

Tools
General mechanic's tool kit: automotive (Appendix E, Item 1)

Materials/Parts
Sealing compound (Appendix B, Item 18)

Manual References
TM 9-2815-237-34P

Equipment Condition
Engine disassembled into subassemblies (para. 2-15).

a. Cleaning

Clean intake manifold (1) or intake manifolds (4) and center manifold (3) in accordance with para. 2-9.

b. Inspection

NOTE

- For general inspection instructions, refer to para. 2-10.
- For turbocharged engines, perform step 1 only.

1. Inspect intake manifold (1) or intake manifolds (4) and center intake manifold (3) for damage. Replace if damaged.
2. Inspect nipple (2) for damage. If damaged, replace nipple (2) using sealing compound.

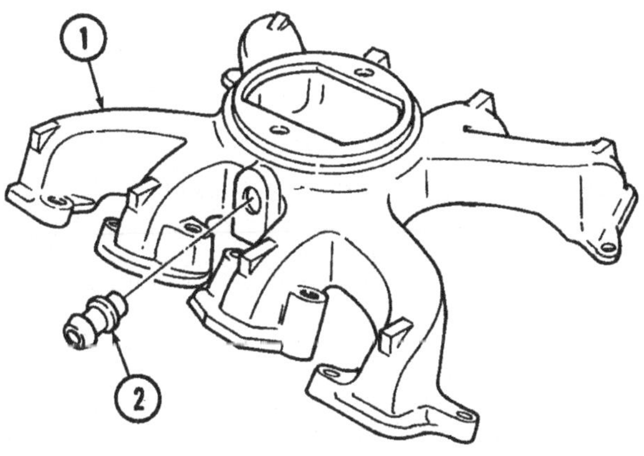

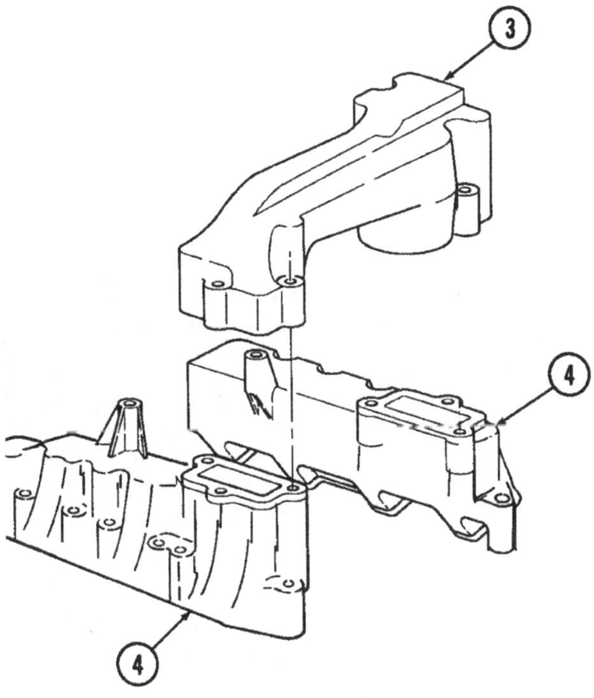

TURBO ENGINE

2-31. FUEL PUMP REPAIR

This task covers:

a. Cleaning **b. Inspection**

INITIAL SETUP:

Tools
General mechanic's tool kit: automotive (Appendix E, Item 1)

Manual References
TM 9-2815-237-34P

Equipment Condition
Engine disassembled into subassemblies (para. 2-15).

a. Cleaning

Clean all components in accordance with para. 2-9.

b. Inspection

NOTE

For general inspection instructions, refer to para. 2-10.

1. Inspect fuel pump (1) for damage. Replace if damaged.
2. Inspect fuel pump lever (2) for damage. Replace fuel pump (1) if damaged.
3. Inspect inlet tube (6) and vent tube (5) for damage. Replace fuel pump (1) if damaged.
4. Inspect adapter plate (3) for damage. Replace if damaged.
5. Inspect fuel pump push rod (4) for damage. Replace if damaged.

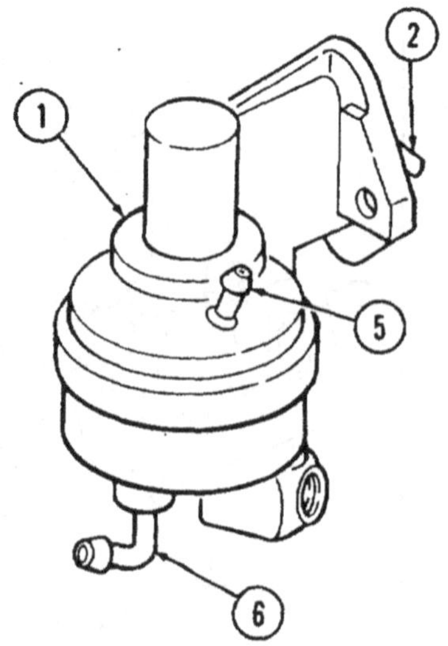

2-32. FUEL INJECTION PUMP REPAIR

This task covers:

a. Disassembly
b. Cleaning
c. Inspection
d. Assembly

INITIAL SETUP:

Tools

General mechanic's tool kit:
automotive (Appendix E, Item 1)

Special Tools

Hex head driver, 5/32-in. (Appendix E, Item 54)
Hex head driver, 1/8-in. (Appendix E, Item 55)
Hex head driver, 5/16-in. (Appendix E, Item 9)
Hex head driver, 1/4-in. (Appendix E, Item 56)
Torx drive (Appendix E, Item 42)
End cap wrench (Appendix E, Item 44)
Holding fixture (Appendix E, Item 46)
Governor weight support (Appendix E, Item 43)
Drive shaft bearing installer (Appendix E, Item 39)
Drive shaft seal installer (Appendix E, Item 40)
Roller-to-roller setting tool (Appendix E, Item 45)
Throttle and shutoff shaft bushing (Appendix E, Item 41)
Pilot tube mandrel (Appendix E, Item 47)
Delivery valve extractor (Appendix E, Item 48)

Material/Parts

Overhaul kit (Appendix D, Item 27)
Seven lockwashers (Appendix D, Item 25)
Two locknuts (Appendix D, Item 24)
Adjusting plug (Appendix D, Item 35)
Valve stop (Appendix D, Item 53)
Cap (Appendix D, Item 4)
Calibrating fluid (Appendix B, Item 4)
Seal lubricant (Appendix, B Item 12)
Sealing compound (Appendix B, Item 18)
Lacquer thinner (Appendix B, Item 20)
or Acetone (Appendix B, Item 2)

Manual References

TM 9-2815-237-34P
TM 9-2320-280-20

Equipment Condition

Remove fuel injector pump (para. 2-15).

General Safety Instructions

- Wear eyeshields when using compressed air.
- Do not allow solvents to get on skin or clothing.
- Compressed air for cleaning purpose will not exceed 30 psi (207 kPa).

NOTE

- The fuel injection pump repair procedures for pre-1990, 1990 and above 6.2L and 6.5L engines are basically the same. Difference in procedures are noted. Refer to para. 1-10 to determine engine model before ordering replacement parts.
- Before disassembly, fuel injection pump should be tested (para. 2-33.c, 2-34.c, or 2-35.c) on calibration stand to determine type of pump malfunction.
- Do not disassemble fuel injection pump if no malfunction is found after preliminary test.
- Work area should be clean, well-ventilated, and free from blowing dirt and dust.

2-32. FUEL INJECTION PUMP REPAIR (Cont'd)

a. Disassembly

CAUTION

Fuel pump body must be thoroughly cleaned before disconnecting any attaching components to prevent foreign particles from entering pump.

1. Clean exterior of fuel injection pump (9) in accordance with para. 2-9.
2. Mount pump (9) in holding fixture and secure fixture in vise.
3. Remove three screws (2), lockwashers (3), and washers (4) from governor control cover (1). Discard lockwashers (3).

NOTE

If fuel is found in pump, discard fuel in an approved container.

4. Remove cover (1) and gasket (13). Discard gasket (13).

NOTE

- Perform steps 5, 6, and 7 for 6.5L engines and 6.2L engines 1990 and above.
- Perform steps 8 and 9 for pre-1990 engines.

5. Remove cap (8) from guide stud (7). Using 1/4-in. hex head driver, remove guide stud (7), seal (5), and nut (6) from min-max governor (12). Discard seal (5) and cap (8).
6. Turn throttle shaft (11) to low idle and remove idle spring (16).
7. Turn throttle shaft (11) to wide open throttle, loosen sleeve (17) on min-max governor (12) and lift min-max governor (12) and inner governor spring (14) out of pump (9) and disengage min-max governor (12) from governor thrust spring (15).
8. Using 1/4-in. hex head driver, remove guide stud (7) and washer (10) from min-max governor (12). Discard washer (10).
9. Turn throttle shaft (11) to wide open throttle and pull min-max governor (12) out of pump (9).

2-32. FUEL INJECTION PUMP REPAIR (Cont'd)

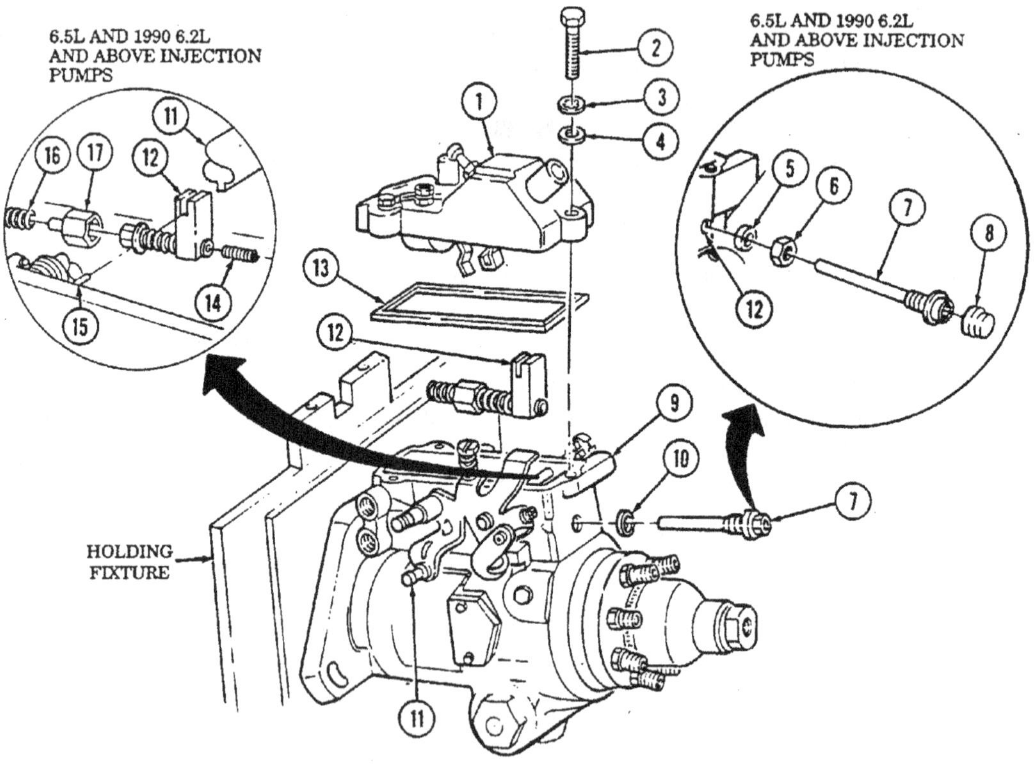

2-32. FUEL INJECTION PUMP REPAIR (Cont'd)

10. Deleted.
11. Remove cam pin (10) and cam (11) from throttle shaft (1).
12. Remove two retaining clips (15) from rocker lever pin (16). Discard clips (15).
13. Remove rocker lever pin (16) and rocker lever (13) from housing (4).
14. Using Torx drive, remove screw (14) from face cam (8) and remove face cam (8) and washer (7) from throttle shaft (1).

NOTE

- Perform steps 15 and 16 for 6.5L engines and for 6.2L engines 1990 and above.
- Perform step 17 for pre-1990 engines.

15. Turn throttle shaft (1) to low idle and release the governor thrust spring (5) from housing boss (6).
16. Remove throttle shaft (1) and governor thrust spring (5) from housing (4), and remove two O-rings (3) and spacer (2) from throttle shaft (1). Discard O-rings (3).
17. Remove throttle shaft (1) from housing (4), and remove two O-rings (3) and spacer (2) from throttle shaft (1). Discard O-rings (3).
18. Hold metering valve (19) and raise governor linkage hook (17) from valve arm pin (18).
19. Hold governor arm (22) forward and pull linkage hook (17) back and detach from governor arm (22).
20. Remove linkage hook (17), governor arm (22), and spring (21) from housing (4).
21. Remove spring (21) from linkage hook (17) and governor arm (22).
22. Remove metering valve (19) from housing (4).
23. Using 1/8-in. hex head driver, remove vent wire socket head screw (20) from housing (4).

2-32. FUEL INJECTION PUMP REPAIR (Cont'd)

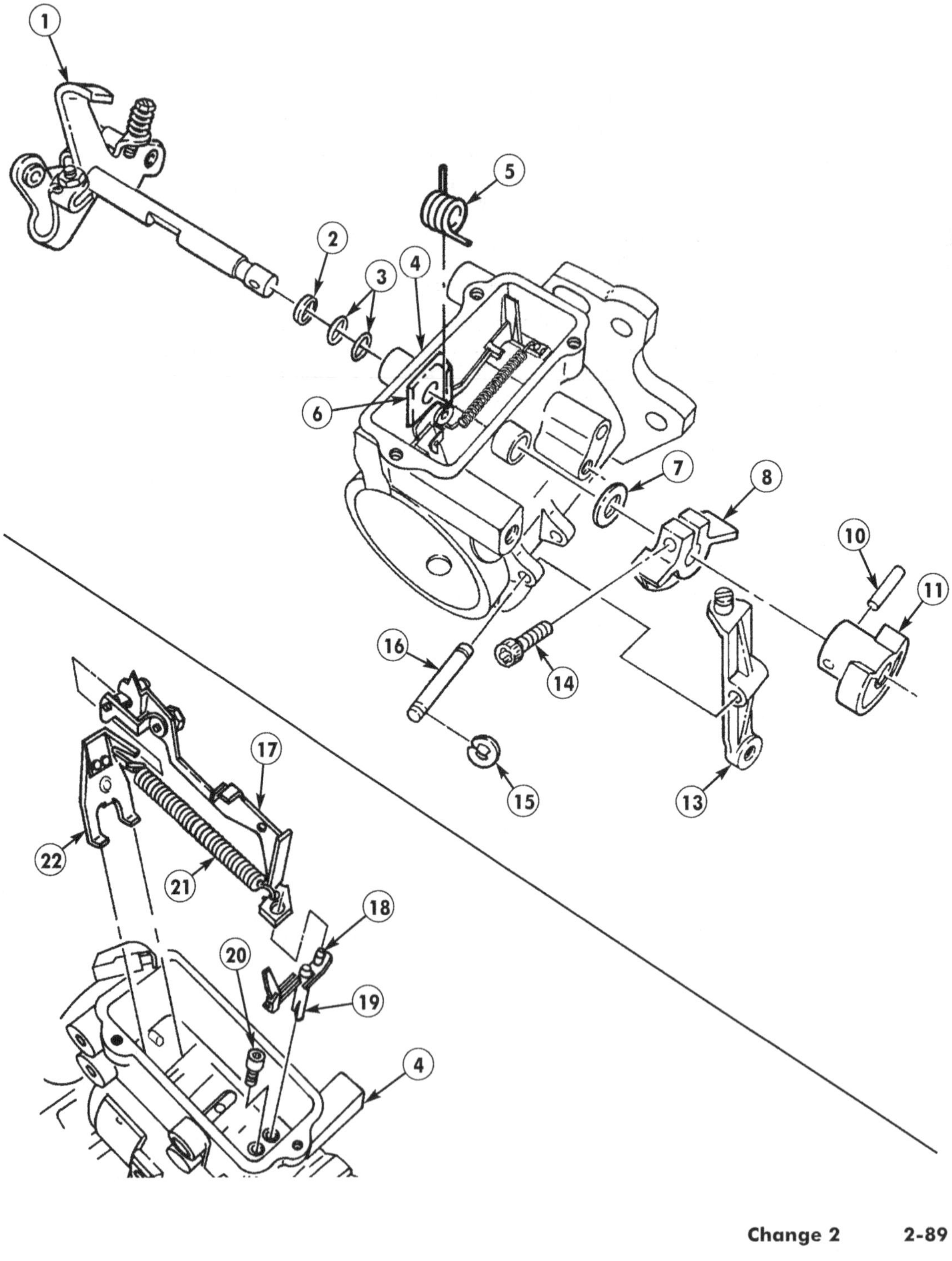

2-32. FUEL INJECTION PUMP REPAIR (Cont'd)

24. Using 5/32-in. hex head driver, remove transfer pump end cap socket-head locking screw (5), plate (4), rubber washer (2), and flat washer (3) from hydraulic head (20). Discard washer (2).
25. Remove inlet tube assembly (8) from fuel pump end cap (9).
26. Using end cap wrench, remove fuel pump end cap (9) from hydraulic head (20).
27. Remove regulator (17) from fuel pump end cap (9).

NOTE

Early production injection pumps have an inlet filter screen, retaining ring, and seal or pressure plate. Late production pumps have a different configuration screen which does not require a retaining ring and seal or pressure plate. When repairing early production pumps, replace the screen with a late production one.

28. Remove inlet filter screen (11) and pressure plate (10), if applicable, from regulator (17).
29. Remove retaining ring (12) and screen (11) from pressure plate (10), if applicable.
30. Remove inlet filter screen seal (16), end plate adjusting plug (13), spring (14), and piston (15) from regulator (17). Discard adjusting plug (13) and seal (16). Tag spring (14) and piston (15) for assembly.
31. Remove transfer pump liner (7), four blades (18), and two springs (6).
32. Remove head-locking capscrews (1) and (21) from housing (22).

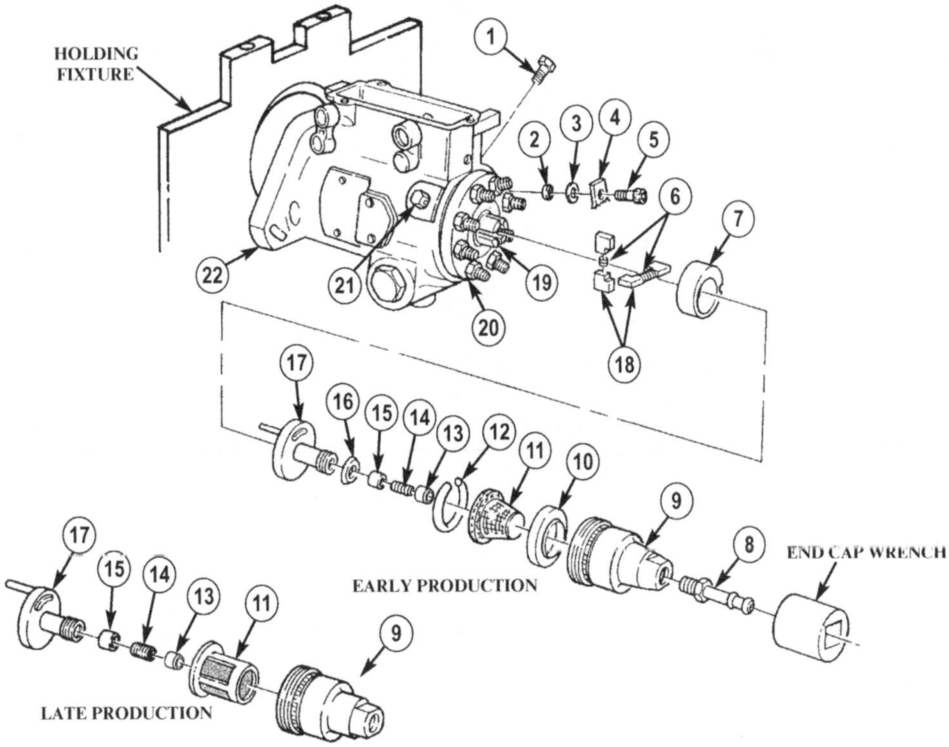

2-32. FUEL INJECTION PUMP REPAIR (Cont'd)

33. Rotate housing (22) and holding fixture in vise so advance plug (28) faces upward.
34. Using 5/16-in. hex head driver, remove head locating screw (27) and O-ring (26). Discard O-ring (26).
35. Remove spring side plug assembly (24) and O-ring (25) from housing (22). Disard O-ring (25).
36. Remove servo advance plunger (23) from spring side plug (24).
37. Using 1/4-in. hex head driver, remove advance plug (28), cam advance pin (30), and O-ring (29) from housing (22). Discard O-ring (29).
38. Remove power side advance piston hole plug (35) and O-ring (34). Discard O-ring (34).
39. Remove servo advance piston assembly (33) from housing (22). Tap piston assembly (33) to remove spring (31) and valve (32). Tag spring (31) for assembly.

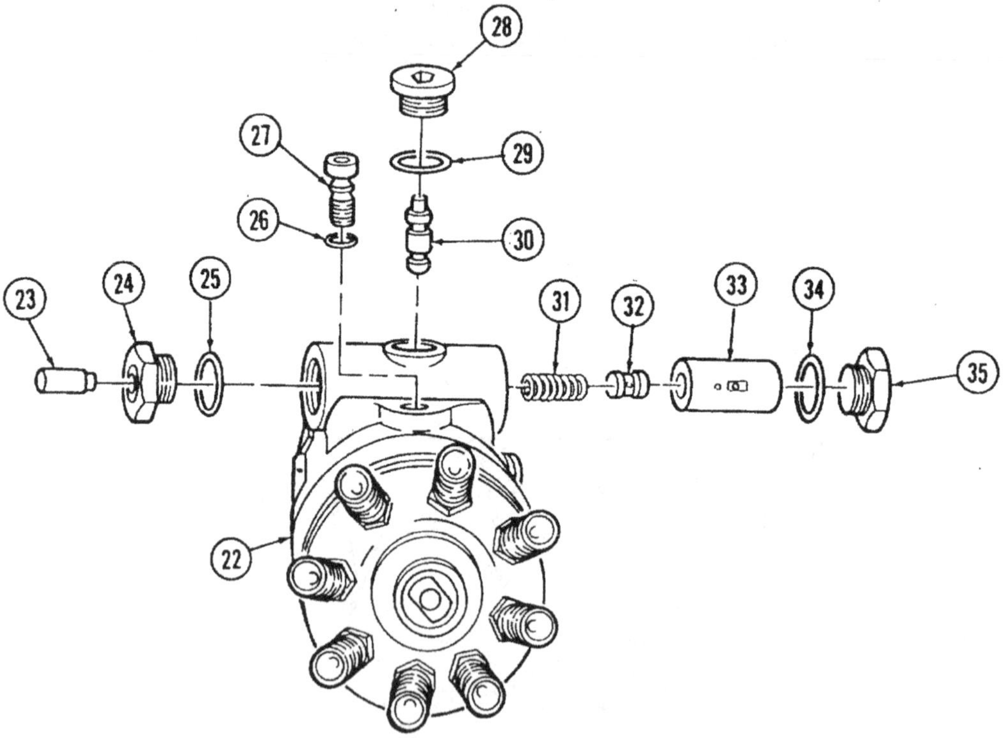

2-32. FUEL INJECTION PUMP REPAIR (Cont'd)

40. Rotate housing (9) and holding fixture right side up.
41. Remove head locking capscrew (1).
42. Remove hydraulic head assembly (7) from housing (9).
43. Remove six weights (8), washer (3), and governor thrust sleeve (2) from hydraulic head (7).
44. Remove liner locating ring (4) and two rotor retainers (5) from hydraulic head (7).
45. Remove transfer pump end cap seal (6) from hydraulic head (7). Discard seal (6).

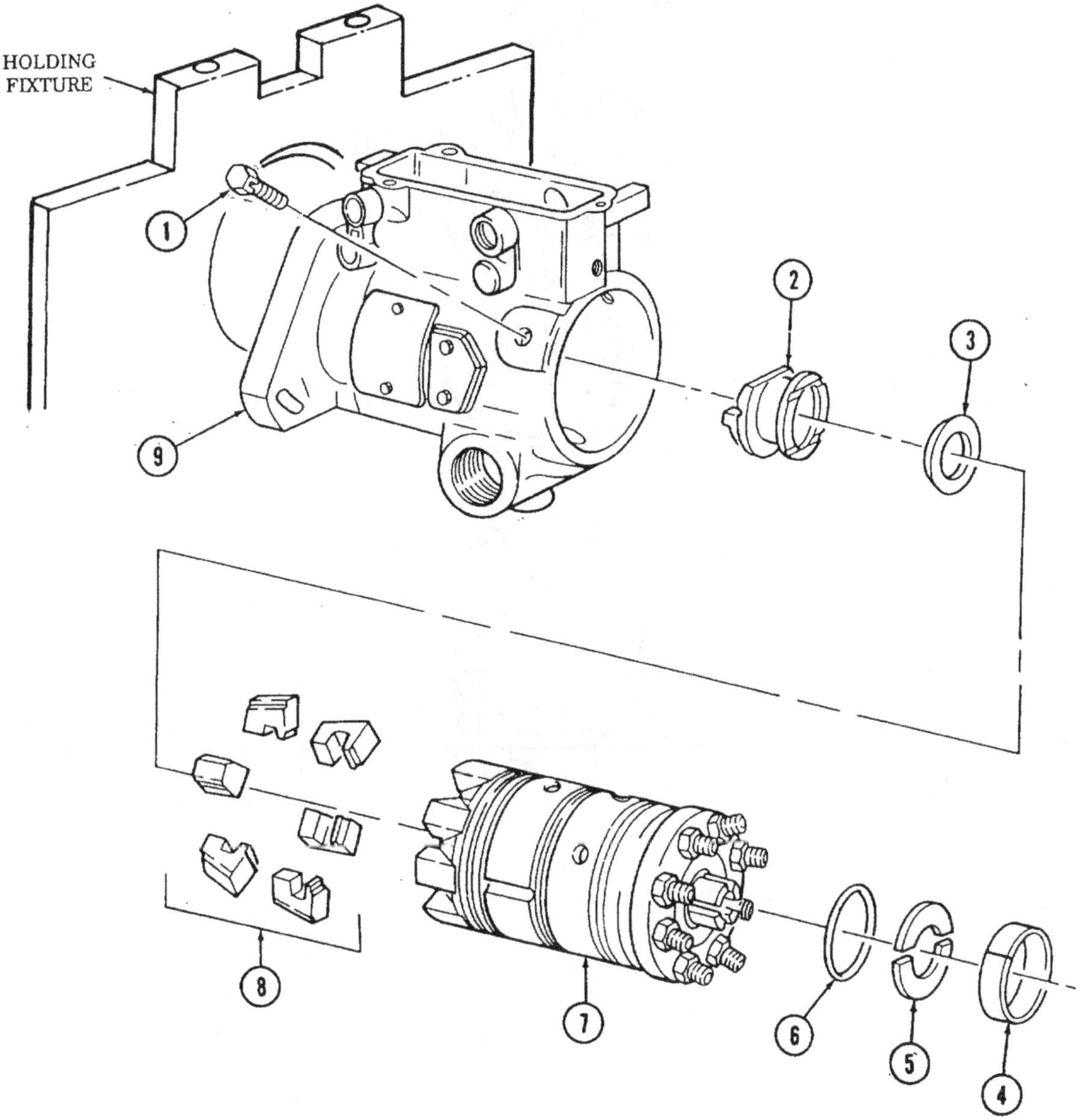

2-32. FUEL INJECTION PUMP REPAIR (Cont'd)

46. Remove snapring (10) and weight retainer (11) from rotor (13).

NOTE

Note direction of arrow on cam ring indicating rotation of pump for assembly.

47. Remove cam ring (12) from hydraulic head (7).

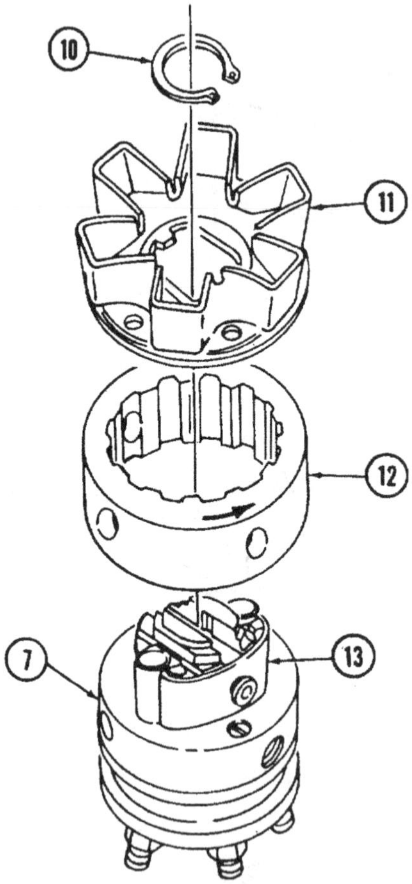

2-32. FUEL INJECTION PUMP REPAIR (Cont'd)

48. Remove two cam roller shoes (2) and two cam rollers (1) from rotor (4).
49. Remove leaf spring adjusting screw (7) and leaf spring (6) from rotor (4).
50. Remove two plungers (3) from rotor (4).
51. Install governor weight support in vise.

CAUTION

- Do not handle precision ground surface of rotor to avoid contamination.
- Handle all parts carefully with clean hands wet with calibrating fluid to avoid contamination.

52. Remove rotor (4) from hydraulic head (5) and place rotor (4) in governor weight support. Using 5/32-in. hex head driver, remove delivery valve stop plug (12).
53. Remove head seal (8) from hydraulic head (5). Discard head seal (2).

NOTE

If delivery valve cannot be removed from rotor, delivery valve extractor must be used.

54. Remove valve stop (11), spring (10), and delivery valve (9) from rotor (4). Discard valve stop (11) and tag spring (10) for assembly.

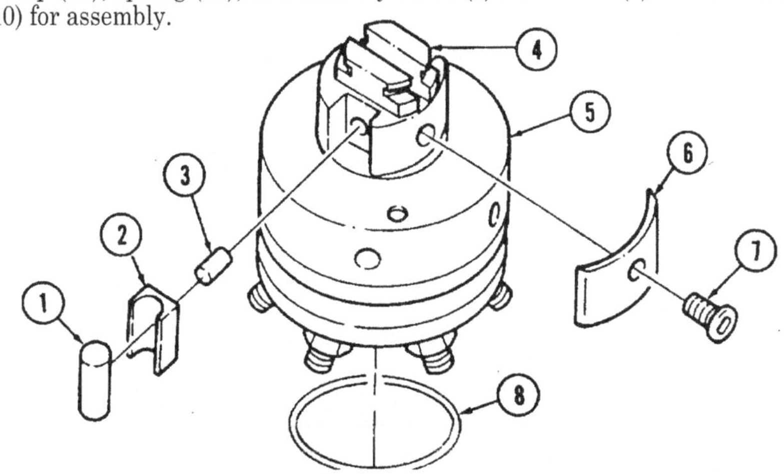

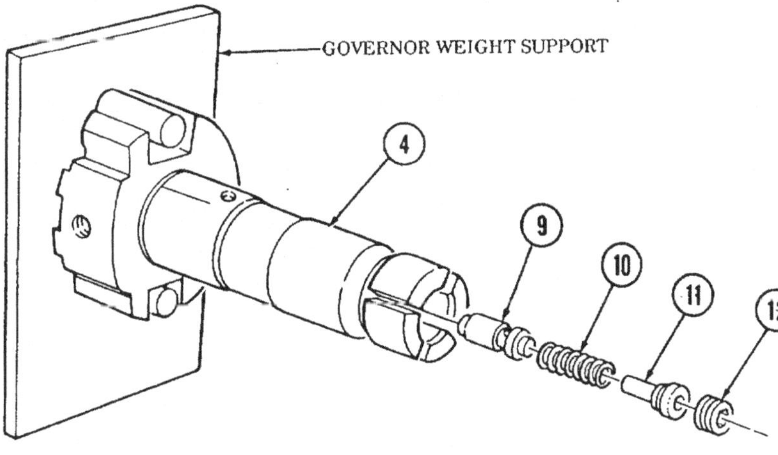

2-32. FUEL INJECTION PUMP REPAIR (Cont'd)

55. Remove drive shaft snapring (18) and drive shaft (15) from housing (17).
56. Remove two black seals (22) and red seal (16) from drive shaft (15). Discard seals (22) and (16).
57. Remove thrust button (13) and spring (14) from drive shaft (15) and tag spring (14) for assembly.
58. Remove two screws (20), timing cover (21), and gasket (19) horn housing (17). Discard gasket (19).
59. Remove housing (17) from holding fixture.

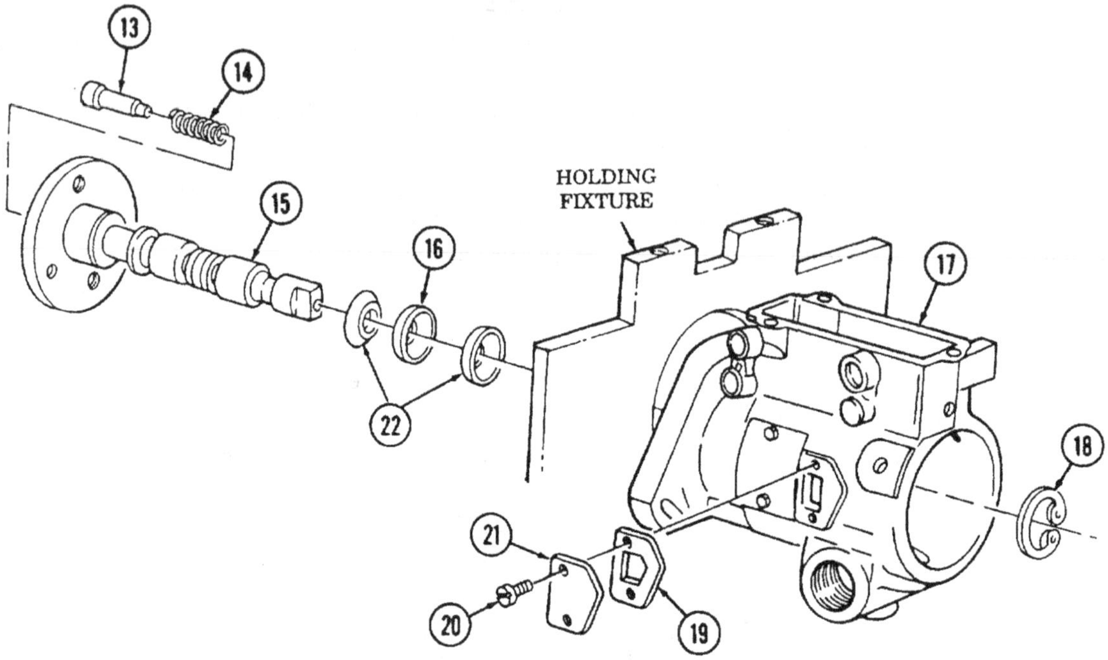

2-32. FUEL INJECTION PUMP REPAIR (Cont'd)

60. Remove terminal (2), washer (3), and shell (4) from stud (9).
61. Remove nut (5), lockwasher (6), washer (7), and fiber washer (8) from stud (9). Discard lockwasher (6).
62. Remove locknut (14), ground strap (13), lockwasher (6), nut (5), lockwasher (6), washer (7), and fiber washer (8) from stud (11). Discard locknut (14) and lockwashers (6).
63. Remove electrical shutoff solenoid (10) from cover (12).
64. Remove check valve (1) and O-ring (15) from cover (12). Discard O-ring (15).

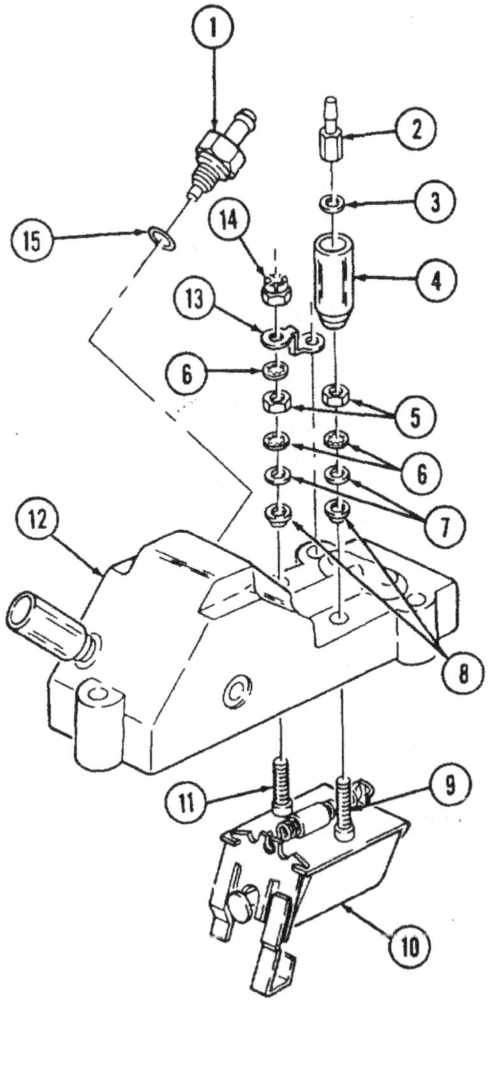

2-32. FUEL INJECTION PUMP REPAIR (Cont'd)

65. Remove terminal (2), washer (3), and shell (16) fromstud (20).
66. Remove nut (5), lockwasher (6), washer (7), and fiber washer (8) from stud (20). Discard lockwasher (6).
67. Remove locknut (14), nut (5), lockwasher (6), washer (7), and fiber washer (8) from stud (17). Discard lockwasher (6) and locknut (14).
68. Remove cold advance solenoid (18) and plunger (19) from cover (12).
69. Remove plunger (19) from cold advance solenoid (18).

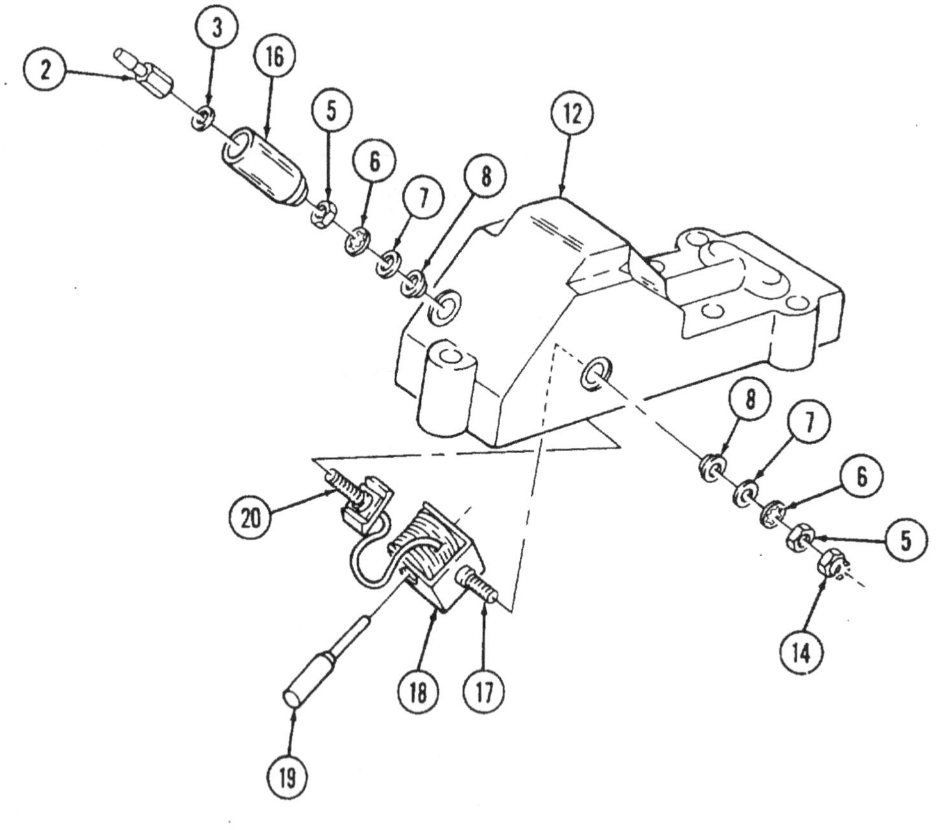

2-32. FUEL INJECTION PUMP REPAIR (Cont'd)

b. Cleaning

CAUTION

- Do not handle precision ground surface of rotor to avoid contamination.
- Handle all parts carefully with clean hands wet with calibrating fluid to avoid contamination.
- Do not immerse hydraulic head in solvent or sealant damage may result.

1. Clean all fuel injection pump components in accordance with instructions in para. 2-9.
2. Rinse hydraulic head (1) and rotor (2) in calibrating fluid.

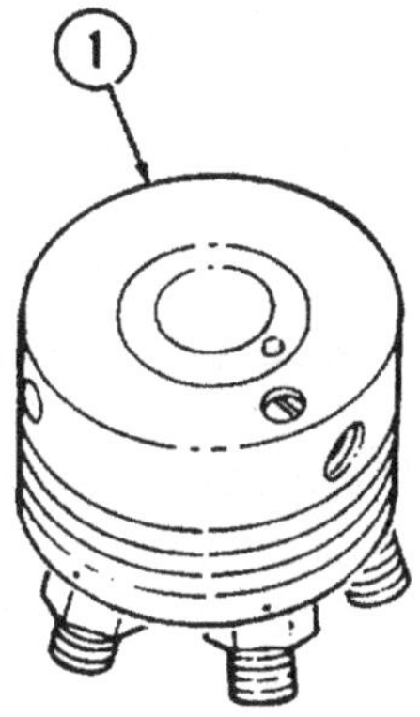

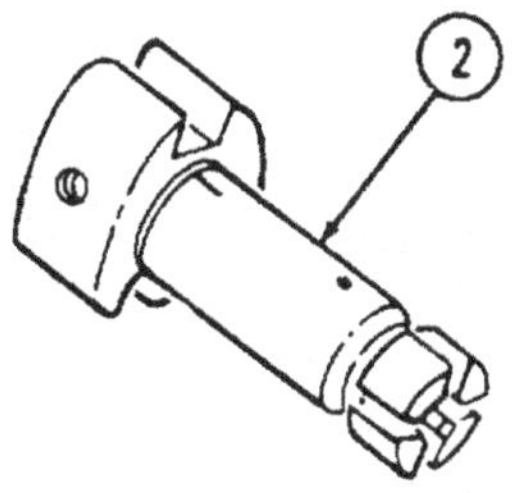

c. Inspection

NOTE

For general inspection instructions, refer to para. 2-10.

1. Inspect housing (5) for damage. Replacc injection pump if damaged.
2. Inspect pilot tube (4) in housing (5) for scoring or damage. If scored or damaged, perform steps 3 through 7, if not, go to step 8.
3. Remove pilot tube (4) and drive shaft bearing (3) from housing (5) using pilot tube mandrel. Discard pilot tube (4).
4. Inspect drive shaft bearing (3) for roughness or damage. Discard if rough or damaged.

2-32. FUEL INJECTION PUMP REPAIR (Cont'd)

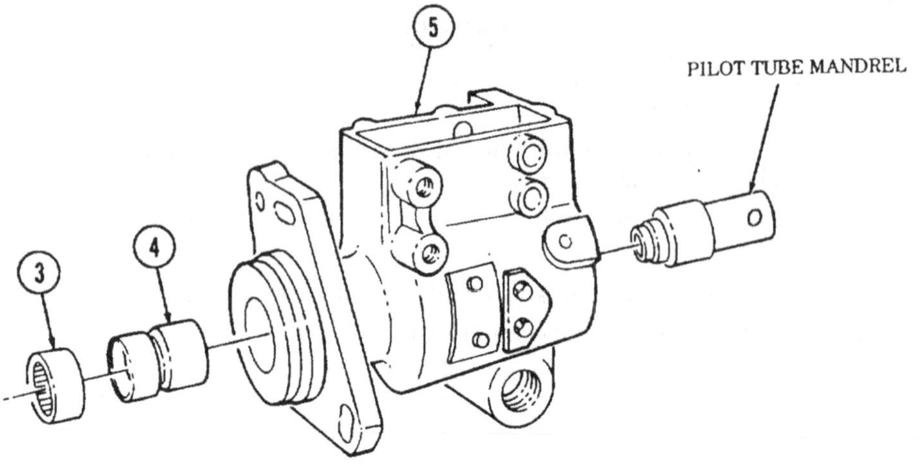

5. Apply sealing compound to outside diameter of pilot tube (4).
6. Using pilot tube mandrel, install pilot tube (4) in housing (5) until groove (6) in pilot tube (4) aligns with vent hole (7) in housing (5).
7. Using drive shaft bearing installer, install drive shaft bearing (3) in housing (5). Go to step 11.
8. Inspect drive shaft bearing (3) for roughness or damage. If rough or damaged, perform steps 9 and 10. If not, go to step 11.
9. Remove drive shaft bearing (3) from housing (5) and discard bearing (3).
10. Using drive shaft bearing installer, install driveshaft bearing (3) in housing (5).

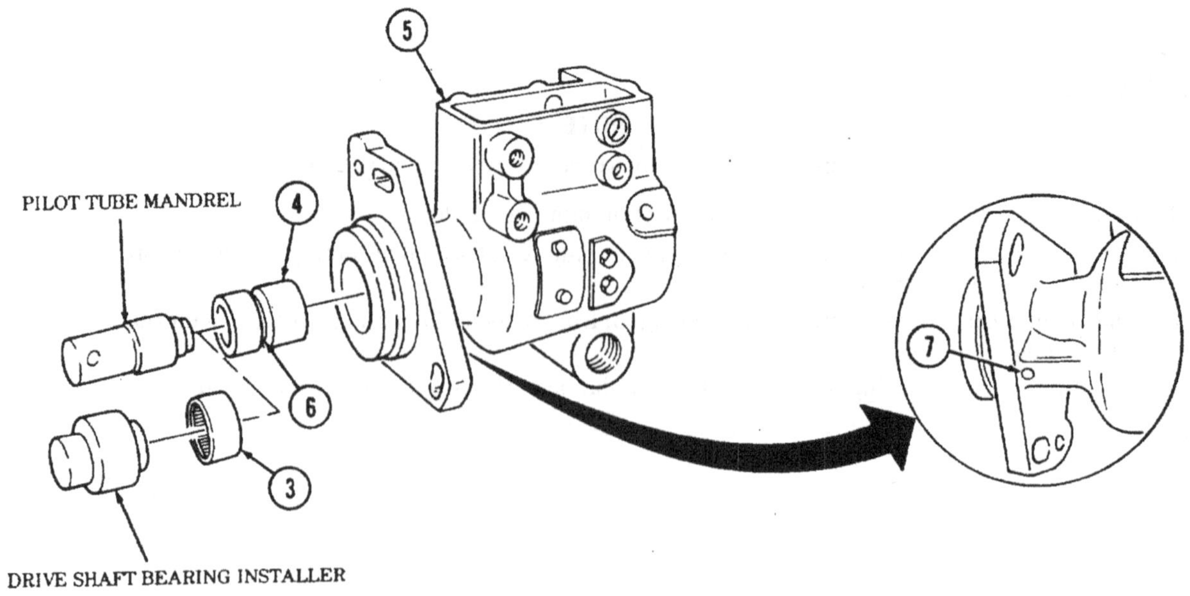

2-32. FUEL INJECTION PUMP REPAIR (Cont'd)

11. Inspect throttle shaft bushings (1) for damage. If damaged, perform steps 12 through 20 using throttle and shutoff shaft bushing installation kit. If not, go to step 21.
12. Install brace into housing (2) to prevent distortion of housing (2) during bushing (1) removal.
13. Install rod through bushings (1).
14. Install bushing guide, two washers, and nuts.
15. Tighten nuts until bushing (1) is removed from housing (2). Remove tools and discard bushing (1).

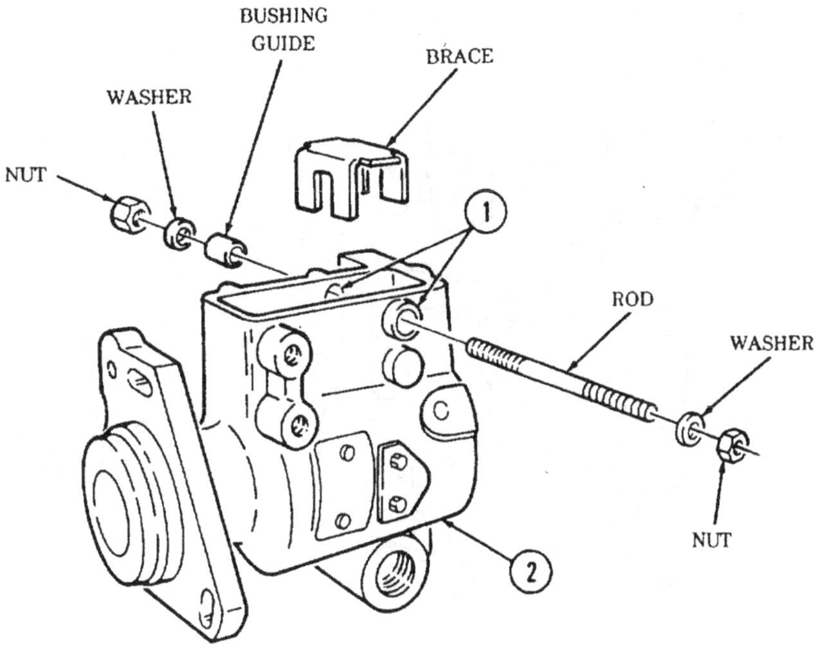

2-32. FUEL INJECTION PUMP REPAIR (Cont'd)

16. Install brace and rod into housing (2).
17. Apply sealing compound to outside diameter of bushing (1).
18. Install bushing (1), bushing guide, two washers, and nuts on rod.
19. Tighten nuts until bushing (1) is past flush with bushing boss (3) on housing (2). Remove tools from housing (2).
20. Repeat steps 12 through 19 for opposite bushing, if required.

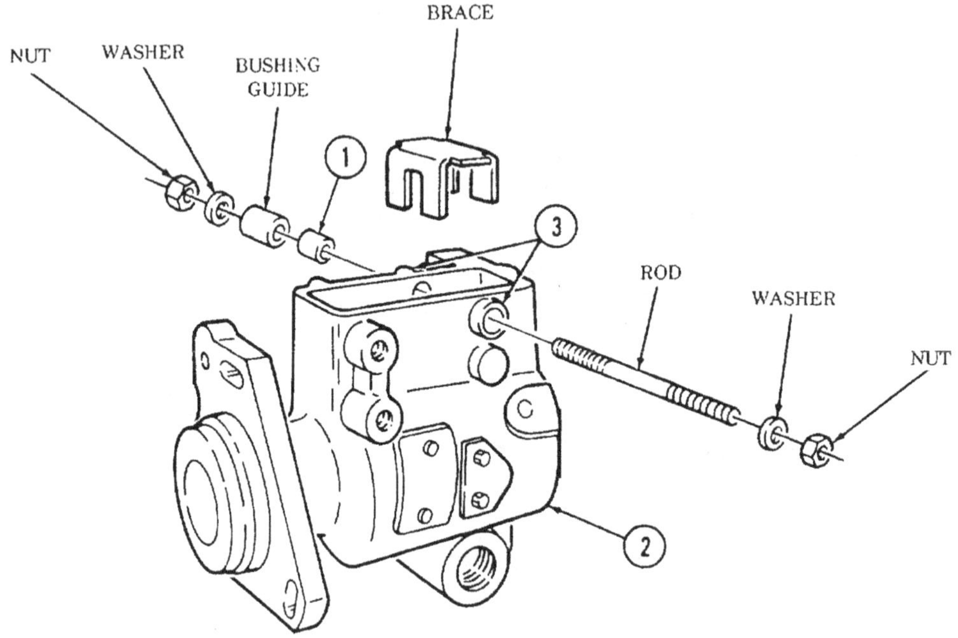

2-32. FUEL INJECTION PUMP REPAIR (Cont'd)

21. Inspect idle spring (2) on min-max governor (1) for corrosion or damage. Replace if corroded or damaged.

NOTE

Perform steps 22 and 23 for 6.5L engines and 6.2L engines 1990 and above.

22. Inspect inner governor spring (7) for corrosion or damage. Replace if corroded or damaged.
23. Inspect governor thrust spring (8) for corrosion or damage. Replace if corroded or damaged.
24. Inspect min-max spring (3) for corrosion or damage. If corroded or damaged, perform steps 25 through 29. If not, go to step 30.
25. Remove idle spring (2) from pushrod (5).
26. Measure distance "A" from throttle block (4) to end of pushrod (5) for assembly.
27. Put throttle block (4) in vise and remove pushrod (5), spring (3), and washer (6). Discard spring (3).
28. Install washer (6) and spring (3) on pushrod (5) and install into throttle block (4).
29. Tighten pushrod (5) until measurement taken in step 26 is obtained.

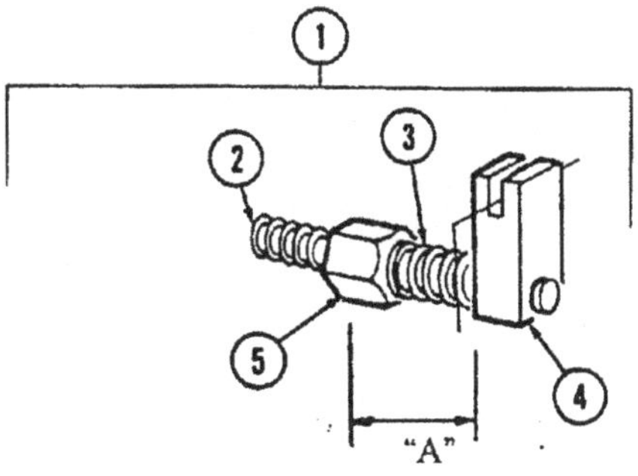

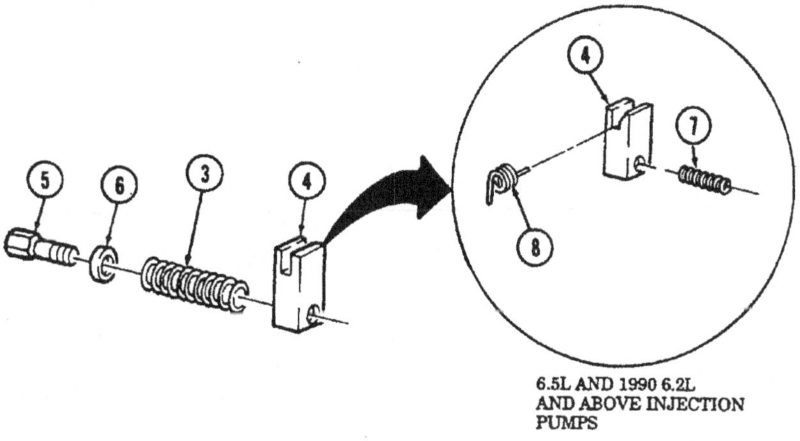

2-32. FUEL INJECTION PUMP REPAIR (Cont'd)

30. Inspect regulatar (10) for damage. Replace if damaged.

NOTE

When removing roll pin, note which hole from where pin is removed, It is marked "C" or "CC".

31. Inspect roll pin (9) in regulator (10) for damage. Replace if damaged.

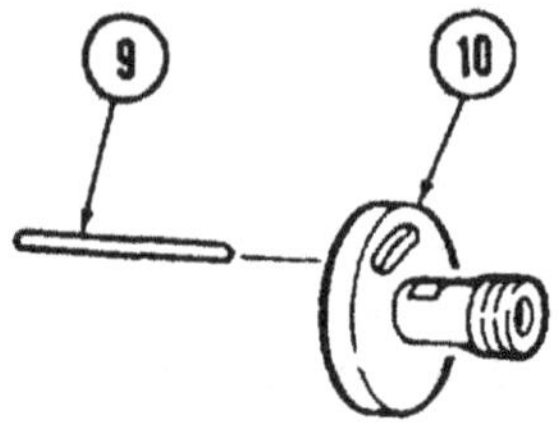

32. Inspect arm spring (11) for damage. Replace if damaged.

CAUTION

Metering valves are available in standard and oversize. If replacing metering valve, check for number "SB-336" on top of hydraulic head. If number is present, oversize metering valve must be used or injection pump may malfunction.

33. Inspect metering valve (12) and arm (13) for damage. If damaged, perform steps 34 and 35. If not, go to step 36.
34. Place valve (12) in holding fixture hole and remove arm (13) from valve (12).
35. Install arm (13) on valve (12).

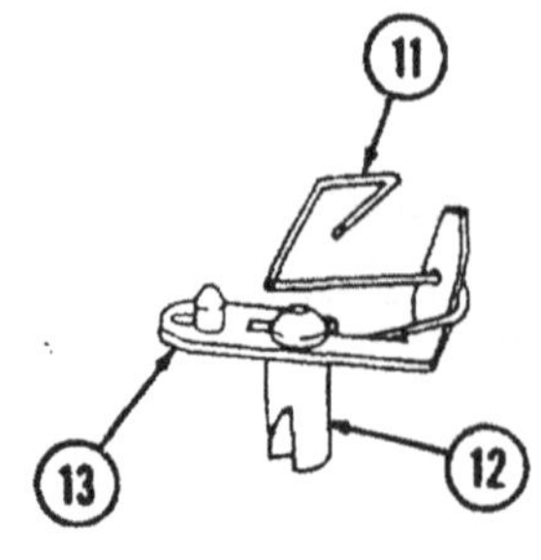

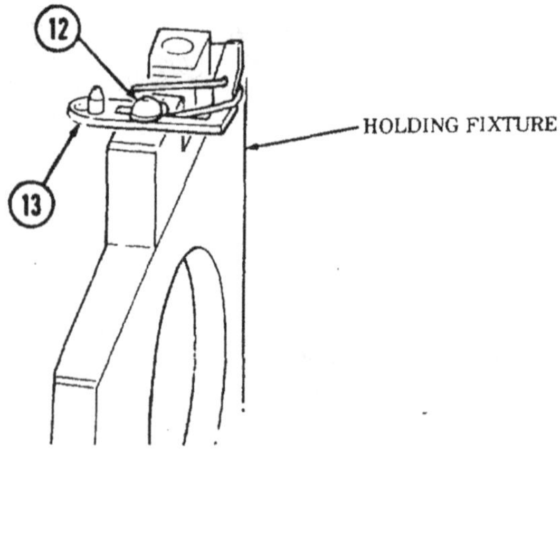

2-32. FUEL INJECTION PUMP REPAIR (Cont'd)

36. Using a micrometer, measure drive shaft (2) at point "A" to determine wear. Inspect drive shaft (2) seal area for nicks, burrs, scratches, or damage. Replace drive shaft (2) if measurement at point "A" is less than 0.305 in. (7.75 mm) or if any nicks, burrs, scratches, or damage is present.
37. Inspect dowel pin (1) in drive shaft (2) for damage. If damaged, replace dowel pin (1). Apply sealing compound to replacement dowel pin (1).

CAUTION

Hydraulic head and rotor are matched sets and must be replaced as an assembly or injection pump damage may result.

38. Inspect fittings (4) in hydraulic head (3) for damage. Replace hydraulic head (3) and rotor (13) if any fittings (4) are damaged.

CAUTION

Do not handle precision ground surface of rotor to prevent contamination.

39. Inspect hydraulic head (3) and rotor (13) for damage. Replace both if either is damaged.
40. Inspect charging ports (11) and discharge port (12) on rotor (13) for chipped edges. Replace hydraulic head (3) and rotor (13) if any chipped edges are apparent.
41. Inspect rotor (13) for scratches. Replace hydraulic head (3) and rotor (13) if scratches are present.

WARNING

Do not allow solvents to get on skin or clothing. Solvents can burn easily and give off harmful vapors. Keep away from open fire and use in a well-ventilated area. If solvent gets on skin or clothing, wash immediately. Failure to follow these instructions could cause injury or death.

NOTE

It maybe necessary to interchange or reverse plunger position when installing plungers into rotor.

42, While holding rotor (13) under clean calibrating fluid, insert plungers (10) into bores of rotor (13). With thumb and forefinger over roller shoe (9) slots, tilt from side to side to determine freedom of movement. If the plungers (10) are sticking, but not visibly damaged, clean both plungers (10) and bore with a soft brush and lacquer thinner or acetone. Check freedom of movement again. If plungers (10) still stick, replace them with the next smaller size ("A" is the largest, "D" is the smallest). If the smallest plungers (10) stick in rotor (13), replace hydraulic head (3) and rotor (13).

43. Inspect leaf spring contact area (14) and tang slot (7) on rotor (13) for excessive wear. Replace hydraulic head (3) and rotor (13) if leaf spring contact area (14) or tang slot (7) are excessively worn.
44. Inspect cam rollers (8) for freedom of rotation in their roller shoes (9). Inspect cam rollers (8) and roller shoes (9) for abrasive wear patterns. Replace cam rollers (8) and roller shoes (9) if cam rollers (8) do not rotate freely or if either shows signs of abrasive wear patterns.
45. Inspect top edge of each roller shoe (9) for chipping or excessive wear. Replace cam rollers (8) and roller shoes (9) if chipped or excessively worn.
46. Inspect leaf spring (6) for excessive wear. Replace leaf spring (6) if excessively worn.
47. Inspect leaf spring screw (5) for tightness in rotor (13). Replace leaf spring screw (5) if screw (5) does not thread tightly into rotor (13).

2-32. FUEL INJECTION PUMP REPAIR (Cont'd)

NOTE

Machining marks between cam lobes should not be considered damage.

48. Inspect cam lobes (16) on cam ring (15) for spalling or flaking out. Replace cam ring (15) if cam lobes (16) are spalled or flaked.
49. Inspect all transfer pump blades (17) and transfer pump liner (18) for scoring or damage. Replace all transfer pump blades (17) and transfer pump liner (18) if any are scored or damaged.
50. Using a micrometer, measure transfer pump blades (17) at point "A" to determine wear. If any transfer blade (17) is smaller than 0.538 in. (13.67 mm), replace all transfer pump blades (17) and transfer pump liner (18).
51. Inspect governor weights (21) for excessive wear at heel (19) and toe (20) Replace governor weights (21) if excessively worn.
52. Inspect delivery valve reaction cuff (22) for chipping or erosion. Replace delivery valve (23) if reaction cuff (22) is chipped or eroded.
53. Inspect vent wire (25) in vent wire screw (24) for freedom of movement. If vent wire (25) is stuck, replace vent wire screw (24).

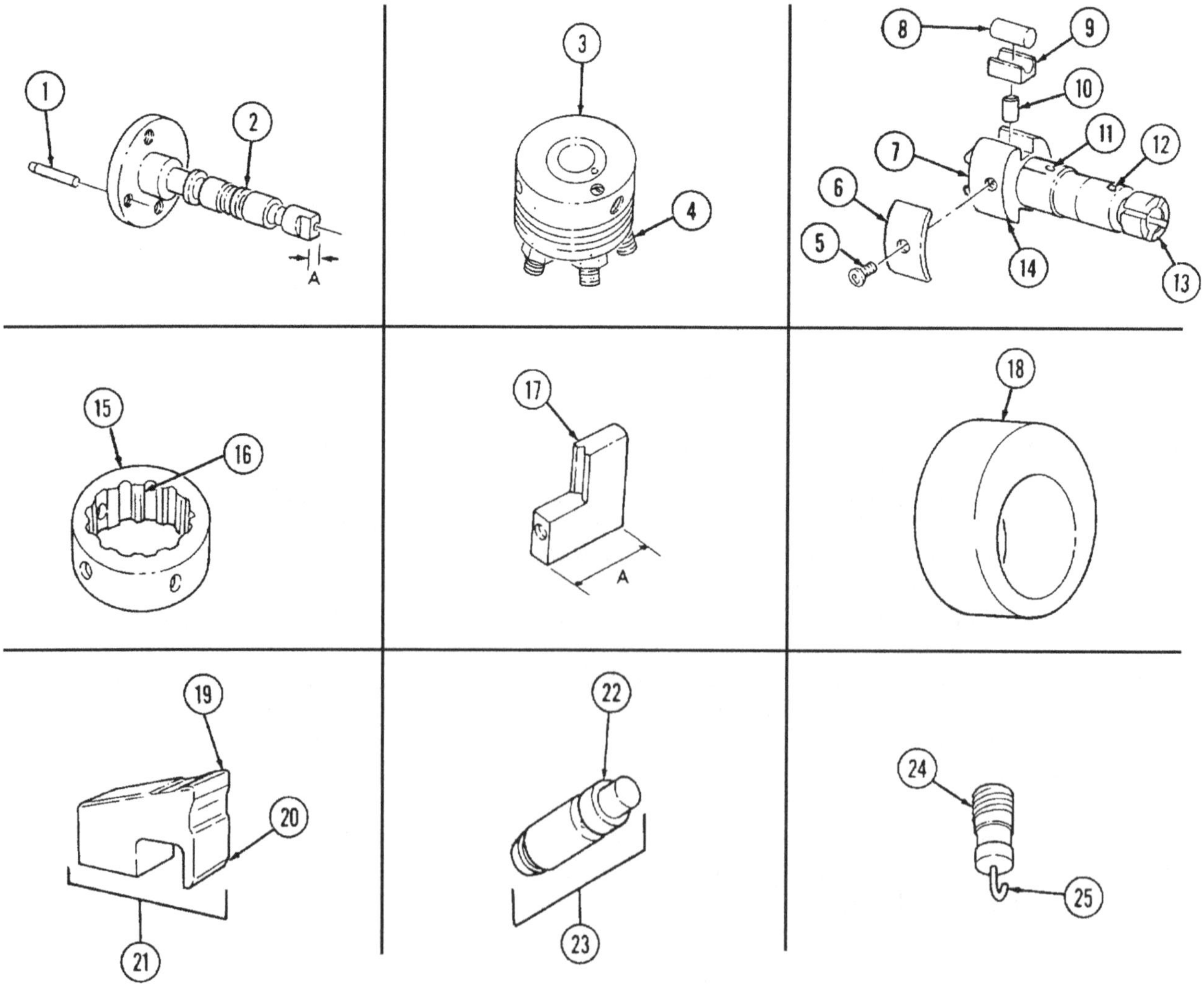

2-32. FUEL INJECTION PUMP **REPAIR**(Cont'd)

54. Inspect throttle shaft (3) for distortion or damage. Replace if distorted or damaged.

NOTE

Perform step 55 only if throttle shaft was replaced.

55. Install low idle screw (1), spring (2), max travel screw (4), and nut (5) in throttle shaft (3).
56. Inspect rubber connectors (6) for deterioration, cracks, or damage. Replace if cracked, deteriorated, or damaged.
57. Inspect pump cover (7) for damage. Replace if damaged.
58. Inspect check valve (8) for damage. Replace if damaged.
59. Inspect cold start advance solenoid (10) for damage. Replace if damaged.
60. Inspect insulators (9) and plunger (11) for damage, Replace if damaged,
61. Inspect insulators (13) and electrical shutoff solenoid (12) for damage. Replace if damaged.
62. Refer to para. 2-10 for general inspection instructions for all other injection pump parts.

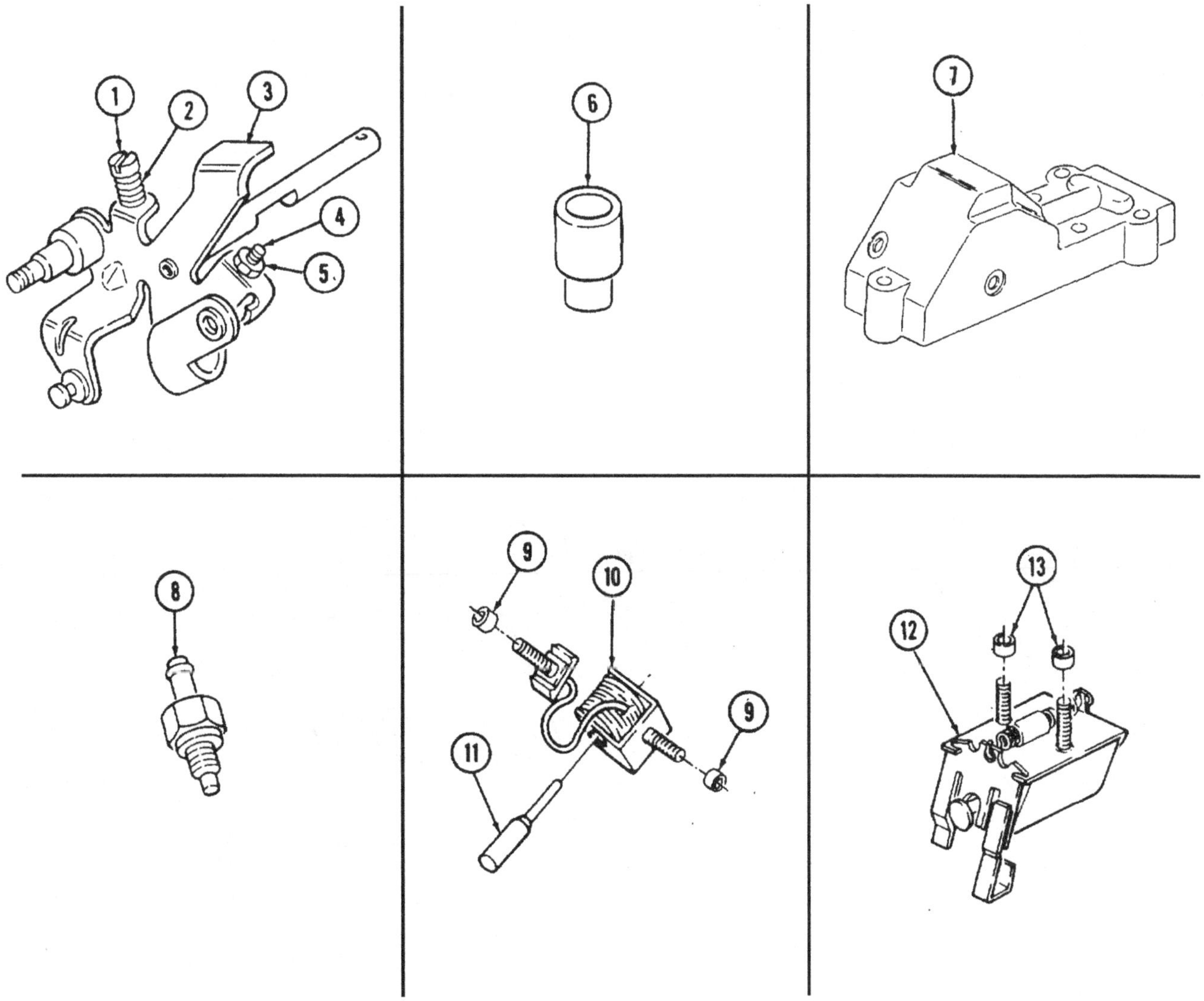

2-32. FUEL INJECTION PUMP REPAIR (Cont'd)

d. Assembly

CAUTION

Handle all parts carefully with clean hands wet with calibrating fluid to avoid contamination.

1. Mount injection pump housing (20) in holding fixture and secure with vise.
2. Lubricate seals (17), (18), and (19) with seal lubricant and install black seal (17) in groove (25), red seal (18) in groove (24) and black seal (19) in groove (23) using drive shaft seal installer.
3. Install thrust button spring (15) and thrust button (14) in drive shaft assembly (16).

NOTE

Note position of timing mark on end of drive shaft tang when installing drive shaft assembly. This is necessary to align rotor to drive shaft.

4. Install drive shaft assembly (16) in housing (20).
5. Install snapring (21) in groove (22) on drive shaft assembly (16).

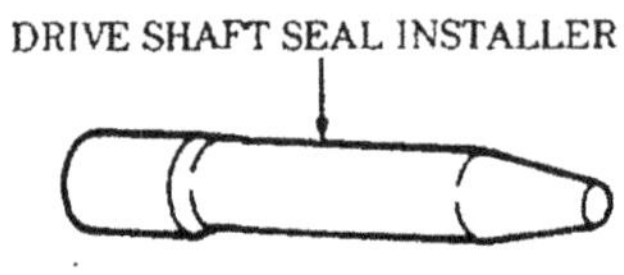

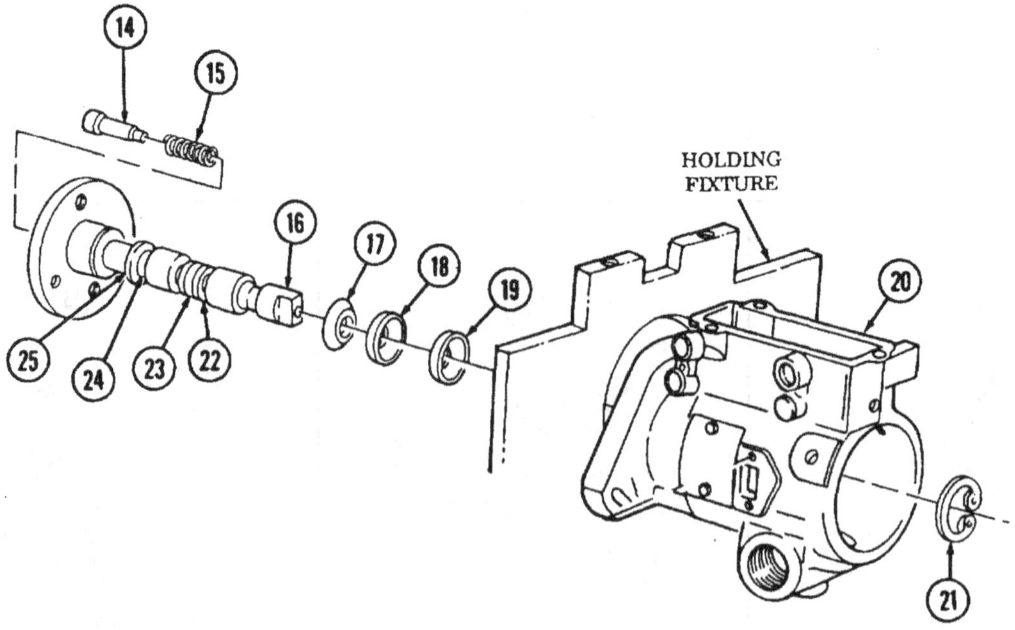

2-32. FUEL INJECTION PUMP REPAIR (Cont'd)

CAUTION

- Do not handle precision ground surface of rotor to avoid contamination.
- Handle all parts carefully with clean hands wet with calibrating fluid to avoid contamination.

6. Mount governor weight support in vise and rinse rotor (1) in calibrating fluid and install in support.

NOTE

- Delivery valves are available in standard and oversize. If replacing delivery valve, check for marking "OS" on side of rotor. If "OS" marking is present, oversize delivery valve must be used.
- Oversize delivery valves have a black marking at the base of delivery valve.

7. Install delivery valve (2) in rotor (1).
8. Install valve spring (3) and valve stop (4) in rotor (1).

CAUTION

Excessive tightening of stop plug will cause seizure of hydraulic head and rotor.

9. Using 5/32-in. hex head driver, install stop plug (5) in rotor (1) and tighten to 85-90 lb-in. (10 N•m).

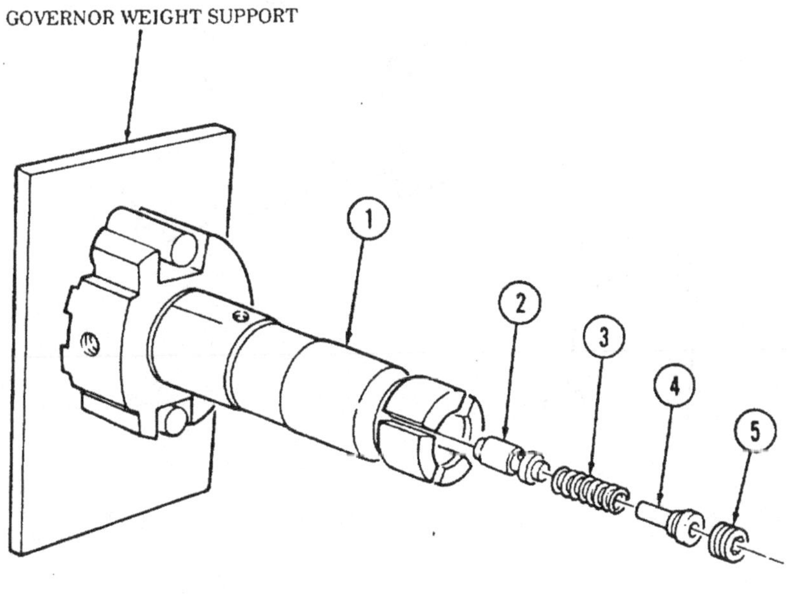

2-32. FUEL INJECTION PUMP REPAIR (Cont'd)

10. Install leaf spring (9) and adjusting screw (10) on rotor (1).
11. Remove rotor (1) from governor weight support and submerge in calibrating fluid.

NOTE

- Plungers of any given nominal diameter are graded in four select fit sizes: A, B, C, and D. The rotor is etched with the letter indicating bore size. This mark is found on the base of the rotor.
- If plunger replacement is required, check the size designation on the rotor and use the plunger of corresponding part number as indicated by letter designation on rotor.

12. Install two plungers (8) in rotor (1).
13. Assemble two cam roller shoes (7), and cam rollers (6), and install in rotor (1).

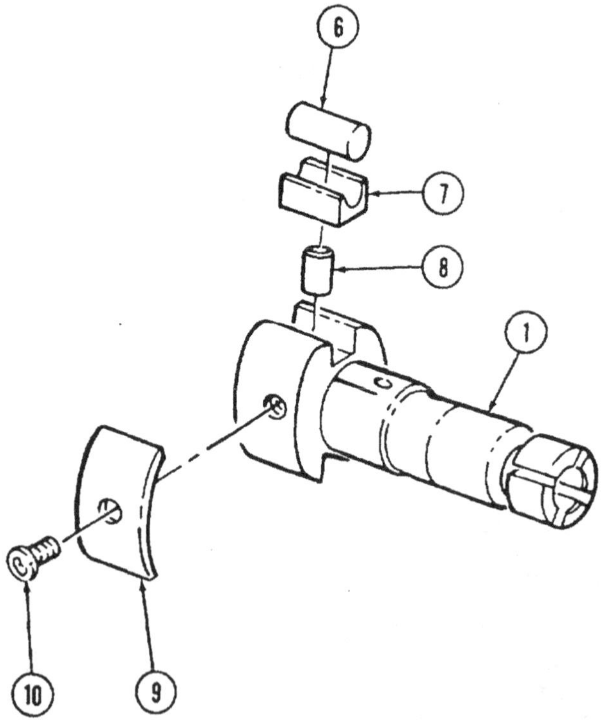

2-32. FUEL INJECTION PUMP REPAIR [Cont'd]

14. Install roller-to-roller setting tool in soft-jaw vise.
15. Install roller (1) in roller-broiler setting tool.

WARNING

Always wear eyeshields when using compressed air. Failure to wear eyeshields may result in eye injury.

16. Apply compressed air regulated to 40-100 psi (276-689 kPa) to roller-broiler setting tool.

NOTE

Increase roller-to-roller dimension 0.006 in. (0.152 mm) if using oversize cam ring.

17. Measure roller-to-roller dimension and compare to pump specification 1.981 in. (50.31 mm).

NOTE

Roller-to-roller dimension is a completely accurate maximum fuel adjustment and should not differ from pump specification.

18. Turn leaf spring adjusting screw (2) clockwise to increase or counterclockwise to decrease roller-to-roller dimension.
19. Rotate rotor (1) until one cam roller (3) aligns with roller-to-roller setting tool dial indicator plunger.
20. Loosen indicator dial retaining screw and set preload by sliding indicator down until plunger depresses on cam roller (3) 0.015 in. (0.381 mm). Tighten screw.
21. Rotate dial on indicator to "zero" indicator on high point of cam roller (3).

NOTE

If cam roller centrality is greater than 0.008 in. (0.203 mm), cam shoes can be interchanged or replaced to achieve tolerance. Centrality must be rechecked after each time components are changed.

22. Measure cam roller centrality by rotating rotor (1) in either direction until high point of other cam roller (3) depressed dial indicator plunger. Indicator reading should be 0.000 in. to 0.008 in. (0.000 to 0.203 mm).
23. Rinse hydraulic head (4) with calibrating fluid.
24. Remove rotor (1) from roller-to-roller setting tool and rinse with calibrating fluid.
25. Install rotor (1) in hydraulic head (4).

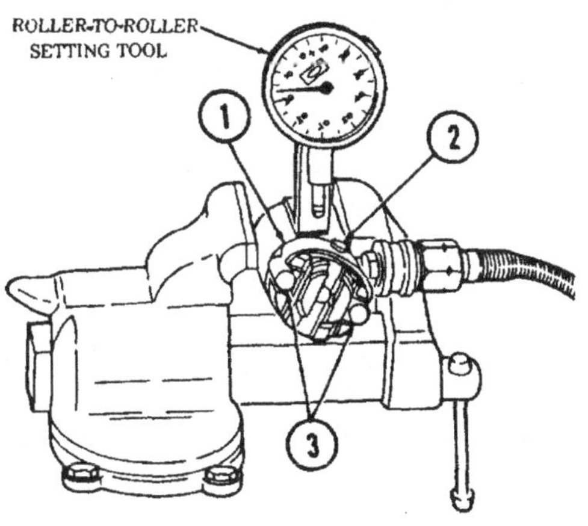

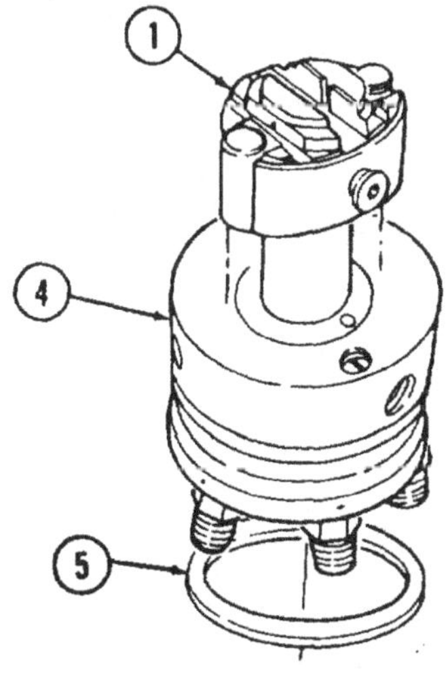

2-32. FUEL INJECTION PUMP REPAIR (Cont'd)

26. Install head seal (5) on rotor (4).
27. Install cam ring (8) on hydraulic head (4) with directional arrow on ring showing counterclockwise rotation of pump rotor (1).
28. Install weight retainer (7) on rotor (1), align with check ball, and secure with snapring (6).

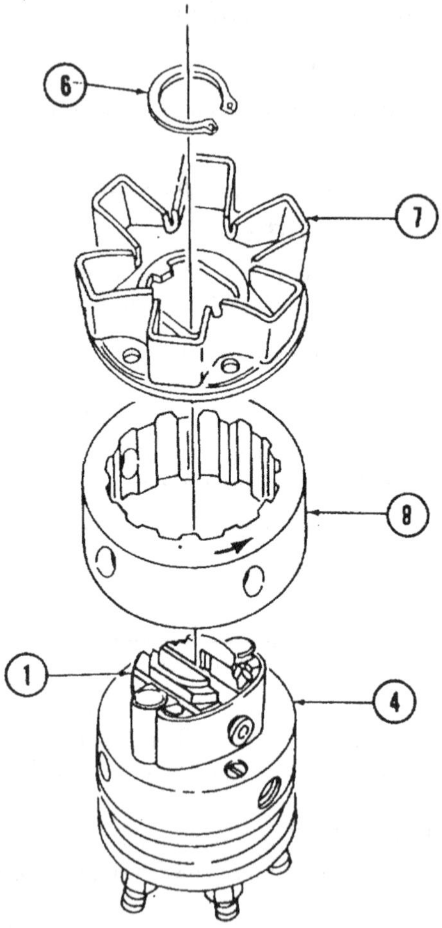

2-32. FUEL INJECTION PUMP REPAIR (Cont'd)

29. Install transfer pump end cap seal (2) in hydraulic head (1).

CAUTION

Use care when handling rotor weight retainer and hydraulic head before retainers are installed. Rotor can fall out and cause damage to components.

30. Install two rotor retainers (3) with cut out portions meeting over roll pin hole in hydraulic head (1).
31. Install liner locating ring (4) on hydraulic head (1) with split 90° from roll pin hole.
32. Install transfer pump liner (5) so slot aligns with roll pin hole in hydraulic head (1).

CAUTION

Blades must not be cocked during installation. Sharp edge of liner can score blade ends.

NOTE

- Transfer pump blades are available in standard and oversize. Oversize blades are marked in black and must be used whenever possible.
- Oversize and standard transfer pump blades do not have to be used as matched sets.

33. Install two springs (6) in four transfer pump blades (7) and install transfer pump blades (7) in rotor (8).

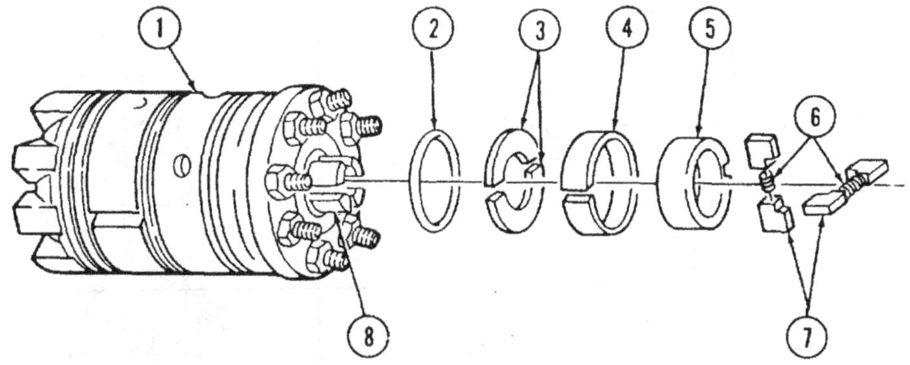

2-32. FUEL INJECTION PUMP REPAIR (Cont'd)

CAUTION

Adjusting plug must not be installed past flush with end of regulator or pump damage will result.

34. Install piston (10), spring (13), and adjusting plug (11) in regulator (9). Tighten adjusting plug (11) until flush with end of regulator (9).

NOTE

Early production injection pumps have an inlet filter screen, retaining ring, and seal or pressure plate. Late production pumps have a different configuration screen which does not require a retaining ring and seal or pressure plate. When repairing early production pumps, replace the screen with a late production one.

35. Deleted
36. Install filter screen (12) on regulator (9).
37. Lightly coat filter screen assembly (14) and threads of transfer end cap (16) with seal lubricant.
38. Install filter screen assembly (14) and end cap (16) on hydraulic head (1). Apply slight pressure on top of end cap (16) and rotate counterclockwise until a slight click is heard.
39. Turn end cap (16) clockwise until snug.
40. Install inlet tube assembly (15) on end cap (16).

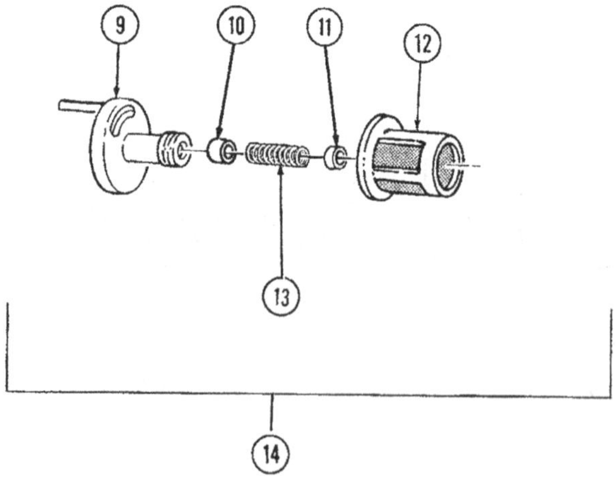

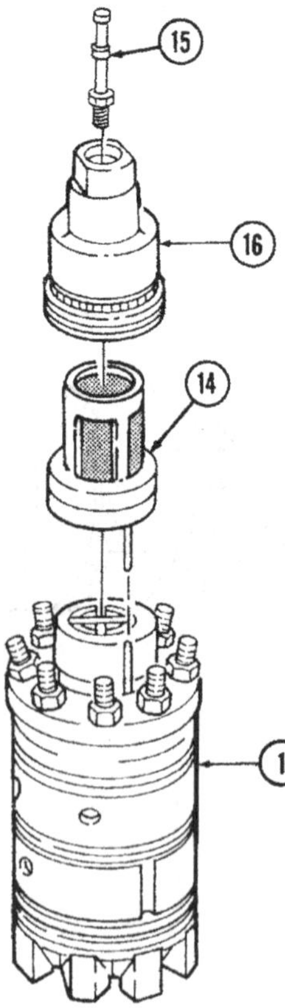

2-32. FUEL INJECTION PUMP REPAIR (Cont'd)

41. Install six governor weights (3) in weight retainer (4).

NOTE

Governor weights must work freely after installation.

42. Install thrust washer (2) and thrust sleeve (1) into lower slots of governor weights (3).

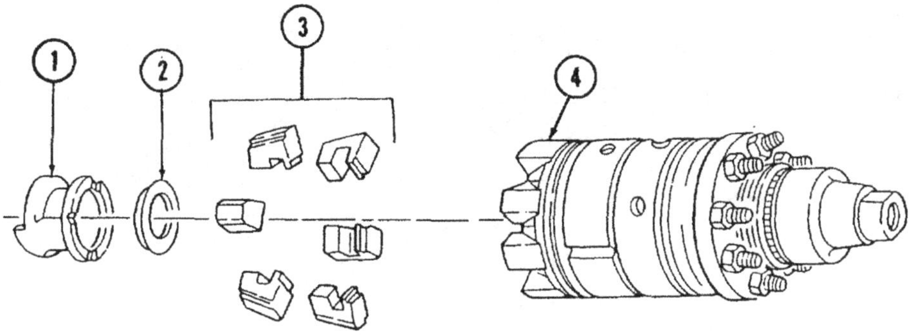

2-32. FUEL INJECTION PUMP REPAIR (Cont'd)

43. Align timing mark (11) on drive shaft (5) and timing mark (6) on rotor (7).

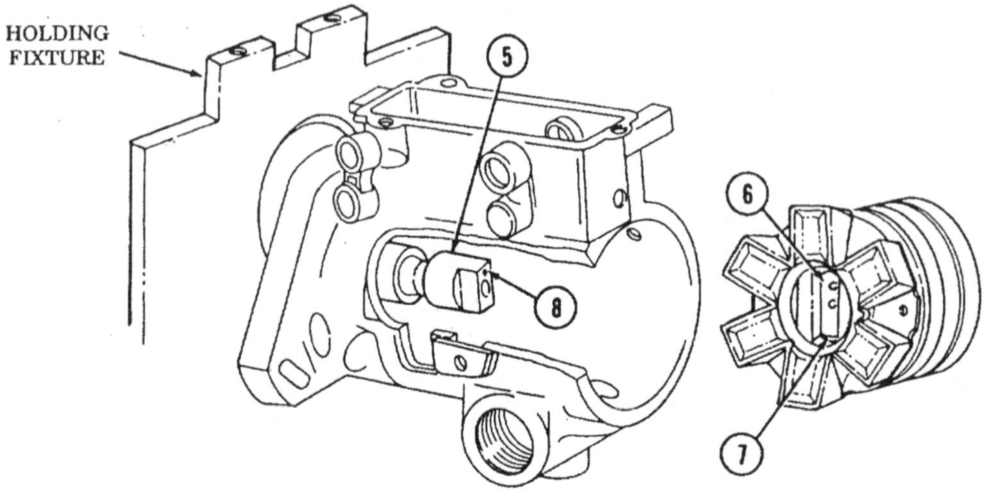

44. Apply light coat of seal lubricant to inside edge of housing (12).
45. Rotate cam ring (9) so index mark (11) is aligned with timing cover access hole (13).

CAUTION

Hydraulic head must only be installed far enough to install head locking screws, or leakage and damage to hydraulic head O-ring seal will result.

46. Install hydraulic head (10) in housing (12) with slight rotary motion.
47. Rotate head (10) and align head locking capscrew holes with holes in housing (12).
48. Install two head locking capscrews (8) finger tight.

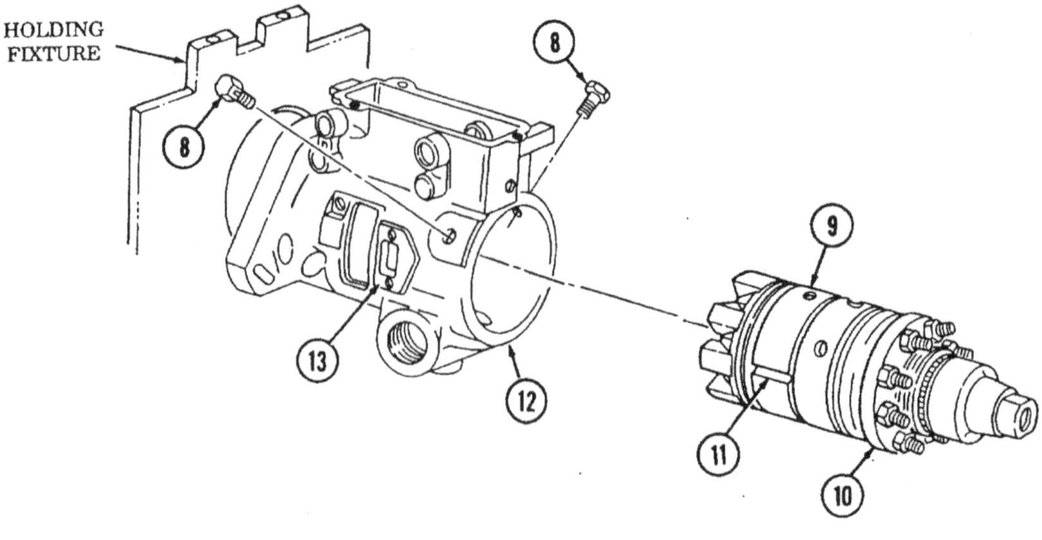

2-32. FUEL INJECTION PUMP REPAIR (Cont'd)

49. Rotate housing (11) and holding fixture in vise so advance plug (8) faces up.
50. Install O-ring (6) on head locating screw (7).
51. Lightly coat head locating screw (7) with seal lubricant and install in housing (11). Tighten screw (7) to 180-220 lb-in. (20-25 N•m) using 5/16-in. hex head driver.
52. Install servo advance piston (14) in housing (11) with notch and two small holes facing head locating screw (7).
53. Install O-ring (13) and power side advance piston hole plug (12) in housing (11). Tighten plug (12) to 215-265 lb-in. (24-30 N•m).

NOTE

Rotate cam ring to align hole for cam advance pin in housing.

54. Install cam advance pin (10).
55. Install O-ring (9) and advance plug (8) in housing (11). Tighten advance plug (8) with 1/4-in. hex head driver to 75-100 lb-in. (9-11 N•m).
56. Install servo advance valve (5) on spring (4) and install servo advance valve (5) in servo advance piston (14).
57. Install servo advance plunger (1) and O-ring (3) on spring side plug (2).
58. Install spring side plug assembly (2) in housing (11) so servo advance plunger (1) fits in spring (4). Tighten sprinig side plug assembly (2) to 215-265 lb-in. (24-30 N•m).

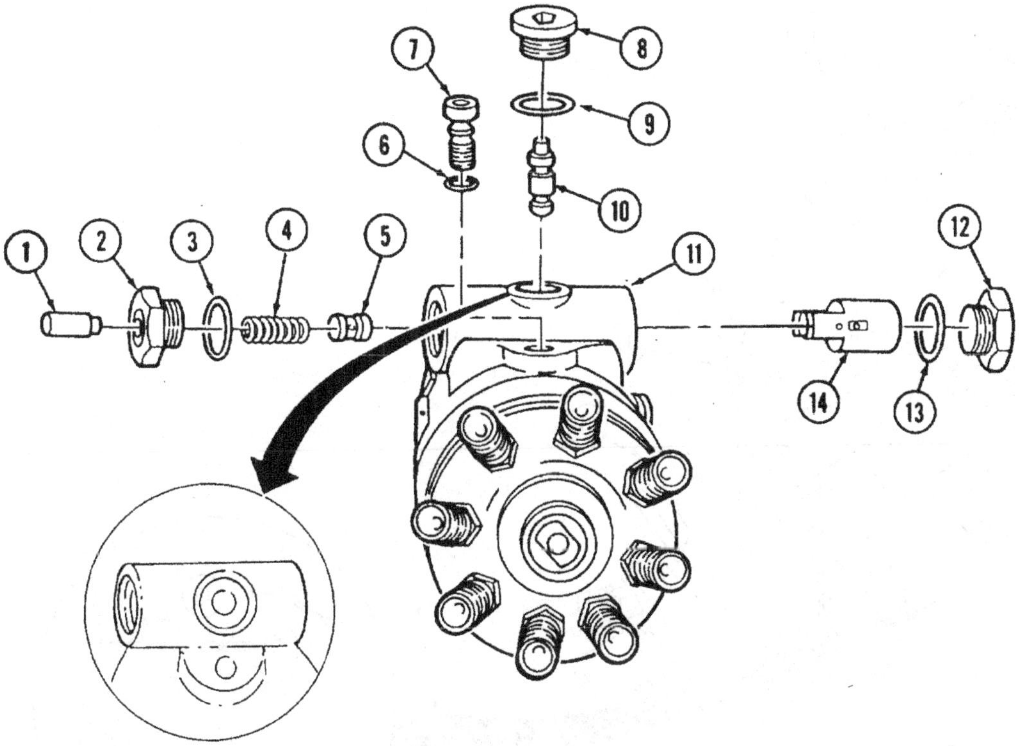

2-32. FUEL INJECTION PUMP REPAIR (Cont'd)

59. Rotate housing (11) and holding fixture 180°

NOTE

The vent wire screw is available in different sizes (identification numbers are etched on each) to adjust return oil flow. If original screw is to be replaced, use the same size. The final determination will be made during calibration.

60. Install vent wire screw (18). Tighten to 25-30 lb-in. (3 N•m) with 1/8-in. hex head driver.
61. Install metering valve assembly (17) in housing (11).
62. Connect linkage hook (15) to governor arm (20).
63. Install spring (19) on governor arm (20) and linkage hook (15).

CAUTION

Governor arm must be installed correctly or injection pump will malfunction.

64. Install governor arm (20) in housing (11) so that feet on governor arm (20) fit in slots in thrust sleeve (21).
65. Install linkage hook (15) on valve arm pin (16).

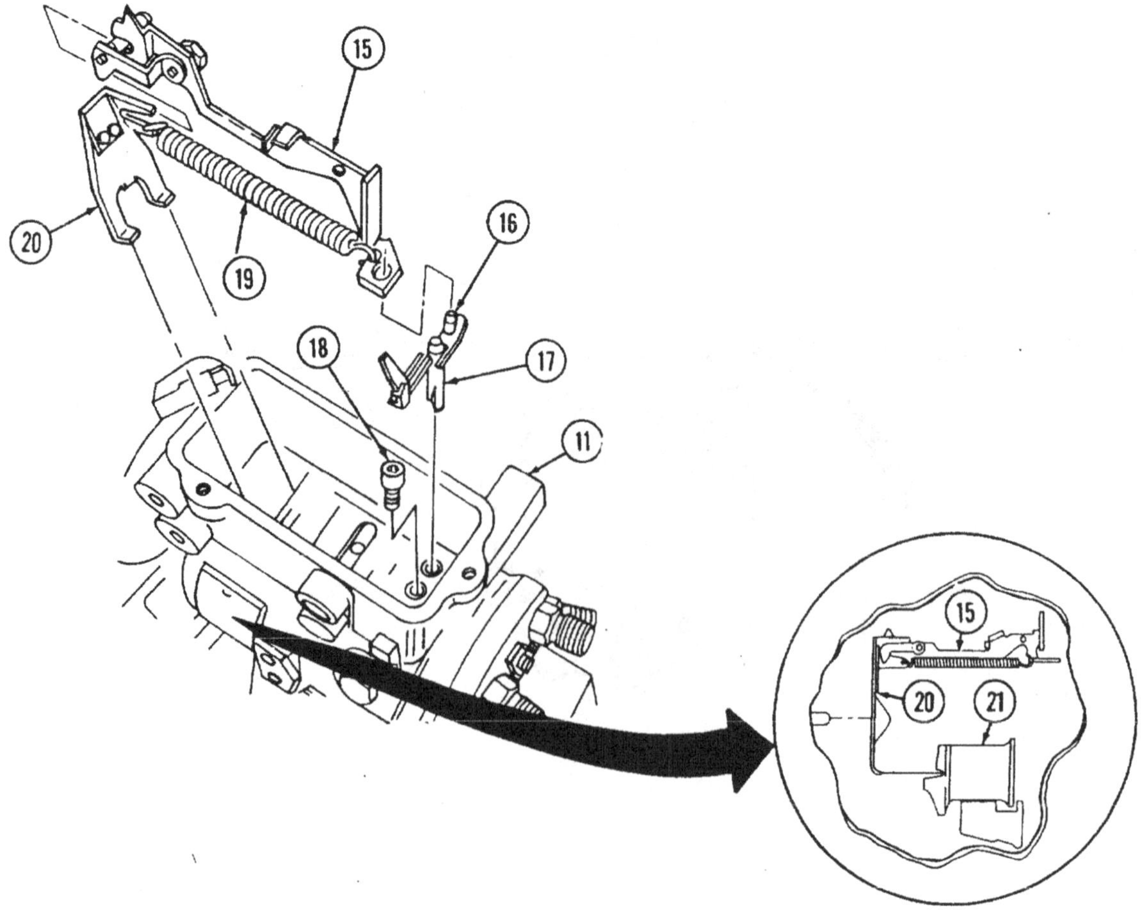

2-32. FUEL INJECTION PUMP REPAIR (Cont'd)

NOTE

- Perform step 66 for pre-1990 engines.
- Perform step 67 for 6.5L engines and 6.2L engines 1990 and above.

66. Install spacer (2) and two O-rings (3) on throttle shaft (1) and install throttle shaft (1) in housing (4).
67. Install spacer (2) and two O-rings (3) on throttle shaft (1). Position governor thrust spring (5) inside housing (4) and insert throttle shaft (1) through housing (4) and thrust spring (5).
68. Install mylar washer (6) and face cam (7) on throttle shaft (1) with screw (13). Finger tighten screw (13).
69. Install rocker lever (12) on housing (4) with rocker lever pin (15) and two retaining clips (14).
70. Put 0.005-in. (0.127 mm) feeler gauge between throttle shaft mylar washer (6) and housing boss (16).
71. Put throttle shaft (1) in idle position. Squeeze throttle shaft (1) and face cam (7) tightly toward each other and rotate face cam (7) so lower roller rests approximately in middle of low idle step. Tighten screw (13) to 30 lb-in. (3 N·m) with Torx drive and remove feeler gauge.
72. Align holes in kickdown cam (10) and throttle shaft (1).
73. Install kickdown cam (10) on throttle shaft (1) with retaining pin (9).
74. Deleted.

2-32. FUEL INJECTION PUMP REPAIR (Cont'd)

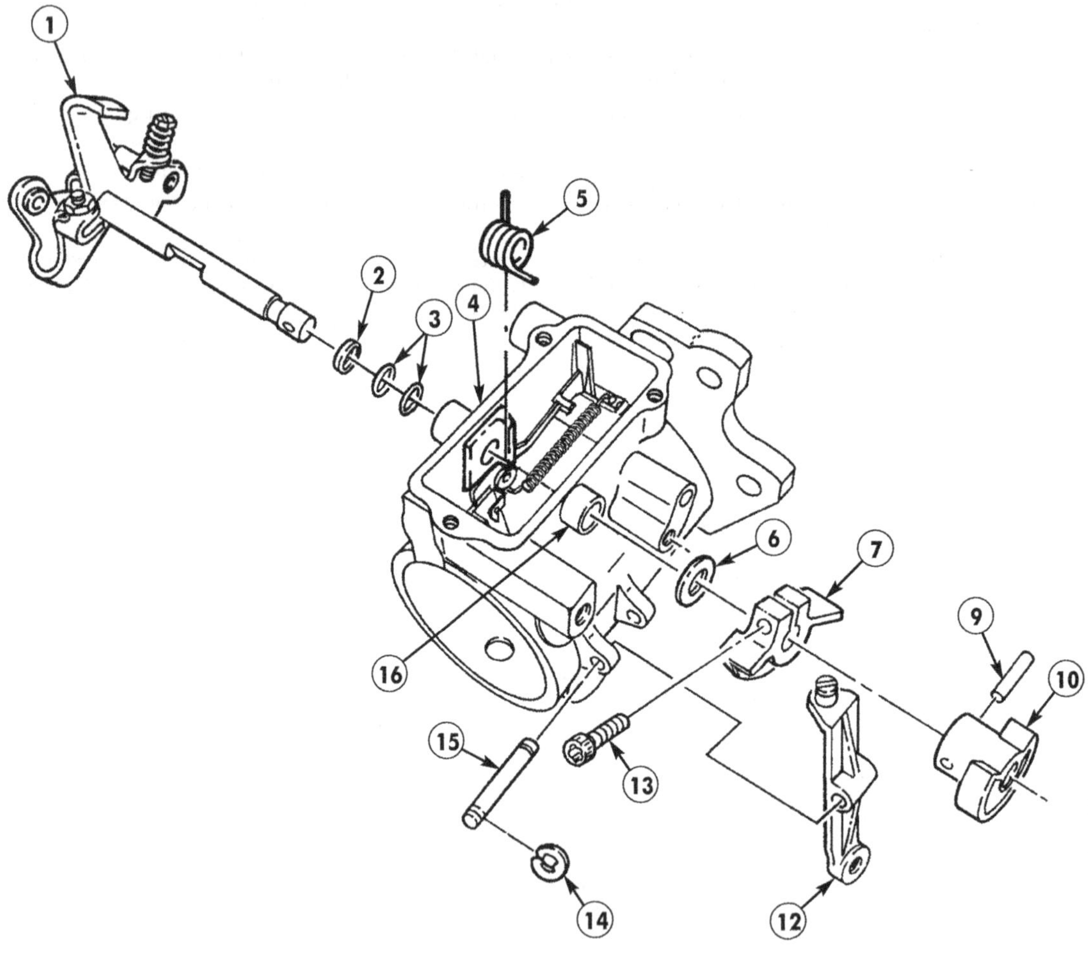

2-32. FUEL INJECTION PUMP REPAIR (Cont'd)

CAUTION

Tightly wound coils on idle spring must be installed facing throttle block, or injection pump may malfunction.

75. Install idle spring (1) on pushrod (2).

NOTE

- Perform step 76 for pre-1990 6.2L engines.
- Perform steps 77 through 79 for 6.5L engines and 6.2L engines 1990 and above.

76. With throttle shaft (8) in wide open position, install min-max governor (4) on notch of throttle shaft (8). Rotate throttle shaft (8) to low idle position and lower min-max governor (4) into housing (5).
77. Insert inner governor spring (9) into pushrod (2).
78. With throttle shaft (8) in wide open position, install min-max governor (4) on notch of throttle shaft (8) and position tang (12) of thrust spring (11) in groove on throttle block (3). Rotate throttle shaft (8) to low idle position and lower min-max governor (4) into housing (5).
79. Hook tang (10) of thrust spring (11) in groove (13) of housing (5).

CAUTION

Guide stud must be installed between metering valve arm and spring. Incorrect installation may cause injection pump to malfunction.

NOTE

- Perform steps 80 and 81 for vehicles with 6.5L and 1990 and above 6.2L engines.
- Perform step 82 for vehicles with pre-1990 6.2L engines.

80. Install nut (17) and seal (16) on guide stud (7). Install guide stud (7) in housing (5) between metering valve arm (15) and spring (14) and into min-max governor (4). Tighten guide stud (7) to 80-90 lb-in. (9-10 N·m) using 1/4-in. hex head driver.
81. Install cap (18) on guide stud (7).
82. Install guide stud (7) and washer (6) in housing (5) between metering valve arm (15) and spring (14) and into min-max governor (4). Tighten guide stud (7) to 80-90 lb-in. (9-10 N·m) using 1/4-in. hex head driver.

2-32. FUEL INJECTION PUMP REPAIR (Cont'd)

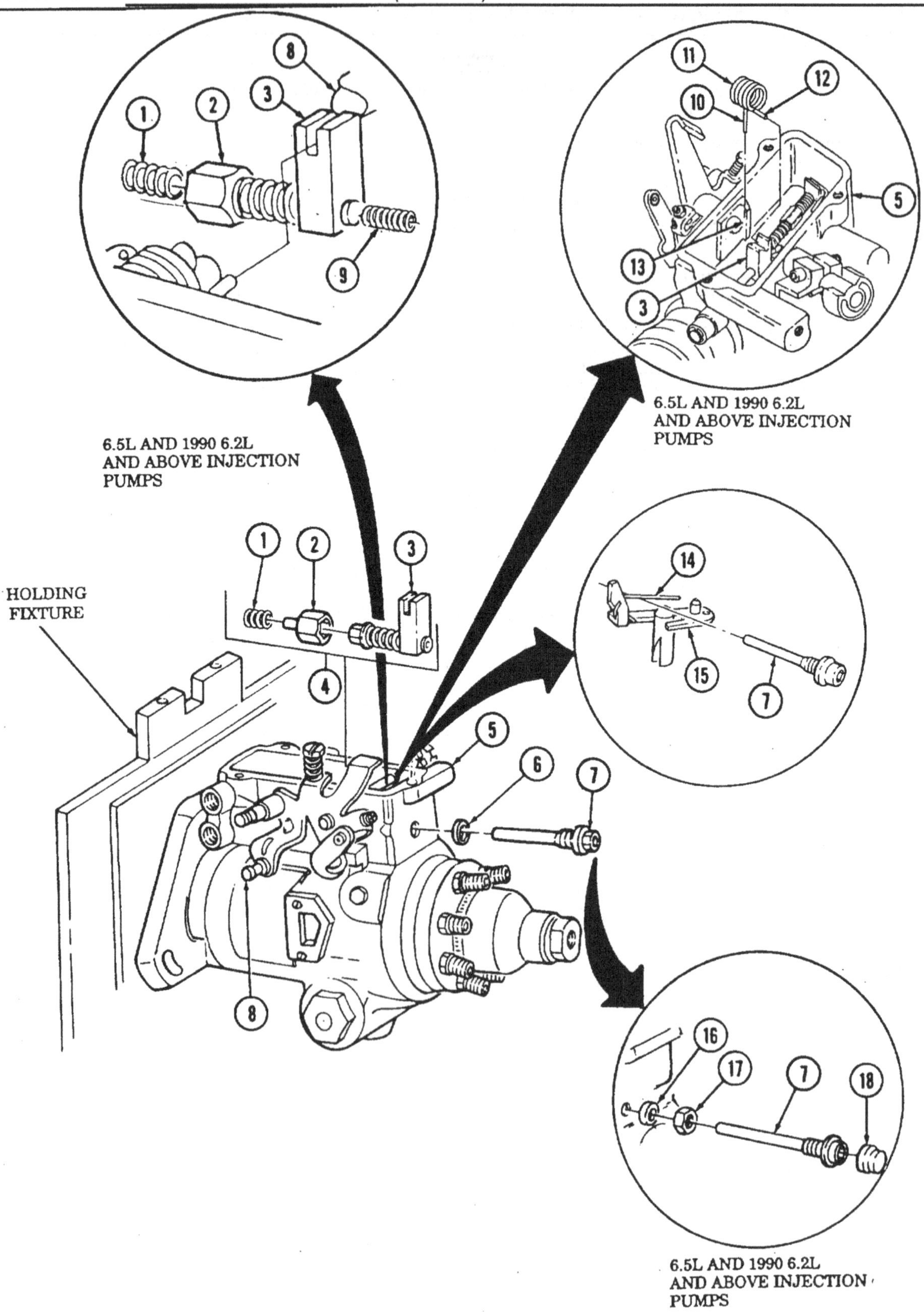

2-32. FUEL INJECTION PUMP REPAIR (Cont'd)

83. Install O-ring (15) and check valve (1) in cover (2).
84. Install plunger (10) in cold start advance solenoid (9) and stud (11) in hole in rear of cover (2).
85. Install cold start advance solenoid (9) in cover (2) so small tip of plunger (10) fits in port of check valve (1) and stud (8) fits through hole inside of cover (2).
86. Install fiber washer (3), washer (4), lockwasher (5), nut (6), and locknut (7) on stud (8).
87. Install fiber washer (3), washer (4), lockwasher (5), and nut (6) on stud (11).
88. Install shell (12), washer (13), and terminal (14) on stud (11).

2-32. FUEL INJECTION PUMP REPAIR (Cont'd)

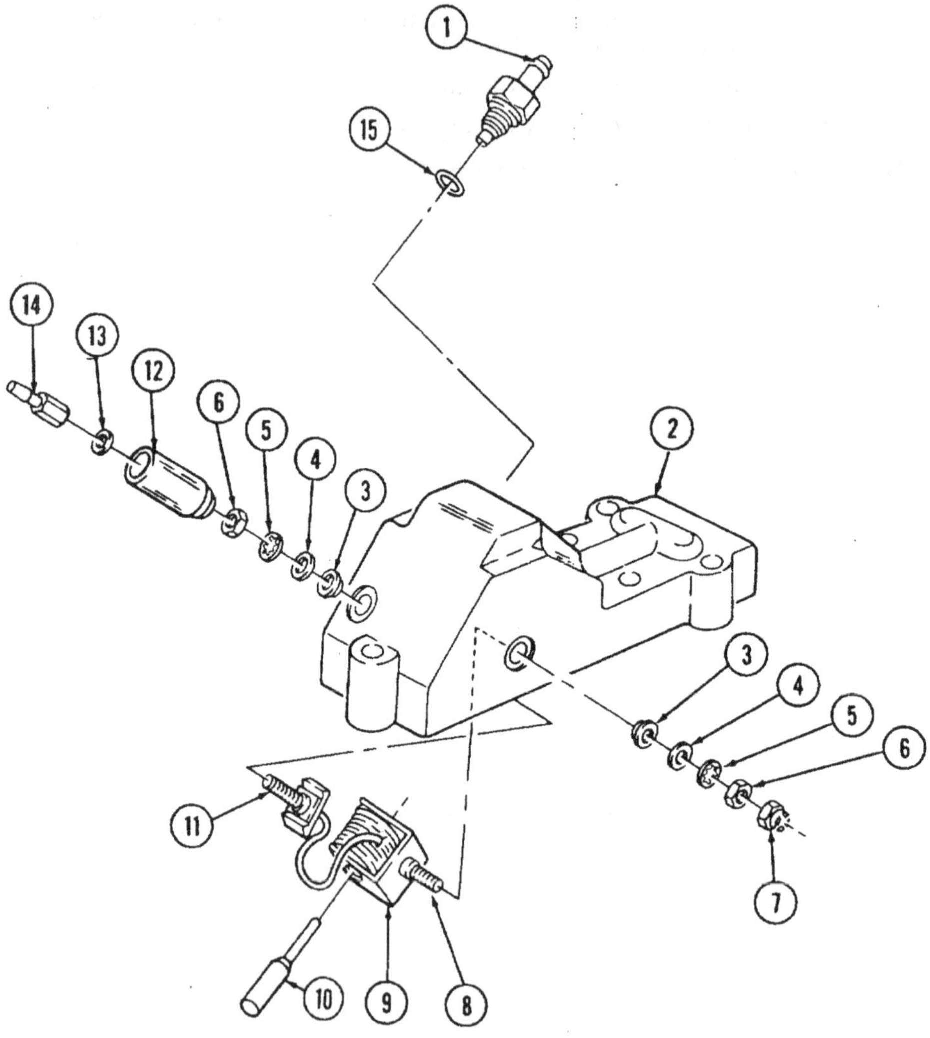

2-32. FUEL INJECTION PUMP REPAIR (Cont'd)

89. Install electrical shutoff solenoid (9) in cover (11).
90. Install fiber washer (7), washer (6), lockwasher (5), nut (4), lockwasher (12), ground strap (13) and locknut (14) on stud (10).
91. Install fiber washer (7), washer (6), lockwasher (5), and nut (4) on stud (8).
92. Install shell (3), washer (2), and terminal (1) on stud (8).

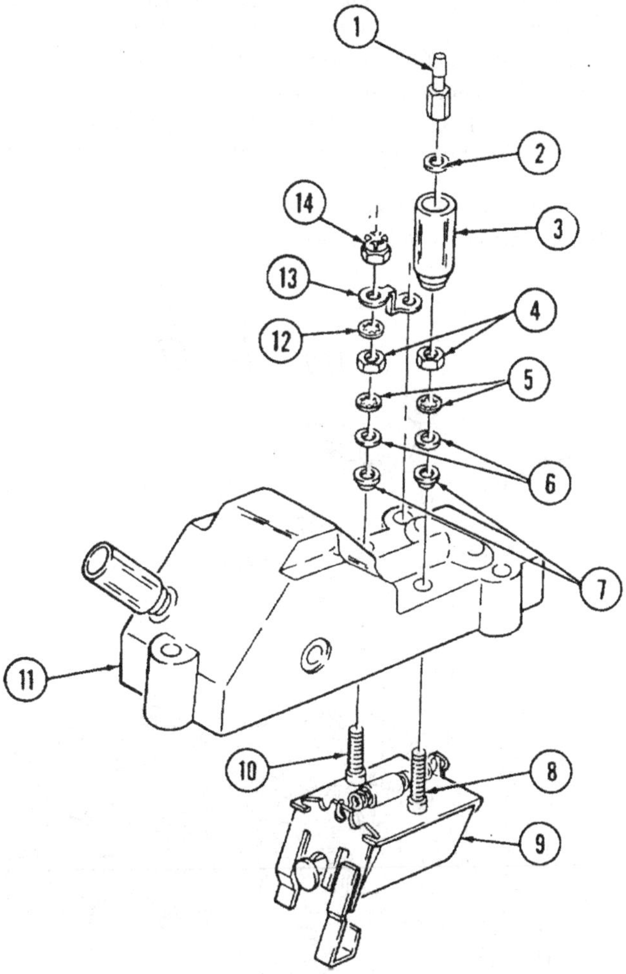

2-32. FUEL INJECTION PUMP REPAIR (Cont'd)

93. Install washer (15) between ground strap (16) and cover (11).

CAUTION

When installing cover on housing, cover must be positioned on housing ahead of threaded holes and slid rearward to align holes in cover with threaded holes in housing. Incorrect installation of cover may cause pump to malfunction.

94. Install gasket (31) and cover (11) on housing (30) and secure with three screws (17), lock-washers (18), and washers (19). Tighten screws (17) to 35-45 lb-in. (4-5 N ·m).
95. Install timing cover (28) and gasket (27) on housing (30) and secure with two screws (29).
96. Using end cap wrench, tighten end cap (26) to 30-37 lb-ft (41-50 N ·m).
97. Install transfer pump end cap locking screw (25), plate (24), flat washer (23), and rubber washer (22) in hydraulic head (21). Tighten screw (25) to 70-80 lb-in. (8-9 N ·m) using 5/32-in. hex head driver.
98. Tighten two head locking capscrews (20) to 180-220 lb-in. (20-25 N ·m).
99. Remove pump (30) from holding fixture and vise.

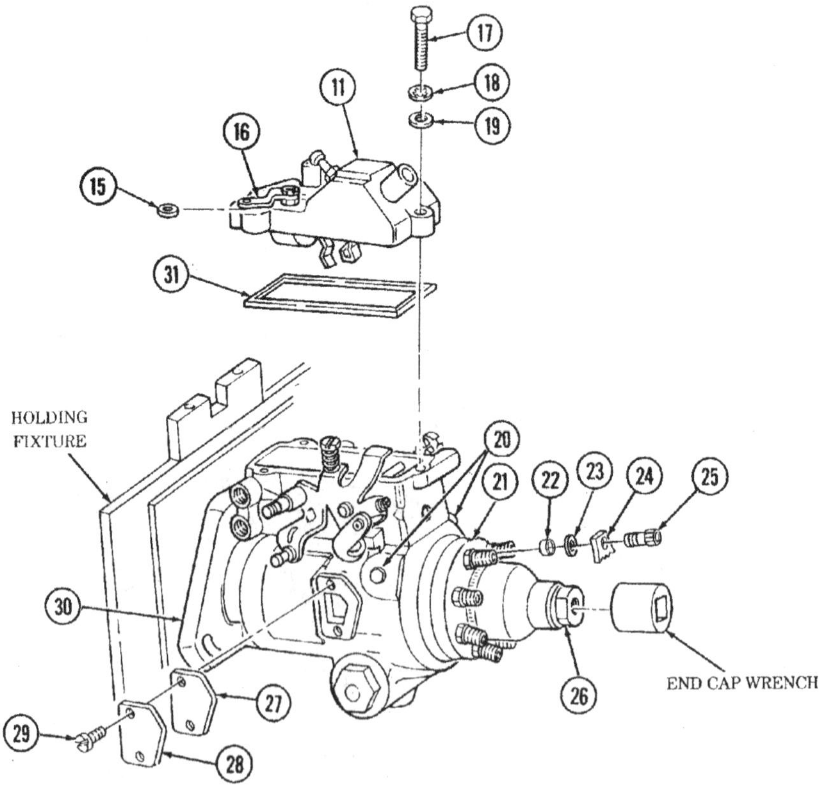

FOLLOW-ON TASKS: • Calibrate fuel injection pump (paras. 2-33, 2-34, or 2-35).
• Install fuel injector pump (para. 2-41).

2-33. PRE-1990 (6.2L) FUEL INJECTION PUMP (DB2829-4523) CALIBRATION

This task covers:

a. Test Bench Requirements
b. Injection Pump Information
c. Injection Pump Check (Prior to Service)
d. Injection Pump Settings (Following Pump Repair)

INITIAL SETUP:

Tools

General mechanic's tool kit automotive (Appendix E, Item 1)

Special Tools

Roller-to-roller setting tool (Appendix E, Item 45)
Linkage gap tool (Appendix E, Item 49)
Automatic advance indicator (Appendix E, Item 50)
Throttle lever gauge (Appendix E, Item 51)
Protractor (Appendix E, Item 52)
Air timing gauge (Appendix E, Item 53)
Holding fixture (Appendix E, Item 46)
Torx drive (Appendix E, Item 42)

Materials/Parts

Calibrating fluid (Appendix B, Item 4)
Sealing compound (Appendix B, Item 19)

Manual References

TM 9-2815-237-34P

Test Equipment

Digital tachometer
Voltage source (variable)
Pressure gauge (0-160 psi, 1 psi increments)
Pressure gauge (0-30 psi)
Vacuum gauge (0-30 in. Hg)
Flowmeter with three-way valve
Temperature gauge (0°-250°F)
Zero-backlash coupling device
Calibrating nozzles and lines
Orifice 13211
Calibration stand

NOTE

The following calibration procedure applies to pre-1990 6.2L fuel injection pumps. Refer to para. 2-34 for calibration on 1990 6.2L and above pumps and para. 2-35 for 6.5L pumps.

a. Test Bench Requirements

The fuel injection pump requires a test bench capable of at least 2250 rpm shaft speed. The test bench must be equipped with the following items:

(a) A zero-backlash coupling device.

(b) A digital tachometer.

(c) A variable voltage source.

(d) Test gauges:

(1) 0-160 psi (0-1103 kPa) pressure gauge calibrated in 1 psi increments to measure transfer Pump pressure. This gauge should be located as close as possible to the pump, with an in-line shutoff valve located at the pump.

(2) 0-30 psi (0-206.9 kPa) pressure gauge to measure housing pressure.

(3) 0-30 in. Hg (0-101.3 kPa) vacuum gauge to measure transfer pump lift. This gauge should be in the supply line to the pump, with a shutoff valve between the gauge and the oil source.

(4) A flowmeter for measuring return oil should be used with a three-way valve to permit the flowmeter to be in use only during the return oil check.

(5) 0°-250°F (-17.8°-121. 1°C) temperature gauge, located at the pump inlet, to monitor inlet temperature.

2-33. PRE-1990 (6.2L) FUEL INJECTION PUMP (DB2829-4523) CALIBRATION (Cont'd)

(e) Calibrating nozzles and injection lines:

(1) Eight 0.020-in. (0.5 mm) orifice plate nozzles set at 1700 ± 25 psi (11,721 ± 172 kPa).

(2) Eight 0.093-in. (2.36 mm) inside diameter by 25-in. (63.5 cm) long injection lines.

(f) Calibrating fluid conforming to SAE J967, maintaining a temperature of 110°-115°F (43°-46°C) at the pump inlet over all speed and load ranges.

(g) Orifice fitting in the inlet line between the supply pump and shutoff valve, to maintain constant fluid circulation and temperature.

b. Injection Pump Information

NOTE

All speeds are in engine rpm unless otherwise noted.

1. Injection pump operating speed is half of engine speed.
2. Injection pump rotation is counterclockwise.
3. Injection pump is timed to the outlet port in the 10 o'clock position, when viewed from the transfer pump end.

c. Injection Pump Check (Prior to Service)

1. Before mounting the pump on the test stand, check the drive shaft for freedom of rotation in the housing. Pump should be mounted securely in the appropriate adapter, and the intermediate coupling checked for freedom of movement.
2. Connect supply and return lines securely.
3. Install high pressure injection lines. Use copper gaskets or steel washers where required. Leave fuel line connections at the pump and nozzles loose.
4. Apply 17.6 V to electric shutoff solenoid and operate pump at 2000 rpm wide-open throttle (WOT) for 10 minutes to bring to operating temperature and clear air from system.
5. Tighten all connections securely. Check connections for leaks while operating.
6. Check fuel delivery at 650 rpm (low idle). Fuel delivery should be 10-18 mm^3/stroke. If fuel delivery is not within the range specified, adjust low idle screw. Turn screw clockwise to decrease or counterclockwise to increase fuel delivery.
7. Check housing pressure at 650 rpm, while applying 17.6 V to housing pressure cold advance solenoid. Housing pressure should be 0-1 psi (0-7 kPa).
8. Check electric shutoff solenoid for pull-in with 17.6 V maximum with throttle lever in low idle position at 400 rpm.
9. Fuel delivery measurement:

(a) In order to minimize variations between readings due to residual oil in the graduates, the following timed cycle should be used:

(1) Draw

(2) Settle 30 seconds

(3) Drain 30 seconds

(4) Draw

(5) Settle 30 seconds

(6) Read graduates

(7) Repeat cycle and average readings

2-33. PRE-1990 (6.2L) FUEL INJECTION PUMP (DB2829-4523) CALIBRATION (Cont'd)

(b) Energize electric shutoff solenoid and check delivery against the chart below.

RPM	THROTTLE POSITION	MM³/ STROKE"	CAM MOVEMENT (DEGREES)	HOUSING PRESSURE
150	WOT	28 min.		
400	WOT	47 min.		
400	WOT***	4 max.		
650	Low idle	10-18	1.5 min.	
650	Low idle		2.75 min.	0-1 psi (0-6.9 kPa)**
1500		21.5-23.5	2.25-4.75	
2000	WOT	51.0-55.0	0.5-2.5	
3200	WOT		4.25-7.25	
3200	Low idle		10.0 max.	
3600	WOT	44.5 min.		
3900	WOT	42.5 min.		
4200		8 max.		

* Maximum cylinder variation should be ± 6 mm³ from the average flow of all cylinders.
** With housing pressure cold advance solenoid energized.
*** With electric shutoff solenoid de-energized.

10. Disconnect pump and mount in holding fixture.
11. Using air timing gage, check pump timing. Set to -0.75 to +1.25 degrees.
12. Refer to para. 2-7, Engine Troubleshooting, to diagnose any malfunctions encountered during checkout.

d. Injection Pump Settings (Following Pump Repair)

1. Roller-to-roller dimension must be 1.981 ± 0.0005 in. (50.32 ± 0.013 mm). With oversize cam ring installed, roller-to-roller dimension will be 1.987 ± 0.0005 in. (50.47 ± 0.013 mm). Maximum eccentricity allowed is 0.008 in. (0.20 mm) TIR.
2. Linkage gap must be 0.130-0.175 in. (3.30-4.45 mm).

(a) Mount the pump horizontally in holding fixture and remove governor cover.

(b) Hold the throttle lever in the wide-open throttle (WOT) position. Rotate pump drive shaft counterclockwise until a click is heard as the rounded contact points on the governor arm engage the slots in the thrust sleeve. Continue to rotate the drive shaft until the gap between the governor arm and pump housing is minimal.

(c) Using linkage gap tool, check the clearance limits between the rear of the shutoff shaft and the vertical tab on the linkage hook.

(d) If adjustment is required, loosen governor linkage hook adjustment screw and extend the linkage hook to its maximum open length (throttle lever must be in WOT).

(e) Insert the appropriate step of the linkage gap tool between the vertical hook tab and the throttle shaft, with the step facing the shaft. The tool must be held vertically and parallel to the linkage hook tab.

(f) With the linkage hook pin seated firmly in the governor arm slot and the governor arm in the minimal gap position (step 2.(b)), shorten the linkage hook assembly until the face of the vertical tab is flush against the tool. Tighten adjusting screw to 12-15 lb-in. (1 N•m).

(g) Recheck linkage gap and adjust if necessary.

2-33. PRE-1990 (6.2L) FUEL INJECTION PUMP (DB2829-4523) CALIBRATION (Cont'd)

3. Set the throttle lever in correct low idle position prior to calibration using throttle lever gauge. The low idle screw must beat 34 ± 2 degrees as measured between vertical centerline of throttle shaft bushing bore and throttle connection stud.
4. Flush the pump with calibrating oil to remove metal chips and possible contamination.
5. Before mounting the pump on the test stand, check the drive shaft for freedom of rotation in the housing. Pump should be mounted securely in the appropriate adapter, and the intermediate coupling checked for freedom of movement.
6. Connect supply and return lines securely.
7. Install high pressure injection lines. Use copper gaskets or steel washers where required. Leave fuel line connections at the pump and nozzles loose.
8. Apply 17.6 V to electric shutoff solenoid and operate pump at 2000 rpm (WOT) for 10 minutes to bring to operating temperature and clear air from system.
9. Tighten all connections securely. Check connections for leaks while operating.
10. Set transfer pump pressure at 2000 rpm (WOT) for 70-76 psi (482.7-524 kPa); test stand boost should be set at 5 ± 0.5 psi (34.5 ± 3.4 kPa).
11. Set return oil flow to 225-375 cc/minute at 2000 rpm (WOT).
12. Measure housing pressure:

    (a) Check housing pressure at 650 rpm, with throttle shaft in low idle position, pressure should be 8-12 psi (55-83 kPa).

    (b) Energize housing pressure cold advance solenoid with 17.6 V, housing pressure should be 0-1 psi (0-6.9 kPa).

13. With governor cover removed and throttle shaft in low idle position, set min-max governor assembly for 14.0 ± 2 mm^3/stroke at 650 rpm (fuel reading with cover installed).
14. Check electric shutoff solenoid for pull-in with 17.6 V maximum with throttle lever in low idle position at 400 rpm.
15. Transfer pump lift should be 18 in. Hg (60.8 kPa) at 400 rpm.
16. Mechanical/Light Load Advance Setting

    (a) Check housing pressure at 650 rpm, with throttle shaft in low idle position, pressure should be 8-12 psi (55-83 kPa).

    (b) With throttle in a WOT position, rotate face cam to its maximum lift position in contact with the cam roller.

    (c) At 2000 rpm, adjust trimmer screw for 1.5 degrees cam advance.

    (d) Set throttle for 22.5 ± 1 mm^3/stroke at 1500 rpm and rotate face cam to obtain 3.5 degrees cam advance.

    (e) Tighten face cam screw to 28-32 lb-in. (3-4 N•m) and check face cam retention.

        (1) Put 0.005-in. (0.127 mm) feeler gauge between throttle shaft mylar washer and housing

        (2) Put throttle shaft in idle position. Squeeze throttle shaft and face cam tightly toward each other and rotate face cam as lower roller rests approximately in middle at low idle step. Tighten screw to 30 lb-in. (3 N•m) with torx drive and remove feeler gauge.

17. Set total throttle travel, With throttle in low idle position, align zero degree mark on protractor with center of rib on rocker lever. Move throttle to WOT and adjust maximum travel screw to 80-84 degrees.

2-33. PRE-1990 (6.2L) FUEL INJECTION PUMP (DB2829-4523) CALIBRATION (Cont'd)

NOTE

Any changes to pump or settings after this point will require retesting of pump, starting with step 16.

18. Fuel delivery measurement:

(a) In order to minimize variations between readings due to residual oil in the graduates, the following timed cycle should be used:

(1) Draw
(2) Settle 30 seconds
(3) Drain 30 seconds
(4) Draw
(5) Settle 30 seconds
(6) Read graduates
(7) Repeat cycle and average readings

(b) Check points:

(1) At 1500 rpm (WOT) de-energize electric shutoff solenoid. Delivery should be 4 mm^3/stroke maximum.

(2) Re-energize electric shutoff solenoid and check delivery against the chart below.

RPM	THROTTLE POSITION	MM^3/ STROKE*	CAM MOVEMENT (DEGREES)	PRESSURE	
				TRANSFER PUMP	HOUSING
150	WOT	28 min.		12 psi (83 kPa) min.	
400	WOT	47 min.			
400	WOT***	4 max.			
650	Low idle	12-16	1.5 min.		
650	Low idle		2.75 min.		0-1 psi (0-7 kPa)**
1500		21.5-23.5	2.5-4.5		8-12 psi (55-83 kPa)
2000	WOT	51.0-55.0	0.5-2.5		8-12 psi (55-83 kPa)
3200	WOT		4.25-6.75		8-12 psi (55-83 kPa)
3200	Low idle		10.0 max.		8-12 psi (55-83 kPa)
3600	WOT	46 min.			
3900	WOT	44 min.			
4200	WOT	8 max.		135 psi (931 kPa) max.	

* Maximum cylinder variation should be ± 6 mm^3 from the average flow of all cylinders.
** With housing pressure cold advance solenoid energized.
*** With electric shutoff solenoid de-energized.

19. Disconnect pump and mount in holding fixture.

20. Using air timing gauge, check pump timing. Set to +0.25 degree* 0.50 degree.

21. Tighten all fasteners to specifications, refer to para. 2-32d.

22. Apply sealing compound as follows:

(a) Apply one drop at interface of maximum travel screw and locknut.

CAUTION

Do not allow sealing compound to enter gap between face cam and housing. Binding of the throttle shaft may result.

(b) Apply one drop to threaded end of face cam screw where it protrudes from face cam.

(c) Apply one drop to interface of servo advance adjusting screw and rocker lever, making sure sealing compound does not enter hex recess of screw.

2-34. 1990 (6.2L) FUEL INJECTION PUMP (DB2829-4879) CALIBRATION

This task covers:

a. Test Bench Requirements
b. Injection Pump Information
c. Injection Pump Check (Prior to Service)
d. Injection Pump Settings (Following Pump Repair)

INITIAL SETUP:

Tools
General mechanic's tool kit:
automotive (Appendix E, Item 1)

Special Tools
Roller-to-roller setting tool (Appendix E, Item 45)
Linkage gap tool (Appendix E, Item 49)
Automatic advance indicator (Appendix E, Item 50)
Throttle lever gauge (Appendix E, Item 51)
Protractor (Appendix E, Item 52)
Air timing gauge (Appendix E, Item 53)
Holding fixture (Appendix E, Item 46)
Torx drive (Appendix E, Item 42)

Materials/Parts
Calibrating fluid (Appendix B, Item 4)
Sealing compound (Appendix B, Item 19)

Manual References
TM 9-2815-237-34P

Test Equipment
Digital tachometer
Voltage source (variable)
Pressure gauge (0-160 psi, 1 psi increments)
Pressure gauge (0-30 psi)
Vacuum gauge (0-30-in. Hg)
Flowmeter with three-way valve
Temperature gauge (0°-250°F)
Zero-backlash coupling device
Calibrating nozzles and lines
Orifice 13211
Calibration stand

NOTE

The following calibration procedure applies to 1990 6.2L and above fuel injection pumps. Refer to para. 2-33 for calibration on pre-1990 6.2L pumps and para 2-35 for 6.5L pumps.

a. Test Bench Requirements

The fuel injection pump requires a test bench capable of at least 2250 rpm shaft speed. The test bench must be equipped with the following items:

(a) A zero-backlash coupling device.

(b) A digital tachometer.

(c) A variable voltage source.

(d) Test gauges:

(1) 0-160 psi (0-1103 kPa) pressure gauge calibrated in 1 psi increments to measure transfer pump pressure. This gauge should be located as close as possible to the pump, with an in-line shutoff valve located at the pump.

(2) 0-30 psi (0-206.9 kPa) pressure gauge to measure housing pressure.

(3) 0-30 in. Hg (0-101.3 kPa) vacuum gauge to measure transfer pump lift. This gauge should be in the supply line to the pump, with a shutoff valve between the gauge and the oil source.

(4) A flowmeter for measuring return oil should be used with a three-way valve to permit the flowmeter to be in use only during the return oil check.

(5) 0°-250°F (-17.8°-121.1°C) temperature gauge, located at the pump inlet, to monitor inlet temperature.

2-34. 1990 (6.2L) FUEL INJECTION PUMP (DB2829-4879) CALIBRATION (Cont'd)

(e) Calibrating nozzles and injection lines:

(1) Eight 0.020-in. (0.5 mm) orifice plate nozzles set at 1700 ± 25 psi (11,721 ± 172 kPa).

(2) Eight 0.093-in. (2.36 mm) inside diameter by 25 in. (63.5 cm) long injection lines.

(f) Calibrating fluid conforming to SAE J967, maintaining a temperature of 110°-115°F (43°-46°C) at the pump inlet over all speed and load ranges.

(g) Orifice fitting in the inlet line between the supply pump and shutoff valve, to maintain constant fluid circulation and temperature.

b. Injection Pump Information

NOTE

All speeds are in engine rpm unless otherwise noted.

1. Injection pump operating speed is half of engine speed.
2. Injection pump rotation is counterclockwise.
3. Injection pump is timed to the outlet port in the 10 o'clock position, when viewed from the transfer pump end.

c. Injection Pump Check (Prior to Service)

1. Before mounting the pump on the test stand, check the drive shaft for freedom of rotation in the housing. Pump should be mounted securely in the appropriate adapter, and the intermediate coupling checked for freedom of movement.
2. Connect supply and return lines securely.
3. Install high pressure injection lines. Use copper gaskets or steel washers where required. Leave fuel line connections at the pump and nozzles loose.
4. Apply 17.6 V to electric shutoff solenoid and operate pump at 2000 rpm wide-open throttle (WOT) for 10 minutes to bring to operating temperature and clear air from system.
5. Tighten all connections securely. Check connections for leaks while operating.
6. Check fuel delivery at 650 rpm (low idle). Fuel delivery should be 10-18 mm^3/stroke. If fuel delivery is not within the range specified, adjust low idle screw. Turn screw clockwise to decrease or counter-clockwise to increase fuel delivery.
7. Check housing pressure at 650 rpm, while applying 17.6 V to housing pressure cold advance solenoid. Housing pressure should be 0-1 psi (0-7 kPa) and 8-12 psi (55-83 kPa) when de-energized.
8. Check electric shutoff solenoid for pull-in with 17.6 V maximum with throttle lever in low idle position at 400 rpm.
9. Fuel delivery measurement

(a) In order to minimize variations between readings due to residual oil in the graduates, the following timed cycle should be used:

(1) Draw

(2) Settle 30 seconds

(3) Drain 30 seconds

(4) Draw

(5) Settle 30 seconds

(6) Read graduates

(7) Repeat cycle and average readings

2-34. 1990 (6.2L) FUEL INJECTION PUMP (DB2829-4879) CALIBRATION (Cont'd)

(b) Energize electric shutoff solenoid and check delivery against the chart below.

RPM	THROTTLE POSTION	MM^3/ STROKE*	CAM MOVEMENT (DEGREES)	PRESSURE	
				TRANSFER PUMP	HOUSING
150	WOT	28 min.		12 psi (83 kPa) min.	0-12 psi (0-83 kPa)
400	WOT	45 min.			
400	WOT**	4 max.			
650	Low idle	10-18			8-12 psi (55-83 kPa)
650	Low idle		1.25 min.		0-1 psi (0-7 kPa)
1500		21.5-23.5	1.0-3.5		
2000	WOT	56 max.		60-62 psi (414-427 kPa)	
2400	WOT		0.5-3.0		
3200	Low idle		9.0-10.0		
3500	WOT	48-53	4.5-6.5		
3600	WOT	47 min.			
3800	WOT	33 min.			
4050	WOT	15 *max.*		125 psi (862 kPa) max.	

* Maximum cylinder variation should be ± 6 mm^3 from the average flow of all cylinders.
** With housing pressure cold advance solenoid energized.

10. Disconnect pump and mount in holding fixture.
11. Using air timing gauge, check pump timing. Set to -0.5 to +1.5 degrees.
12. Total throttle travel from low idle to wide-open throttle should be 80-84 degrees.
13. Refer to para. 2-7, Engine Troubleshooting, to diagnose any malfunctions encountered during checkout.

d. Injection Pump Settings (Following Pump Repair)

1. Roller-to-roller dimension must be 1.975 ± 0.0005 in. (50.17 ± 0.013 mm). With oversize cam ring installed, roller-to-roller dimension will be 1.981 ± 0.0005 in. (50.32 ± 0.013 mm). Maximum eccentricity allowed is 0.008 in. (0.20 mm) TIR.
2. Linkage gap must be 0.130-0.175 in. (3.30-4.45 mm).
   (a) Mount the pump horizontally in holding fixture and remove governor cover.
   (b) Hold the throttle lever in the wide-open throttle (WOT) position. Rotate pump drive shaft counterclockwise until a click is heard as the rounded contact points on the governor arm engauge the slots in the thrust sleeve. Continue to rotate the drive shaft until the gap between the governor arm and pump housing is minimal.
   (c) Using linkage gap tool, check the clearance limits between the rear of the shutoff shaft and the vertical tab on the linkage hook.
   (d) If adjustment is required, loosen governor linkage hook adjustment screw and extend the linkage hook to its maximum open length (throttle lever must be in WOT).
   (e) Insert the appropriate step of the linkage gap tool between the vertical hook tab and the throttle shaft, with the step facing the shaft. The tool must be held vertically and parallel to the linkage hook tab.
   (f) With the linkage hook pin seated firmly in the governor arm slot and the governor arm in the minimal gap position (step 2.(b)), shorten the linkage hook assembly until the face of the vertical tab is flush against the tool. Tighten adjusting screw to 12-15 lb-in. (1 N•m).
   (g) Recheck linkage gap and adjust if necessary.

2-34. 1990 (6.2L) FUEL INJECTION PUMP (DB2829-4879) CALIBRATION (Cont'd)

3. Set the throttle lever in correct low idle position prior to calibration using throttle lever gauge. The low idle screw must beat 34± 2 degrees as measured between vertical centerline of throttle shaft bushing bore and throttle connection stud.
4. Flush the pump with calibrating oil to remove metal chips and possible contamination.
5. Before mounting the pump on the test stand, check the drive shaft for freedom of rotation in the housing. Pump should be mounted securely in the appropriate adapter, and the intermediate coupling checked for freedom of movement.
6. Connect supply and return lines securely.
7. Install high pressure injection lines. Use copper gaskets or steel washers where required. Leave fuel line connections at the pump and nozzles loose.
8. Apply 17.6 V to electric shutoff solenoid and operate pump at 2000 rpm (WOT) for 10 minutes to bring to operating temperature and clear air from system.
9. Tighten all connections securely. Check connections for leaks while operating.
10. Set transfer pump pressure at 2000 rpm (WOT) for 60-62 psi (414-428 kPa); test stand boost should be set at 5 ± 0.5 psi (34.5 ± 3.4 kPa).
11. Set return oil flow to 225-375 cc/minute at 2000 rpm (WOT).
12. Measure housing pressure:
    (a) Check housing pressure at 650 rpm, with throttle shaft in low idle position, pressure should be 8-12 psi (55-83 kPa).
    (b) Energize housing pressure cold advance solenoid with 17.6 V. Housing pressure should be 0-1 psi (0-7 kPa).
13. With governor cover removed and throttle shaft in low idle position, set min-max governor assembly for 7-9 mm ^{3}/stroke at 650 rpm (fuel reading with cover installed).

NOTE

If low idle screw adjustment is required, refer to step 3. Adjustment is ± 1 turn from the point in step 3.

14. At 650 rpm, adjust low idle screw for 13-15 mm ^{3}/stroke for the remainder of the setting and checking procedures.
15. During and at the completion of all settings, ensure face cam to throttle shaft end play measures 0.004-0.006-in. (0.10-0.15 mm) between throttle shaft spacer and housing.
16. At 3500 rpm (WOT), rotate face cam to its maximum lift position in contact with the cam roller.
17. At 3500 rpm (WOT), adjust trimmer screw for 5.5 degrees cam advance and set fuel delivery to 50-51 mm ^{3}/stroke.
18. Set guide stud for 35-37 mm ^{3}/stroke at 3800 rpm (WOT). Tighten guide stud nut to 80-90 lb-in. (9-10 N•m) and check dimension from top of nut to top of guide stud for 0.400-in. (10.2 mm) maximum.
19. Set throttle for 21.5-23.5 mm ^{3}/troke at 1500 rpm and rotate face cam to obtain 2.25 degrees cam advance.
20. Tighten face cam screw to 28-30 lb-in. (3 N•m) and check face cam retention.
    (a) Put 0.005-in. (0.127 mm) feeler gauge between throttle shaft mylar washer and housing boss.
    (b) Put throttle shaft in idle position. Squeeze throttle shaft and face cam tightly toward each other and rotate face cam as lower roller rests approximately in middle at low idle step. Tighten screw to 30 lb-in. (3 N•m) with torx drive and remove feeler gauge.
21. Set total throttle travel. With throttle in low idle position, align zero degree mark on protractor with center of rib on rocker lever. Move throttle to WOT and adjust maximum travel screw to 80-84 degrees.

NOTE

Any changes to pump or settings after this point will require retesting of pump, starting with step 12.

2-34. 1990 (6.2L) FUEL INJECTION PUMP (DB2829-4879) CALIBRATION (Cont'd)

22. Fuel delivery measurement:

    (a) In order to minimize variations between readings due to residual oil in the graduates, the following timed cycle should be used:

    (1) Draw
    (2) Settle 30 seconds
    (3) Drain 30 seconds
    (4) Draw
    (5) Settle 30 seconds
    (6) Read graduates
    (7) Repeat cycle and average readings

    (b) Check points:

    (1) At 1500 rpm (WOT) de-energize electric shutoff solenoid. Delivery should be 4 mm^3/stroke maximum.

    (2) Re-energize electric shutoff solenoid and check delivery against the chart below.

RPM	THROTTLE POSITION	MM^3/ STROKE*	CAM MOVEMENT (DEGREES)	PRESSURE	
				TRANSFER PUMP	HOUSING
150	WOT	28 min.		12 psi (83 kPa) min.	0-12 psi (0-83 kPa)
400	WOT	45 min.			
400	WOT**	4 max.			
400	Low idle***				
650	Low idle	12-16			8-12 psi (55-83 kPa)
650	Low idle		1.25 min.		0-1 psi (0-7 kPa)
1500		21.5-23.5	1.0-3.5		
2000	WOT	56 max.		60-62 psi (414-427 kPa)	
2400	WOT		0.5-3.0		
3200	Low idle		9.0-10.0		
3500	WOT	48-53	4.5-6.5		
3600	WOT	47 min.			
3800	WOT	33 min.			
4050	WOT	15 max.		125 psi (862 kPa) max.	

* Maximum cylinder variation should be ± 6 mm^3 from the average flow of all cylinders.
** With housing pressure cold advance solenoid energized.
*** Check electric shutoff solenoid for pull-in with 17.6 volts maximum.

23. Disconnect pump and mount in holding fixture.
24. Using air timing gauge, check pump timing. Set to +0.25 degree ± 0.50 degree.
25. Tighten all fasteners to specifications, refer to para. 2-32d.
26. Apply sealing compound as follows:

    (a) Apply one drop at interface of maximum travel screw and locknut.

CAUTION

Do not allow sealing compound to enter gap between face cam and housing. Binding of the throttle shaft may result.

    (b) Apply one drop to threaded end of face cam screw where it protrudes from face cam.

    (c) Apply one drop to interface of servo advance adjusting screw and rocker lever, making sure sealing compound does not enter hex recess of screw.

27. Install tamper-proofing cap on guide stud.

2-35. (6.5L) FUEL INJECTION PUMP (DB2831-5149) CALIBRATION

This task covers:

a. Test Bench Requirements
b. Injection Pump Information
c. Injection Pump Check (Prior to Service)
d. Injection Pump Settings (Following Pump Repair)

INITIAL SETUP:

Tools

General mechanic's tool kit:
automotive (Appendix E, Item 1)

Special Tools

Roller-to-roller setting tool (Appendix E, Item 45)
Linkage gap tool (Appendix E, Item 49)
Automatic advance indicator (Appendix E, Item 50)
Throttle lever gauge (Appendix E, Item 51)
Protractor (Appendix E, Item 52)
Air timing gauge (Appendix E, Item 53)
Holding fixture (Appendix E, Item 46)
Torx drive (Appendix E, Item 42)

Materials/Parts

Calibrating fluid (Appendix B, Item 4)
Sealing compound (Appendix B, Item 19)

Manual References

TM 9-2815-237-34P

Test Equipment

Digital tachometer
Voltage source (variable)
Pressure gauge (0-160 psi, 1 psi increments)
Pressure gauge (0-30 psi)
Vacuum gauge (0-30 in. Hg)
Flowmeter with three-way valve
Temperature gauge (0°-250°F)
Zero-backlash coupling device
Calibrating nozzles and lines
Orifice 13211
Calibration stand

NOTE

The following calibration procedure applies to 6.5L fuel injection pumps. Refer to para. 2-34 for calibration on 1990 6.2L and above pumps and para. 2-33 for pre-1990 6.2L pumps.

a. Test Bench Requirements

The fuel injection pump requires a test bench capable of at least 2,250 rpm shaft speed. The test bench must be equipped with the following items:

(a) A zero-backlash coupling device.

(b) A digital tachometer.

(c) A variable voltage source.

(d) Test gauges:

(1) 0-160 psi (0-1103 kPa) pressure gauge calibration 1 psi increments to meaure transfer pump pressure. This gauge should be located as close as possible to the pump, with an in-line shutoff valve located at the pump.

(2) 0-30 psi (0-206.9 kPa) pressure gauge to measure housing pressure.

(3) 0-30 in. Hg (0-101.3 kPa) vacuum gauge to measure transfer pump lift. This gauge should be in the supply line to the pump, with a shutoff valve between the gauge and the oil source.

(4) A flowmeter for measuring return oil should be used with a three-way valve to permit the flowmeter to be in use only during the return oil check.

(5) 0°-250°F (-17.8°-121.1°C) temperature gauge, located at the pump inlet, to monitor inlet temperature.

2-35. (6.5L) FUEL INJECTION PUMP (DB2831-5149) CALIBRATION (Cont'd)

(e) Calibrating nozzles and injection lines:

(1) Eight 0.020-in. (0.5 mm) orifice plate nozzles set at 1700 ± 25 psi (11,721 ± 172 kPa).

(2) Eight 0.098-in. (2.5 mm) inside diameter by 25 in. (63.5 cm) long injection lines.

(f) Calibrating fluid conforming to SAE J967, maintaing a temperature of 110°-115°F (43°-46°C) at the pump inlet over all speed and load ranges.

(g) Orifice fitting in the inlet line between the supply pump and shutoff valve, to maintain constant fluid circulation and temperature.

b. Injection Pump Information

NOTE

All speeds are in engine rpm unless otherwise noted.

1. Injection pump operating speed is half of engine speed.
2. Injection pump rotation is counterclockwise.
3. Injection pump is timed to the outlet port in the 10 o'clock position, when viewed from the transfer pump end.

c. Injection Pump Check (Prior to Service)

1. Before mounting the pump on the test stand, check the drive shaft for freedom of rotation in the housing. Pump should be mounted securely in the appropriate adapter, and the intermediate coupling checked for freedom of movement.
2. Connect supply and return lines securely.
3. Install high pressure injection lines. Use copper gaskets or steel washers where required. Leave fuel line connections at the pump and nozzles loose.
4. Apply 17.6 V to electric shutoff solenoid and operate pump at 2000 rpm wide-open throttle (WOT) for 10 minutes to bring to operating temperature and clear air from system.
5. Tighten all connections securely. Check connections for leaks while operating.
6. Check fuel delivery at 650 rpm (low idle). Fuel delivery should be 10-18 mm^3/stroke. If fuel delivery is not within the range specified, adjust low idle screw. Turn screw clockwise to decrease or counter-clockwise to increase fuel delivery.
7. Check housing pressure at 650 rpm, while applying 17.6 V to housing pressure cold advance solenoid. Housing pressure should be 0-1 psi (0-7 kPa) and 8-12 psi (55-83 kPa) when de-energized.
8. Check electric shutoff solenoid for pull-in with 17.6 V maximum with throttle lever in low idle position at 400 rpm.
9. Fuel delivery measurement

(a) In order to minimize variations between readings due to residual oil in the graduates, the following timed cycle should be used:

(1) Draw

(2) Settle 30 seconds

(3) Drain 30 seconds

(4) Draw

(5) Settle 30 seconds

(6) Read graduates

(7) Repeat cycle and average readings

2-35. (6.5L) FUEL INJECTION PUMP (DB2831-5149) CALIBRATION (Cont'd)

(b) Energize electric shutoff solenoid and check delivery against the chart below.

RPM	THROTTLE POSITION	MM^3/ STROKE*	CAM MOVEMENT (DEGREES)	PRESSURE	
				TRANSFER PUMP	HOUSING
150	Low idle	27 min.			0-12 psi (0-83 kPa)
400	WOT	43 min.			
400	WOT**	4 max.			
700	Low idle	12-16***	0.5 min. ****		8-12 psi (55-83 kPa)
700	Low idle				0-1 psi (0-7 kPa)**
2000	WOT	56 max.			
3200	ADJ	21.5-23.5	3.75-16.25		
3400	WOT	48-53	3.00-5.00		
3500	WOT	43 min.			
3650	WOT	30 min.			
3900	WOT	15 max.			

* Maximum cylinder variation should be ± 6 mm^3 from the average flow of all cylinders.
** With electric shut-off solenoid de-energized.
*** Reset using low idle screw.
**** Minimum of 1.25 degrees more than reading.

10. Disconnect pump and mount in holding fixture.
11. Using air timing gauge, check pump timing. Set to -2.0 to +0.0 degrees.
12. Total throttle travel from low idle to wide-open throttle should be 75-84 degrees.
13. Refer to para. 2-7, Engine Troubleshooting, to diagnose any malfunctions encountered during checkout.

d. Injection Pump Settings (Following Pump Repair)

1. Roller-to-roller dimension must be 1.968 ± .001 in. (49.99 ± .025 mm). With oversize cam ring installed, roller-to-roller dimension will be 1.976 ± .001 in. (50.19 ± .025 mm). Maximum eccentricity allowed is 0.008 in. (0.20 mm) TIR.
2. Linkage gap must be 0.130-0.175 in. (3.30-4.45 mm).
   (a) Mount the pump horizontally in holding fixture and remove governor cover.
   (b) Hold the throttle lever in the wide-open throttle (WOT) position. Rotate pump drive shaft counterclockwise until a click is heard as the rounded contact points on the governor arm engauge the slots in the thrust sleeve. Continue to rotate the drive shaft until the gap between the governor arm and pump housing is minimal.
   (c) Using linkage gap tool, check the clearance limits between the rear of the shutoff shaft and the vertical tab on the linkage hook.
   (d) If adjustment is required, loosen governor linkage hook adjustment screw and extend the linkage hook to its maximum open length (throttle lever must be in WOT).
   (e) Insert the appropriate step of the linkage gap tool between the vertical hook tab and the throttle shaft, with the step facing the shaft. The tool must be held vertically and parallel to the linkage hook tab.
   (f) With the linkage hook pin seated firmly in the governor arm slot and the governor arm in the minimal gap position (step 2.(b)), shorten the linkage hook assembly until the face of the vertical tab is flush against the tool. Tighten adjusting screw to 12-15 lb-in. (1 N•m).
   (g) Recheck linkage gap and adjust if necessary.

2-35. (6.5L) FUEL INJECTION PUMP (DB2831-5149) CALIBRATION (Cont'd)

3. Set the throttle lever in correct low idle position prior to calibration using throttle lever gauge. The low idle screw must beat 34 ± 2 degrees as measured between vertical centerline of throttle shaft bushing bore and throttle connection stud.
4. Flush the pump with calibrating oil to remove metal chips and possible contamination.
5. Before mounting the pump on the test stand, check the drive shaft for freedom of rotation in the housing. Pump should be mounted securely in the appropriate adapter, and the intermediate coupling checked for freedom of movement.
6. Connect supply and return lines securely.
7. Install high pressure injection lines. Use copper gaskets or steel washers where required. Leave fuel line connections at the pump and nozzles loose.
8. Apply 17.6 V to electric shutoff solenoid and operate pump at 1000 rpm (WOT) for 10 minutes to bring to operating temperature and clear air from system.
9. Tighten all connections securely. Check connections for leaks while operating.
10. Set transfer pump pressure at 2000 rpm (WOT) for 60-62 psi (414-427 kPa); test stand boost should be set at 5 ± 0.5 psi (34.5 ± 3.4 kPa).
11. Set return oil flow to 225-375 cc/minute at 2000 rpm (WOT).
12. Measure housing pressure:
    (a) Check housing pressure at 650 rpm, with throttle shaft in low idle position, pressure should be 8-12 psi (55-83 kPa).
    (b) Energize housing pressure cold advance solenoid with 17.6 V. Housing pressure should be 0-1 psi (0-7 kPa).
13. With governor cover removed and throttle shaft in low idle position, set min-max governor assembly for 7-9 mm ^{3}/stroke at 650 rpm (fuel reading with cover installed).

NOTE

If low idle screw adjustment is required, refer to step 3. Adjustment is ± 1 turn from the point in step 3.

14. At 650 rpm, adjust low idle screw for 13-15 mm ^{3}/stroke for the remainder of the setting and checking procedures.
15. During and at the completion of all settings, ensure face cam to throttle shaft end play measures 0.004-0.006-in. (0.10-0.15 mm) between throttle shaft spacer and housing.
16. At 3400 rpm (WOT), rotate face cam to its maximum lift position in contact with the cam roller.
17. Adjust trimmer screw at 3400 rpm (WOT), for 4.0 degrees cam advance and set fuel delivery to 50-51 mm^3/troke.
18. Set guide stud for 35.5-37.5 mm ^{3}/stroke at 3650 rpm (WOT). Tighten guide stud nut to 80-90 lb-in. (9-10 N•m) and check dimension from top of nut to top of guide stud for 0.400-in. (10.2 mm) maximum.
19. Set throttle for 21.5-23.5 mm ^{3}/stroke at 3200 rpm and rotate face cam to obtain 5.10 degrees cam advance.
20- Tighten face cam screw to 28-30 lb-in. (3 N•m) and check face cam retention.
    (a) Put 0.005-in. (0.127 mm) feeler gauge between throttle shaft mylar washer and housing boss.
    (b) Put throttle shaft in idle position. Squeeze throttle shaft and face cam tightly toward each other and rotate face cam as lower roller rests approximately in middle at low idle step. Tighten screw to 30 lb-in. (3 N•m) with torx bit and remove feeler gauge.
21. Set total throttle travel. With throttle in low idle position, align zero degree mark on protractor with center of rib on rocker lever. Move throttle to WOT and adjust maximum travel screw to 75-84 degrees.

NOTE

Any changes to pump or settings after this point will require retesting of pump, starting with step 12.

2-35. (6.5L) FUEL INJECTION PUMP (DB2831-5149) CALIBRATION (Cont'd)

22. Fuel delivery measurement

(a) In order to minimize variations between readings due to residual oil in the graduates, the following timed cycle should be used:

(1) Draw
(2) Settle 30 seconds

(3) Drain 30 seconds
(5) Settle 30 seconds
(6) Read graduates
(7) Repeat cycle and average readings

(b) Check points:

(1) At 1500 rpm WOT) de-energize electric shutoff solenoid. Delivery should be 4 mm^3/stroke maximum.

(2) Re-energize electric shutoff solenoid and check delivery against the chart below.

RPM	THROTTLE POSITION	MM^3/ STROKE*	CAM MOVEMENT (DEGREES)	PRESSURE	
				TRANSFER PUMP	HOUSING
150	Low idle	28 min.		16 psi (110 kPa) min.	0-12 psi (0-83 kPa)
400	WOT	43 min.			
400	WOT**	4 max.			
400	Low idle***				
700	Low idle	12-16	1.0 min.		8-12 psi (55-83 kPa)
700	Low idle		1.25 min. ****		0-1 psi (0-7 kPa)
2000	WOT	56 max.		60-62 psi (414-427 kPa)	
3200	ADJ	21.5-23.5	4.0-6.0		
3400	WOT	49-52	3.50-4.50		
3500	WOT	45 min.			
3650	WOT	31.5 min.			
3900	WOT	15 max.		125 psi (862 kPa) max.	

* Maximum cylinder variation should be ± 6 mm^3 from the average flow of all cylinders.
** With electric shut-off solenoid de-energized.
*** Check electric shutoff solenoid for pull-in with 8.8 volts maximum.
**** Minimum of 1.25 degrees more than reading.

23. Disconnect pump and mount in holding fixture.

24. Using air timing gauge, check pump timing. Set to + 0.25 degree ± 0.50 degree.

25. Tighten all fasteners to specifications, refer to para. 2-32d.

26. Apply sealing compound as follows:

(a) Apply one drop at interface of maximum travel screw and locknut.

CAUTION

Do not allow sealing compound to enter gap between face cam and housing. Binding of the throttle shaft may result.

(b) Apply one drop to threaded end of face cam screw where it protrudes from face cam.

(c) Apply one drop to interface of servo advance adjusting screw and rocker lever, making sure sealing compound does not enter hex recess of screw.

27. Install tamper-proofmg capon guide stud.

2-35.1. (6.5L) FUEL INJECTION PUMP (DB2831-5209) CALIBRATION

This task covers:

a. **Test Bench Requirements**
b. **Injection Pump Information**
c. **Injection Pump Check (Prior to Service)**
d. **Injection Pump Settings (Following Pump Repair)**

INITIAL SETUP:

Tools
General mechanic's tool kit: automotive (Appendix E, Item 1)

Special Tools
Roller-to-roller setting tool (Appendix E, Item 45)
Linkage gap tool (Appendix E, Item 49)
Automatic advance indicator (Appendix E, Item 50)
Vacuum valve block gauge (Appendix E, Item 51)
Protractor (Appendix E, Item 52)
Air timing gauge (Appendix E, Item 53)
Holding fixture (Appendix E, Item 46)
Torx drive (Appendix E, Item 42)

Test Equipment
Digital tachometer
Voltage source (variable)
Pressure gauge (0-160 psi, 1 psi increments)
Pressure gauge (0-30 psi)
Vacuum gauge (0-30 in. Hg)
Flowmeter with three-way valve
Temperature gauge (0-250°F (-17.8-121.1°C))
Zero-backlash coupling device
Calibrating nozzles and lines
Orifice 13211
Calibration stand

Materials/Parts
Calibrating fluid (Appendix B, Item 4)
Sealing compound (Appendix B, Item 19)

Manual References
TM 9-2815-237-34P

NOTE

The following calibration procedure applies to 6.5L detuned engine fuel injection pumps. Refer to para. 2-35 for calibration of 6.5L fuel injection pumps.

a. Test Bench Requirements

The fuel injection pump requires a test bench capable of at least 2250 rpm shaft speed. The test bench must be equipped with the following items:

(a) A zero-backlash coupling device.

(b) A digital tachometer.

(c) A variable voltage source.

(d) Test gauges:

(1) 0-160 psi (0-1103 kPa) pressure gauge calibrated in 1 psi increments to measure transfer pump pressure. This gauge should be located as close as possible to the pump, with an in-line shutoff valve located at the pump.

(2) 0-30 psi (0-206.9 kPa) pressure gauge to measure housing pressure.

(3) 0-30 in. Hg (0-101.3 kPa) vacuum gauge to measure transfer pump lift. This gauge should be in the supply line to the pump, with a shutoff valve between the gauge and the oil source.

(4) A flowmeter for measuring return oil should be used with a three-way valve to permit the flowmeter to be in use only during the return oil check.

(5) 0-250°F (-17.8-121.1°C) temperature gauge, located at the pump inlet, to monitor inlet temperature.

2-35.1. (6.5L) FUEL INJECTION PUMP (DB2831-5209) CALIBRATION (Cont'd)

(e) Calibrating nozzles and injection lines:

(1) Eight 0.020-in. (0.5 mm) orifice plate nozzles set at 1700 ± 25 psi (11,721 ± 172 kPa).

(2) Eight 0.098-in. (2.5 mm) inside diameter by 25 in. (63.5 cm) long injection lines.

(f) Calibrating fluid conforming to SAE J967, maintaining a temperature of 110-115°F (43-46°C) at the pump inlet over all speed and load ranges.

(g) Orifice fitting in the inlet line between the supply pump and shutoff valve, to maintain constant fluid circulation and temperature.

b. Injection Pump Information

NOTE

All speeds are in engine rpm unless otherwise noted.

1. Injection pump operating speed is half of engine speed.
2. Injection pump rotation is counterclockwise.
3. Injection pump is timed to the outlet port in the 10 o'clock position when viewed from the transfer pump end.

c. Injection Pump Check (Prior to Service)

1. Before mounting pump on test stand, check driveshaft for freedom of rotation in the housing. Pump should be mounted securely in the appropriate adapter, and the intermediate coupling should be checked for freedom of movement.
2. Connect supply and return lines securely.
3. Install high-pressure injection lines. Use copper gaskets and/or steel washers where required. Leave fuel line connections at the pump and nozzles loose.
4. Apply 17.6 volts to electric shutoff solenoid and operate pump at 2000 rpm Wide-Open Throttle (WOT) for 10 minutes to bring to operating temperature and clear air from system.
5. Tighten all connections securely. Check connections for leaks while operating.
6. Check fuel delivery at 650 rpm (low idle). Fuel delivery should be 10-18 mm^3/stroke. If fuel delivery is not within the range specified, adjust low-idle screw. Turn screw clockwise to decrease or counterclockwise to increase fuel delivery.
7. Check housing pressure at 650 rpm while applying 17.6 volts to housing pressure cold-advance solenoid. Housing pressure should be 0-1 psi (0-7 kPa) and 8-12 psi (55-83 kPa) when de-energized.
8. Fuel delivery:

   (a) Check electric shutoff solenoid for pull-in with 17.6 volts maximum with throttle lever in low-idle position at 400 rpm.

   (b) Maximum fuel delivery under condition a. should be 9 mm^3/stroke minimum.
9. Fuel delivery measurement:

   (a) In order to minimize variations between readings due to residual oil in the graduates, the following timed cycle should be used:

   (1) Draw
   (2) Settle 30 seconds
   (3) Drain 30 seconds
   (4) Draw
   (5) Settle 30 seconds
   (6) Read graduates
   (7) Repeat cycle and average readings

2-35.1. (6.5L) FUEL INJECTION PUMP (DB2831-5209) CALIBRATION (Cont'd)

(b) Energize electric shutoff solenoid and check delivery against the chart below.

RPM	THROTTLE POSITION	MM STROKE*	CAM MOVEMENT (DEGREES)	HOUSING PRESSURE
150	Low idle	25 min.		0-12 psi (0-83 kPa)
400	WOT	35.5 min.		
400	WOT**	4 max.		
700	Low idle	12-16***	0.5 min.	8-12 psi (55-83 kPa)
700	Low idle		****	0-1 psi (0-7 kPa)**
2000	WOT	51.5 max.		
3200	ADJ	21.5-23.5	3.75-16.25	
3500	WOT	43.5-48.5	3.25-5.25	
3600	WOT	43.5 min.		
3800	WOT	30 min.		
4050	WOT	15 max.		

* Maximum cylinder variation should be ± 6 mm^3 from the average flow of all cylinders.
** With electric shut-off solenoid de-energized.
*** Reset using low-idle screw.
**** Minimum of 1.25 degrees more than reading.

10. Disconnect pump and mount in holding fixture.
11. Using air timing gauge, check pump timing. Set to -2.0 to +0.0 degrees.
12. Total throttle travel from low-idle to WOT should be 80-84 degrees.
13. Refer to para. 2-7, Engine Troubleshooting, to diagnose any malfunctions encountered during checkout.

d. Injection Pump Settings (Following Pump Repair)

1. Roller-to-roller dimension must be 1.964 ± .001 in. (49.89 ± .025 mm). With oversize cam ring installed, roller-to-roller dimension will be 1.972 ± .001 in. (50.09 ± .026 mm). Maximum eccentricity allowed is 0.008 in. (0.20 mm) TIR.
2. Linkage gap must be 0.130-0.1734 in. (3.30-4.40 mm).
   (a) Mount the pump horizontally in holding fixture and remove governor cover.
   (b) Hold the throttle lever in the WOT position. Rotate pump driveshaft counterclockwise until a click is heard as the rounded contact points on the governor arm engage the slots in the thrust sleeve. Continue to rotate driveshaft until gap between governor arm and pump housing is minimal.
   (c) Using linkage gap tool, check clearance limits between rear of shutoff shaft and vertical tab on linkage hook.
   (d) If adjustment is required, loosen governor linkage hook adjustment screw and extend linkage hook to its maximum open length (throttle lever must be in WOT).
   (e) Insert appropriate step of linkage gap tool between vertical hook tab and throttle shaft, with step facing shaft. The tool must be held vertically and parallel to linkage hook tab.
   (f) With linkage hook pin seated firmly in governor arm slot and governor arm in the minimal gap position (step 2.(b)), shorten linkage hook assembly until face of vertical tab is flush against tool. Tighten adjusting screw to 12-15 lb-in. (1 N·m).
   (g) Recheck linkage gap and adjust if necessary.

2-35.1. (6.5L) FUEL INJECTION PUMP (DB2831-5209) CALIBRATION (Cont'd)

3. Set the throttle lever in correct low-idle position prior to calibration using vacuum valve block gauge. The low-idle screw must be at 34 ± 2 degrees as measured between vertical centerline of throttle shaft bushing bore and throttle connection stud.
4. Flush pump with calibrating oil to remove metal chips and possible contamination.
5. Before mounting pump on test stand, check driveshaft for freedom of rotation in housing. Pump should be mounted securely in the appropriate adapter, and the intermediate coupling should be checked for freedom of movement.
6. Connect supply and return lines securely.
7. Install high-pressure injection lines. Use copper gaskets or steel washers where required. Leave fuel line connections at the pump and nozzles loose.
8. Apply 17.6 volts to electric shutoff solenoid and operate pump at 1000 rpm (WOT) for 10 minutes to bring to operating temperature and clear air from system.
9. Tighten all connections securely. Check connections for leaks while operating.
10. Set transfer pump pressure at 2000 rpm (WOT) for 60-62 psi (414-427 kPa). Test stand boost should be set at 5 ± 0.5 psi (34.5 ± 3.4 kPa).
11. Set return oil flow to 225-375 cc/minute at 2000 rpm (WOT).
12. Measure housing pressure:
    (a) Check housing pressure at 650 rpm. With throttle shaft in low idle position, pressure should be 8-12 psi (55-83 kPa).
    (b) Energize housing pressure cold advance solenoid with 17.6 volts. Housing pressure should be 0-1 psi (0-7 kPa).
13. With governor cover removed and throttle shaft in low-idle position, set min.-max. governor assembly for 7-9 mm^3/stroke at 650 rpm (fuel reading with cover installed).

NOTE

If low-idle screw adjustment is required, refer to step 3. Adjustment is ± 1 turn from the point in step 3.

14. At 650 rpm, adjust low-idle screw for 13-15 mm^3/stroke for the remainder of the setting and checking procedures.
15. During and at the completion of all settings, ensure face cam-to-throttle shaft end play measures 0.004-0.006 in. (0.10-0.15 mm) between throttle shaft spacer and housing.
16. At 3500 rpm (WOT), rotate face cam to its maximum lift position in contact with the cam roller.
17. Adjust trimmer screw at 3500 rpm (WOT) for 4.25 degrees cam advance and set fuel delivery to 45.5-46.5 mm^3/stroke.
18. Set guide stud for 46.5 mm^3/stroke at 3800 rpm (WOT). Tighten guide stud nut to 80-90 lb-in. (9-10 N·m) and check dimension from top of nut to top of guide stud for 0.425 in. (10.80 mm) maximum.
19. Set throttle for 21.5-23.5 mm^3/stroke at 3200 rpm and rotate face cam to obtain 5.00 degrees cam advance.
20. Tighten face cam screw to 28 30 lb-in. (3 N·m) and check face cam retention.
    (a) Put 0.005-in. (0.127 mm) feeler gauge between throttle shaft mylar washer and housing boss.
    (b) Put throttle shaft in idle position. Squeeze throttle shaft and face cam tightly toward each other and rotate face cam as lower roller rests approximately in middle at lowest step. Tighten screw to 30 lb-in. (3 N·m) with torx bit and remove feeler gauge.
21. Set total throttle travel. With throttle in low-idle position, align zero degree mark on protractor with center of rib on rocker lever. Move throttle to WOT and adjust maximum travel screw to 80-84 degrees.

NOTE

Any changes to pump or settings after this point will require retesting of pump, starting with step 12.

2-35.1. (6.5L) FUEL INJECTION PUMP (DB2831-5209) CALIBRATION (Cont'd)

22. Fuel delivery measurement:
    (a) In order to minimize variations between readings due to residual oil in the graduates, the following timed cycle should be used:
        (1) Draw
        (2) Settle 30 seconds
        (3) Drain 30 seconds
        (4) Draw
        (5) Settle 30 seconds
        (6) Read graduates
        (7) Repeat cycle and average readings
    (b) Check points:
        (1) At 1500 rpm (WOT), de-energize electric shutoff solenoid. Delivery should be 4 mm^3/stroke maximum.
        (2) Re-energize electric shutoff solenoid and check delivery against the chart below.

RPM	THROTTLE POSITION	MM STROKE*	CAM MOVEMENT (DEGREES)	PRESSURE	
				TRANSFER PUMP	HOUSING
150	Low idle	26 min.		16 psi (110 kPa) min.	0-12 psi (0-83 kPa)
400	WOT	36.5 min.			
400	WOT**	4 max.			
400	Low idle***				
700	Low idle	12-16	1.0 min.		8-12 psi (55-83 kPa)
700	Low idle		****		0-1 psi (0-7 kPa)
2000	WOT	51.5 max.		60-62 psi (414-427 kPa)	
3200	ADJ	21.5-23.5	4.0-6.0		
3500	WOT	44.5-47.5	3.75-4.75		
3600	WOT	44 min.			
3800	WOT	31.5 min.			
4050	WOT	15 max.		125 psi (862 kPa) max.	

* Maximum cylinder variation should be ± 6 mm^3 from the average flow of all cylinders.
** With electric shut-off solenoid de-energized.
*** Reset using low-idle screw.
**** Minimum of 1.25 degrees more than reading.

23. Disconnect pump and mount in holding fixture.
24. Using air timing gauge, check pump timing. Set to +1.00 degree ± 0.50 degree.
25. Tighten all fasteners to specifications. Refer to para. 2-32d.
26. Apply sealing compound as follows:
    (a) Apply one drop at interface of maximum travel screw and locknut.

CAUTION

Do not allow sealing compound to enter gap between face cam and housing. Binding of the throttle shaft may result.

    (b) Apply one drop to threaded end of face cam screw where it protrudes from face cam.
    (c) Apply one drop to interface of servo advance adjusting screw and rocker lever, ensuring sealing compound does not enter hex recess of screw.
27. Install tamper-proof cap on guide stud.

2-35. 6.5L FUEL INJECTION PUMPS (DB2831-5485 OR DB2831-5079) CALIBRATION

This task covers:

a. **Test Bench Requirements**
b. **Injection Pump Information**
c. **Injection Pump Check (Prior to Service)**
d. **Injection Pump Settings (Following Pump Repair)**

INITIAL SETUP:

Tools
General mechanic's tool kit:
automotive (Appendix E, Item 1)

Special Tools
Roller-to-roller setting tool (Appendix E, Item 45)
Linkage gap tool (Appendix E, Item 49)
Automatic advance indicator (Appendix E, Item 50)
Vacuum valve block gauge (Appendix E, Item 51)
Protractor (Appendix E, Item 52)
Air timing gauge (Appendix E, Item 53)
Holding fixture (Appendix E, Item 46)
Torx drive (Appendix E, Item 42)

Test Equipment
Digital tachometer
Voltage source (variable)
Pressure gauge (0-160 psi, 1 psi increments)
Pressure gauge (0-30 psi)
Vacuum gauge (0-30 in. Hg)
Flowmeter with three-way valve
Temperature gauge (0°-250°F)
Zero-backlash coupling device
Calibrating nozzles and lines
Orifice 13211
Calibration stand

Materials/Parts
Calibrating fluid (Appendix B, Item 4)
Sealing compound (Appendix B, Item 19)

Manual References
TM 9-2815-237-34P

NOTE

The following calibration procedure applies to 6.5L turbocharged fuel injection pumps. Refer to para. 2-34 for calibration of 1990 6.2L and above pumps and para. 2-33 for pre-1990 6.2L pumps.

a. Test Bench Requirements

The fuel injection pump requires a test bench capable of at least 2250 rpm shaft speed. The test bench must be equipped with the following items:

(a) A zero-backlash coupling device.
(b) A digital tachometer.
(c) A variable voltage source.
(d) Test gauges:
   (1) 0-160 psi (0-1103 kPa) pressure gauge calibrated in 1 psi increments to measure transfer pump pressure. This gauge should be located as close as possible to the pump, with an in-line shutoff valve located at the pump.
   (2) 0-30 psi (0-206.9 kPa) pressure gauge to measure housing pressure.
   (3) 0-30 in. Hg (0-101.3 kPa) vacuum gauge to measure transfer pump lift. This gauge should be in the supply line to the pump, with a shutoff valve between the gauge and the oil source.
   (4) A flowmeter for measuring return oil should be used with a three-way valve to permit the flowmeter to be in use only during the return oil check.
   (5) 0-250°F (-17.8-121.1°C) temperature gauge, located at the pump inlet, to monitor inlet temperature.

2-35(3). FUEL INJECTION PUMPS (DB2831-5485 OR DB2831-5079) CALIBRATION (Cont'

(e) Calibrating nozzles and injection lines:

(1) Eight 0.020-in. (0.5 mm) orifice plate nozzles set at 1700 ± 25 psi (11,721 ± 172 kPa).

(2) Eight 0.098-in. (2.5 mm) inside diameter by 25 in. (63.5 cm) long injection lines.

(f) Calibrating fluid conforming to SAE J967, maintaining a temperature of 110°-115°F (43°-46°C) at the pump inlet over all speed and load ranges.

(g) Orifice fitting in the inlet line between the supply pump and shutoff valve, to maintain constant fluid circulation and temperature.

b. Injection Pump Information

NOTE

All speeds are in engine rpm unless otherwise noted.

1. Injection pump operating speed is half of engine speed.
2. Injection pump rotation is counterclockwise.
3. Injection pump is timed to the outlet port in the 10 o'clock position when viewed from the transfer pump end.

c. Injection Pump Check (Prior to Service)

1. Before mounting the pump on the test stand, check the drive shaft for freedom of rotation in the housing. Pump should be mounted securely in the appropriate adapter, and the intermediate coupling should be checked for freedom of movement.
2. Connect supply and return lines securely.
3. Install high-pressure injection lines. Use copper gaskets or steel washers where required. Leave fuel line connections at the pump and nozzles loose.
4. Apply 17.6 V to electric shutoff solenoid and operate pump at 2000 rpm wide-open throttle (WOT) for 10 minutes to bring to operating temperature and clear air from system.
5. Tighten all connections securely. Check connections for leaks while operating.
6. Check fuel delivery at 650 rpm (low-idle). Fuel delivery should be 10-18 mm^3/stroke. If fuel delivery is not within the range specified, adjust low-idle screw. Turn screw clockwise to decrease or counter-clockwise to increase fuel delivery.
7. Check housing pressure at 650 rpm while applying 17.6 V to housing pressure cold-advance solenoid. Housing pressure should be 0-1 psi (0-7 kPa) and 8-12 psi (55-83 kPa) when de-energized.
8. Check electric shutoff solenoid for pull-in with 17.6 V maximum with throttle lever in low-idle position at 400 rpm.
9. Fuel delivery measurement:

(a) In order to minimize variations between readings due to residual oil in the graduates, the following timed cycle should be used:

(1) Draw

(2) Settle 30 seconds

(3) Drain 30 seconds

(4) Draw

(5) Settle 30 seconds

(6) Read graduates

(7) Repeat cycle and average readings

2-35 (1). FUEL INJECTION PUMPS (DB2831-5485 OR DB2831-5079) CALIBRATION (Cont'd)

(b) Energize electric shutoff solenoid and check delivery against the chart below:

RPM	THROTTLE POSITION	MM³ STROKE*	CAM MOVEMENT (DEGREES)	PRESSURE	
				TRANSFER PUMP	HOUSING
150	Low-idle	36 min.			0-12 psi (0-83 kPa)
400	WOT	57 min.			
400	WOT**	4 max.			
700	Low-idle	10-18***	3.0 min.		8-12 psi (55-83 kPa)
700	Low-idle		****		0-1 psi (0-7 kPa)**
1800	WOT	65.6-70.5			
2000	WOT				
3200	WOT	58.5 min.	2.0-4.0		
3200	ADJ	21.5-23.5	5.0-7.5		
3200	Low-idle		11.0-12.0		
3650	WOT	32 min.			
3900	WOT	15 max.			

* Maximum cylinder variation should be ± 6 mm^3 from the average flow of all cylinders.
** With electric shut-off solenoid de-energized.
*** Reset using low-idle screw.
**** Minimum of 1.25 degrees more than reading.

10. Disconnect pump and mount in holding fixture.

NOTE

Perform step 12 to check pump timing on the DB2831-5079 fuel injection pump.

11. Using air timing gauge, check pump timing. Set to 1.5 to 3.5 degrees.
12. Using air timing gauge, check pump timing. Set to 1.0 +1.0 degrees.
13. Total throttle travel from low-idle to wide-open throttle should be 75-84 degrees.
14. Refer to para. 2-7, Engine Troubleshooting, to diagnose any malfunctions encountered during checkout.

d. Injection Pump Settings (Following Pump Repair)

1. Roller-to-roller dimension must be 1.975 ± .001 in. (50.19 ± .025 mm). With oversize cam ring installed, roller-to-roller dimension will be 1.882 ± .005 in. (50.34 ± .013 mm). Maximum eccentricity allowed is 0.008 in. (0.20 mm) TIR.
2. Linkage gap must be 0.130-0.175 in. (3.30-4.45 mm).

(a) Mount the pump horizontally in holding fixture and remove governor cover.

(b) Hold the throttle lever in the wide-open throttle (WOT) position. Rotate pump driveshaft counterclockwise until a click is heard as the rounded contact points on the governor arm engage the slots in the thrust sleeve. Continue to rotate driveshaft until gap between governor arm and pump housing is minimal.

(c) Using linkage gap tool, check clearance limits between rear of shutoff shaft and vertical tab on linkage hook.

(d) If adjustment is required, loosen governor linkage hook adjustment screw and extend the linkage hook to its maximum open length (throttle lever must be in WOT).

(e) Insert appropriate step of linkage gap tool between vertical hook tab and throttle shaft, with step facing shaft. The tool must be held vertically and parallel to linkage hook tab.

(f) With linkage hook pin seated firmly in governor arm slot and governor arm in the minimal gap position (step 2.(b)), shorten linkage hook assembly until face of vertical tab is flush against tool. Tighten adjusting screw to 12-15 lb-in. (1 N·m).

(g) Recheck linkage gap and adjust if necessary.

2-35.1. FUEL INJECTION PUMPS (DB2831-5485 OR DB2831-5079) CALIBRATION (Cont'

3. Set the throttle lever in correct low-idle position prior to calibration using vacuum valve block gauge. The low-idle screw must be at 34 ± 2 degrees as measured between vertical centerline of throttle shaft bushing bore and throttle connection stud.
4. Flush the pump with calibrating oil to remove metal chips and possible contamination.
5. Before mounting the pump on the test stand, check the driveshaft for freedom of rotation in the housing. Pump should be mounted securely in the appropriate adapter, and the intermediate coupling checked for freedom of movement.
6. Connect supply and return lines securely.
7. Install high-pressure injection lines. Use copper gaskets or steel washers where required. Leave fuel line connections at the pump and nozzles loose.
8. Apply 17.6 V to electric shutoff solenoid and operate pump at 1000 rpm (WOT) for 10 minutes to bring to operating temperature and clear air from system.
9. Tighten all connections securely. Check connections for leaks while operating.
10. Set transfer pump pressure at 2000 rpm (WOT) for 60-62 psi (414-427 kPa); test stand boost should be set at 5 ± 0.5 psi (34.5 ± 3.4 kPa).
11. Set return oil flow to 225-375 cc/minute at 2000 rpm (WOT).
12. Measure housing pressure:
    (a) Check housing pressure at 650 rpm with throttle shaft in low-idle position. Pressure should be 8-12 psi (55-83 kPa).
    (b) Energize housing pressure cold advance solenoid with 17.6 V. Housing pressure should be 0-1 psi (0-7 kPa).
13. With governor cover removed and throttle shaft in low-idle position, set min-max governor assembly for 7-9 mm^3/stroke at 650 rpm (fuel reading with cover installed).

NOTE

If low-idle screw adjustment is required, refer to step 3. Adjustment is ± 1 turn from the point in step 3.

14. At 650 rpm, adjust low-idle screw for 13-15 mm^3/stroke for the remainder of the setting and checking procedures.
15. During and at the completion of all settings, ensure face cam to throttle shaft end play measures 0.004-0.006-in. (0.10-0.15 mm) between throttle shaft spacer and housing.
16. At 1800 rpm (WOT), rotate face cam to its maximum lift position in contact with the cam roller.
17. Adjust trimmer screw at 1800 rpm (WOT), to advance cam advance and set fuel delivery to 67.5-68.6 mm^3/stroke.
18. Set guide stud for 37.0-39.0 mm^3/stroke at 3650 rpm (WOT). Tighten guide stud nut to 80-90 lb-in. (9-10 N·m) and check dimension from top of nut to top of guide stud for 0.400-in. (10.2 mm) maximum.
19. Set throttle for 22.6 ±0.3 mm^3/stroke at 1800 rpm and rotate face cam to obtain 5.10 degrees cam advance.
20. Tighten face cam screw to 28-30 lb-in. (3 N·m) and check face cam retention.
    (a) Put 0.005-in. (0.127 mm) feeler gauge between throttle shaft mylar washer and housing boss.
    (b) Put throttle shaft in idle position. Squeeze throttle shaft and face cam tightly toward each other and rotate face cam as lower roller rests approximately in middle at low-idle step. Tighten screw to 30 lb-in. (3 N·m) with torx bit and remove feeler gauge.
21. Set total throttle travel. With throttle in low-idle position, align zero degree mark on protractor with center of rib on rocker lever. Move throttle to WOT and adjust maximum travel screw to 75-84 degrees.

NOTE

Any changes to pump or settings after this point will require retesting of pump, starting with step 12.

2-35.1. FUEL INJECTION PUMPS (DB2831-5485 OR DB2831-5079) CALIBRATION (Cont)

22. Fuel delivery measurement:
   (a) In order to minimize variations between readings due to residual oil in the graduates, the following timed cycle should be used:
      (1) Draw
      (2) Settle 30 seconds
      (3) Drain 30 seconds
      (4) Draw
      (5) Settle 30 seconds
      (6) Read graduates
      (7) Repeat cycle and average readings
   (b) Check points:
      (1) At 1500 rpm (WOT), de-energize electric shutoff solenoid. Delivery should be 4 mm^3/stroke maximum.
      (2) Re-energize electric shutoff solenoid and check delivery against the chart below.

RPM	THROTTLE POSITION	MM^3 STROKE*	CAM MOVEMENT (DEGREES)	PRESSURE	
				TRANSFER PUMP	HOUSING
150	Low-idle	37 min.		16 psi (110 kPa) min.	0-12 psi (0-83 kPa)
400	WOT	58 min.			
400	WOT**	4 max.			
400	Low-idle***				
700	Low-idle	12-16	3.5 min.		8-12 psi (55-83 kPa)
700	Low-idle		1.25 min.		0-1 psi (0-7 kPa)
1500	WOT	4 max.			
1500	WOT	68.5-69.5			
2000	WOT	56 max.	****	60-62 psi (414-427 kPa)	
3200	WOT	59.5 min.	2.50-3.60		
3200	ADJ	21.5-23.5	5.25-7.25		
3200	Low-idle		11-12		
3400	WOT	49-52	3.50-4.50		
3500	WOT	45 min.			
3650	WOT	31.5 min.			
3900	WOT	15 max.		125 psi (862 kPa) max.	

* Maximum cylinder variation should be ± 6 mm^3 from the average flow of all cylinders.
** With electric shut-off solenoid de-energized.
*** Check electric shutoff solenoid for pull-in with 8.8 volts maximum.
**** Minimum of 1.25 degrees more than reading.

23. Disconnect pump and mount in holding fixture.

NOTE

Perform step 25 to check pump timing on the DB2831-5079.

24. Using air timing gauge, check pump timing. Set to +0.25 degree ± 0.50 degree.
25. Using air timing gauge, check pump timing. Set to 0 ±0.50 degrees
26. Tighten all fasteners to specifications. Refer to para. 2-32d.

2-35(1.5). FUEL INJECTION PUMPS (DB2831-5485 OR DB2831-5079) CALIBRATION (Cont'

27. Apply sealing compound as follows:
    (a) Apply one drop at interface of maximum travel screw and locknut.

CAUTION

Do not allow sealing compound to enter gap between face cam and housing. Binding of the throttle shaft may result.

    (b) Apply one drop to threaded end of face cam screw where it protrudes from face cam.
    (c) Apply one drop to interface of servo advance adjusting screw and rocker lever, ensuring sealing compound does not enter hex recess of screw.
28. Install tamper-proof cap on guide stud.

2-35. (6.2L or 6.5L) FUEL INJECTION PUMPS (DB2829-4523, DB2829-4879, or DB2831-5149) CALIBRATION USING FUEL INJECTION PUMP TEST STAND (FTIS) MODEL A8022

This task covers:

a. **Mounting Fuel Injection Pump on Test Stand**
b. **Test Stand Setup Procedures**
c. **Removal of Fuel Injection Pump from Test Stand**

INITIAL SETUP:

Tools

General mechanic's tool kit:
automotive (Appendix E, Item 1)

Special Tools

Adapter Kit 74-8033 consisting of:
- Adapter plate (Appendix E, Item 57)
- Advance indicator (Appendix E, Item 58)
- Alignment gauge (Appendix E, Item 59)
- Barring tool (Appendix E, Item 60)
- Connector (Appendix E, Item 61)
- High-pressure lines (Appendix E, Item 62)
- Hose assembly (Appendix E, Item 63)
- Hose assembly (Appendix E, Item 64)
- Male pipe fitting (Appendix E, Item 65)
- Quick-connect plug (Appendix E, Item 66)
- Test adapter (Appendix E, Item 67)
- Test lead, black (Appendix E, Item 68)
- Test lead, red (Appendix E, Item 69)
- Throttle arm positioner assembly (Appendix E, Item 70)
- Tubing (Appendix E, Item 71)

Test Equipment

Test stand, model A8022

Materials/Parts

Three 5/16-18x1 hex-head capscrews (Appendix D, Item 55)
Three 3/8-16x1 hex-head capscrews (Appendix D, Item 56)
Calibration fluid (Appendix B, Item 4)

Manual References

TM 9-2815-237-34P
Instruction manual 74-9225

NOTE

The following calibration procedures apply to fuel injection pumps DB 2829-4523, DB 2829-4879, or DB 2831-5149 only.

a. Mounting Fuel Injection Pump on Test Stand

NOTE

Prior to installing fuel injection pump to drive coupling, ensure alignment gauge is installed.

1. Insert alignment gauge into test stand drive coupling.
2. Install fuel injection pump (1) on adapter plate (6) and secure with three 5/16-in. washers (3) and 5/16-18x1 hex-head capscrews (2).
3. Position fuel injection pump (1), with flange mounting bracket (5) installed, to test stand drive coupling.
4. Using barring tool, rotate driveshaft (12) until tapped holes on pump-mounted drive hub (7) are aligned with slots in test stand drive coupling.
5. Install three 3/8-in. washers (8), 3/8-in. lockwashers (9), and 3/8-16x1 hex-head capscrews (10) in slots and mounting holes on test stand drive coupling and pump-mounted drive hub (7). Remove barring tool.
6. Tighten two 3/8-16x1 hex-head capscrews (4) that secure flange mounting bracket (5) to bedplate (11). Remove alignment gauge.

2-35. (6.2L or 6.5L) FUEL INJECTION PUMPS (DB2829-4523, DB2829-4879, or DB2831-5149) CALIBRATION USING FUEL INJECTION PUMP TEST STAND (FTIS) MODEL A8022 (Cont'd)

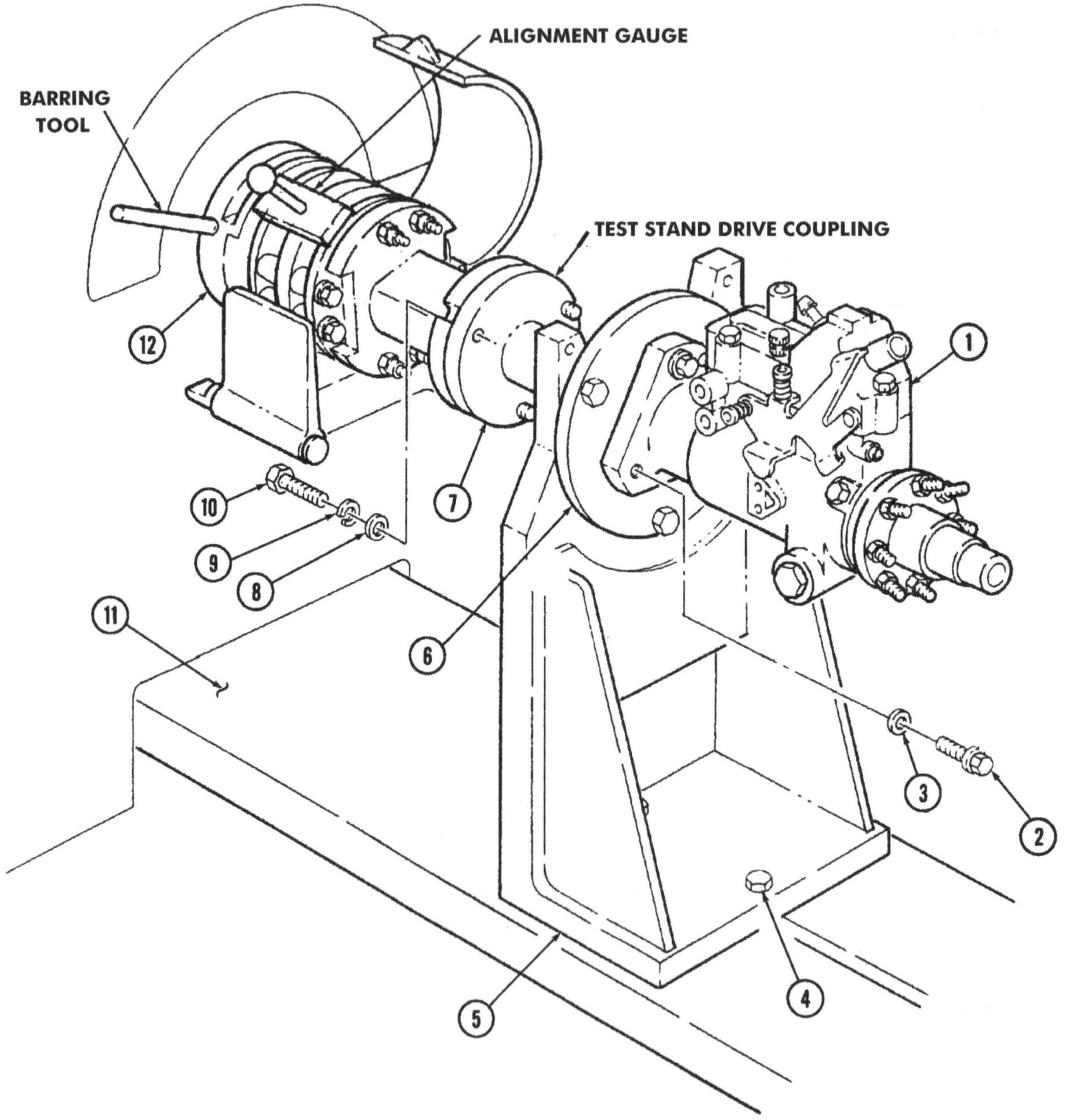

2-35. (6.2L or 6.5L) FUEL INJECTION PUMPS (DB2829-4523, DB2829-4879, or DB2831-5149) CALIBRATION USING FUEL INJECTION PUMP TEST STAND (FTIS) MODEL A8022 (Cont'd)

7. Install test adapter and connector on fuel inlet port (2) of fuel injection pump (1).
8. Connect one end of hose assembly to connector and insert other hose end into calibration fluid supply quick-connect on front panel of test stand (3).
9. Remove screw (8), tab (7), washer (6), and O-ring (5) from fuel injection pump transfer port (4).
10. Position test adapter on fuel injection pump transfer test port (4) and secure into position with existing O-ring (5), tab (7), and washer (6) removed in step 9.
11. Install quick-connect plug into test adapter.
12. Connect one end of rubber hose assembly (9) to quick-connect plug and other hose assembly end to auxiliary pressure quick-connect on front panel of test stand (3).
13. Install male pipe fitting on fuel injection pump (1).
14. Connect one end of tubing to male pipe fitting on fuel injection pump (1) and other end to calibration fluid push-on connector on front panel of test stand (3).

CAUTION

When installing high-pressure delivery lines on fuel injection pump, use two wrenches to prevent damage to discharge fittings.

15. Install eight high-pressure delivery lines on discharge fittings (10) of fuel injection pump (1).

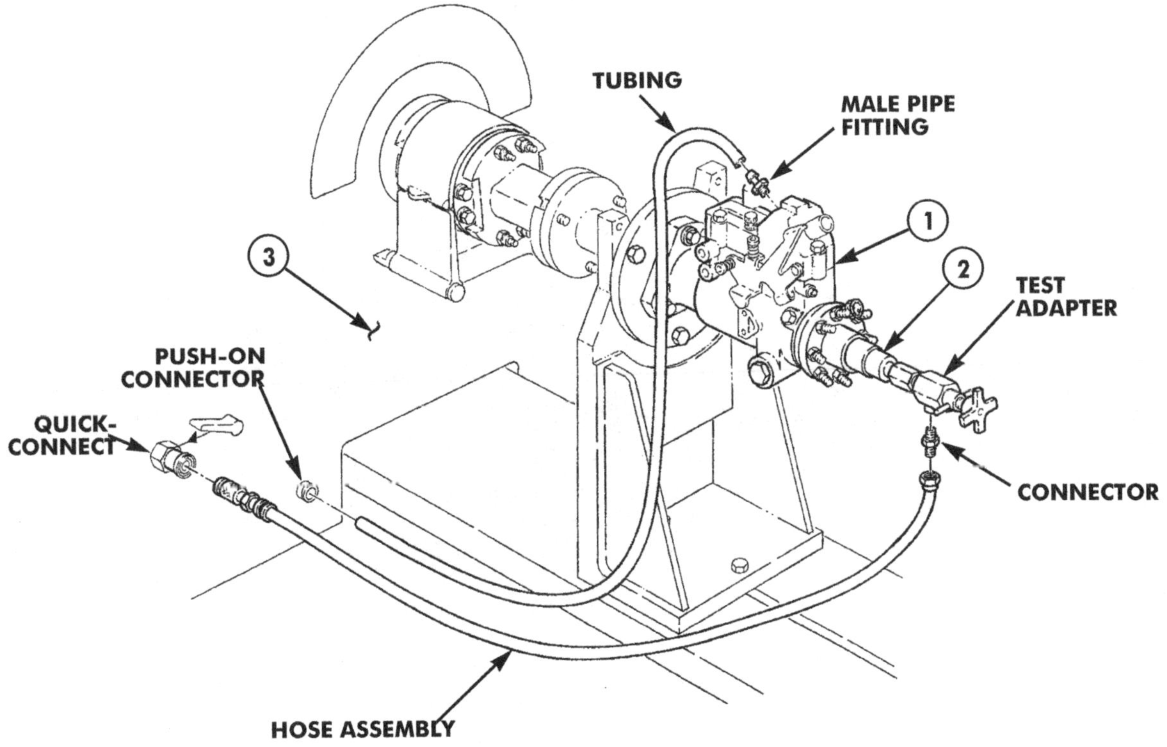

2-35. (6.2L or 6.5L) FUEL INJECTION PUMPS (DB2829-4523, DB2829-4879, or DB2831-5149) CALIBRATION USING FUEL INJECTION PUMP TEST STAND (FTIS) MODEL A8022 (Cont'd)

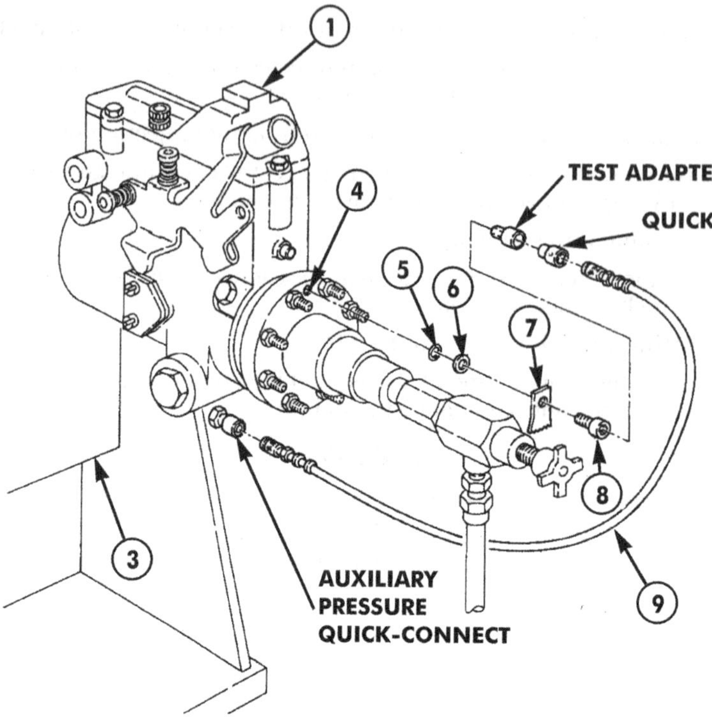

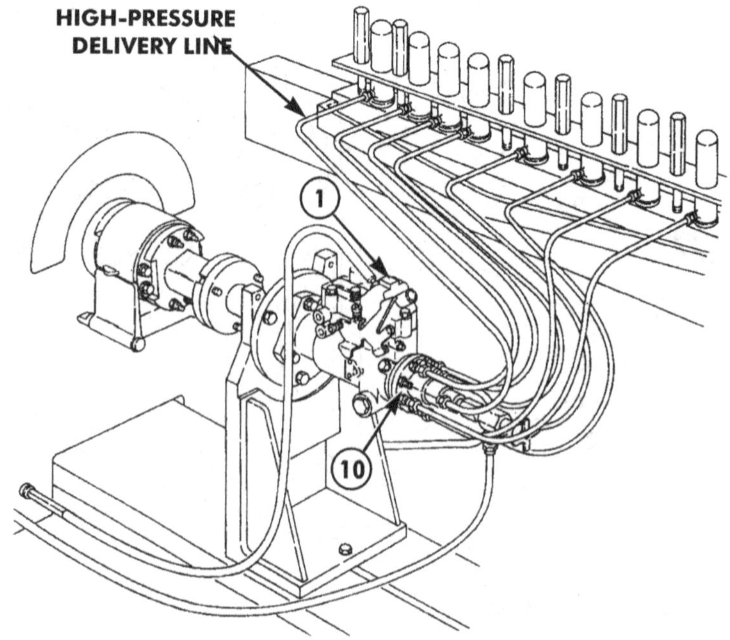

2-35. (6.2L or 6.5L) FUEL INJECTION PUMPS (DB2829-4523, DB2829-4879, or DB2831-5149) CALIBRATION USING FUEL INJECTION PUMP TEST STAND (FTIS) MODEL A8022 (Cont'd)

16. Position throttle arm positioner assembly (3) on bedplate (7) and throttle lever (1).

NOTE

When installing advance indicator, ensure the flat on pointer end engages into the cam ring slot on fuel injection pump.

17. Remove two screws (6), timing cover (5), and gasket (4) from fuel injection pump (2) and install advance indicator with two screws (6).
18. Remove screw plug from advance indicator block and install quick-connect on advance indicator block (8).
19. Attach one end of rubber hose assembly to quick-connect and the other hose assembly end to auxiliary pressure quick-connect plug on front panel of test stand.
20. Connect positive red test lead to DC power supply socket marked (+) on test stand, and attach the alligator clip end of red test lead to the electric shutoff solenoid terminal (9) on fuel injection pump (2).
21. Connect black lead to DC black socket marked “P” on test stand and connect alligator clip end of black lead to ground terminal (10) on fuel injection pump (2).
22. After installation and hookup of the fuel injection pump has been completed, inspect pump, all hoses, and related hardware connections before beginning pump calibration. Review operation test procedures for the FTIS prior to calibrating the fuel injection pump.

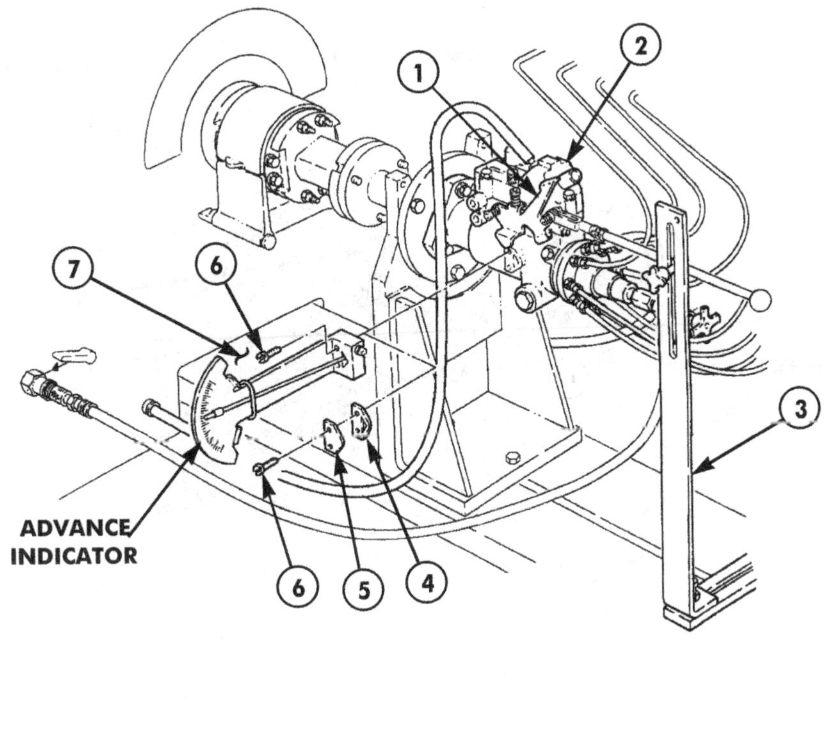

2-35. (6.2L or 6.5L) FUEL INJECTION PUMPS (DB2829-4523, DB2829-4879, or DB2831-5149) CALIBRATION USING FUEL INJECTION PUMP TEST STAND (FTIS) MODEL A8022 (Cont'd)

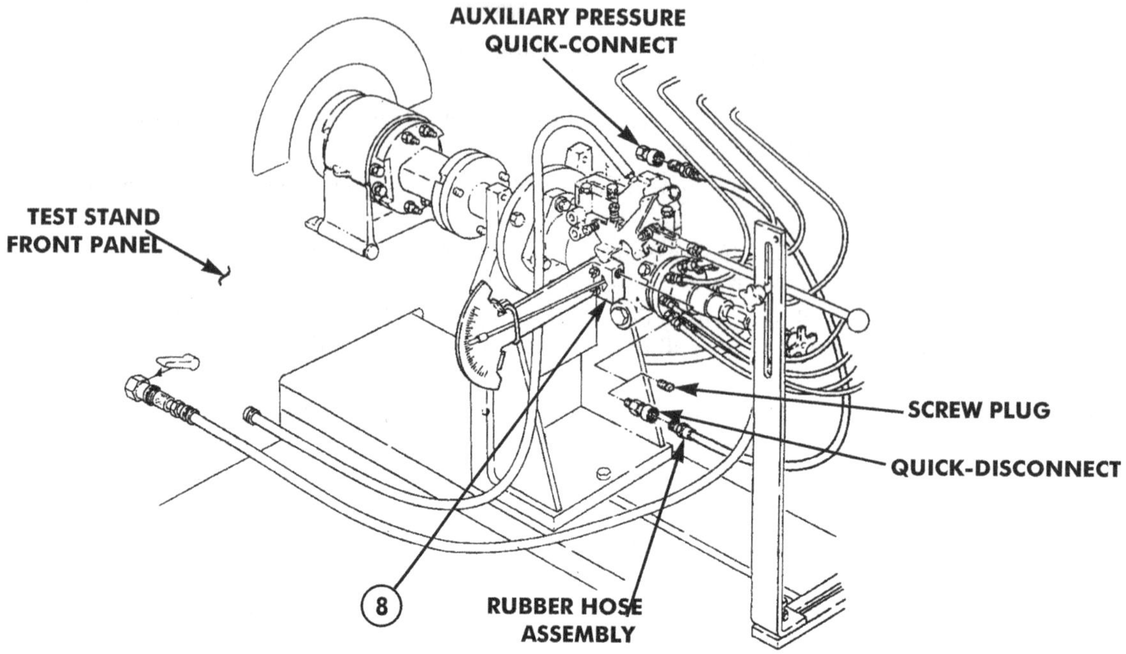

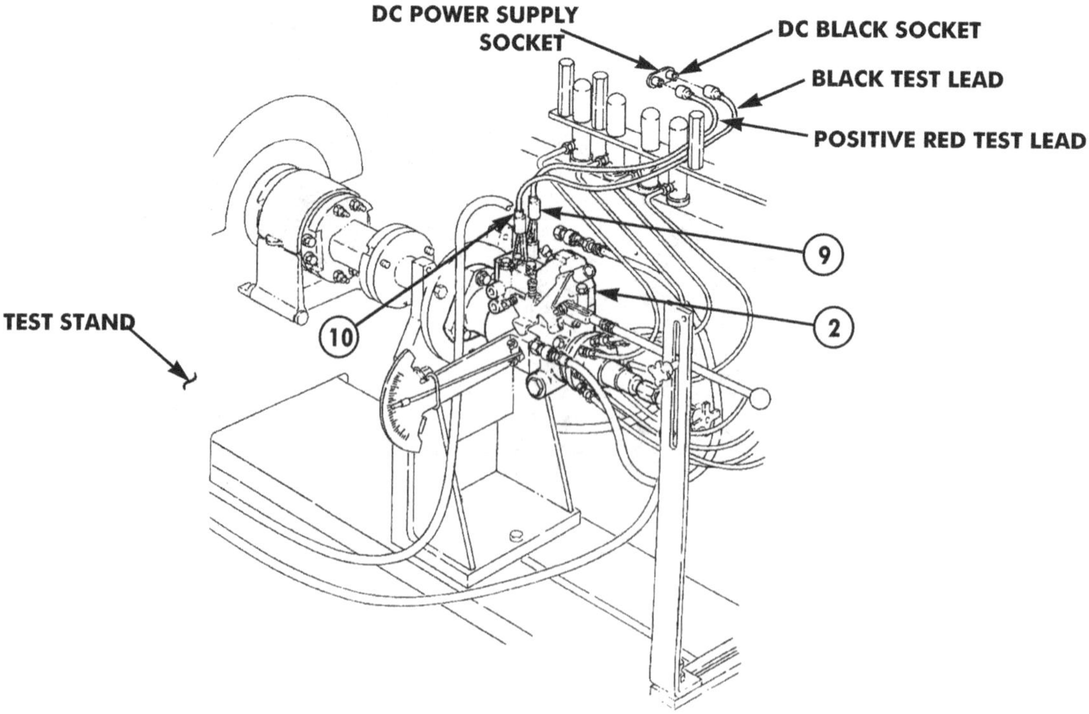

2-35. (6.2L or 6.5L) FUEL INJECTION PUMPS (DB2829-4523, DB2829-4879, or DB2831-5149) CALIBRATION USING FUEL INJECTION PUMP TEST STAND (FTIS) MODEL A8022 (Cont'd)

b. Test Stand Setup Procedures

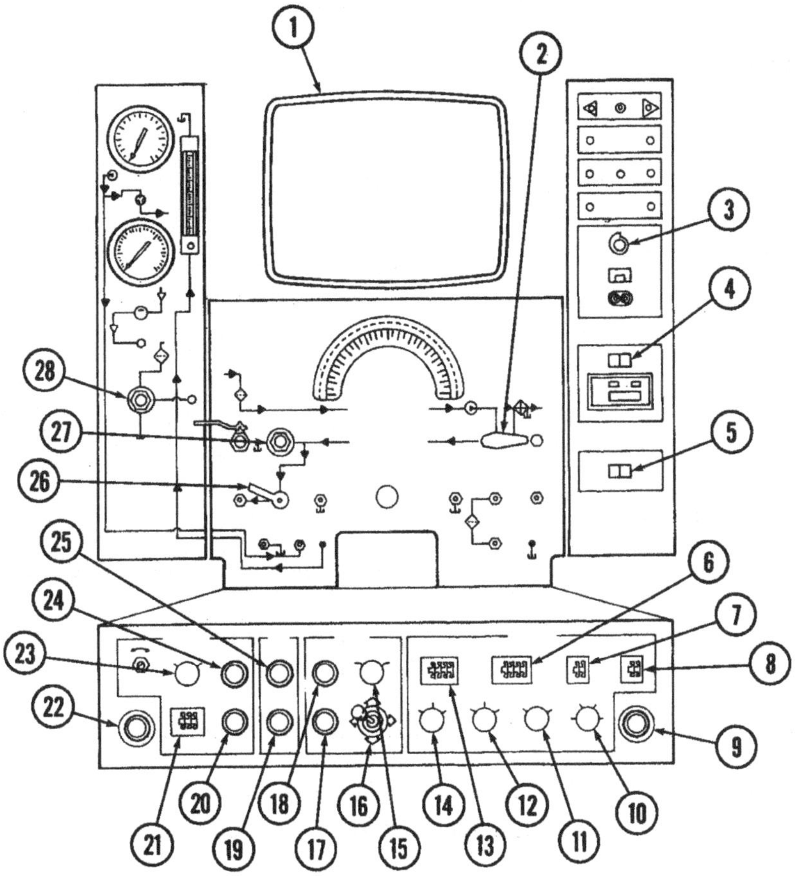

1. Cathode Ray Tube (CRT) Monitor
2. Injection Pump/Auxiliary Selector
3. DC Supply Selector
4. Printer Switch
5. Accumulator Switch
6. Low-Limit Thumbwheel
7. Outlet Number
8. Number of Outlets
9. Stop
10. Mode Selector
11. Ratio Switch
12. Units Switch
13. Hi-Limit Thumbwheel
14. Stroke
15. Pump Rotation Selector
16. Speed Select Control
17. Main Drive Off
18. Main Drive On
19. Lube Oil System Off
20. Calibration Fluid System Off Pushbutton
21. Nominal Temperature Thumbwheel
22. Stop
23. Reservoir Bulkhead Selector
24. Calibration Fluid System On Pushbutton
25. Lube Oil System On
26. Calibration Fluid System
27. Calibration Fluid Pressure
28. Lube Pressure

2-35. (6.2L or 6.5L) FUEL INJECTION PUMPS (DB2829-4523, DB2829-4879, or DB2831-5149) CALIBRATION USING FUEL INJECTION PUMP TEST STAND (FTIS) MODEL A8022 (Cont'd)

1. Set test stand in accordance with the following checklist and instructions in steps 2 through 25.

ITEM	SETTING
Bypass Valve	Open Position
Calibration Fluid System	On
Fluid Selector Valve	Injection Pump
Accumulator Switch	Right
Fluid Shutoff Valve	On
Mode Selector	Auto
Ratio Switch	1:2
Units Switch	mm^3/stroke
Stroke	50
Throttle	WOT
Number of Outlets	8
Outlet Number	Between 1 and 8
Low-Limit Thumbwheel	10.0
Hi-Limit Thumbwheel	Limit Thumbwheel 40.0
Pump Rotation Selector	Counterclockwise
Temperature Select	Bulkhead
Nominal Temperature Thumbwheel	44.7
Master Power Switch	On
DC Supply Selector	17.6 Volts Energize Electric Shutoff Solenoid (ESO)
Lube Pressure	4.5 to 5.5 psi
Lube Oil System	Pushbutton On
Main Drive	Pushbutton On
Printer Switch	On
Speed Select Control	Increase/Decrease Control

2. Set fuel injection pump per calibration chart specifications and settings in para. 2-33, 2-34, or 2-35.
3. Secure throttle positioner arm in Wide Open Throttle (WOT) position.
4. Turn calibration fluid pressure regulator fully counterclockwise.
5. Press calibration fluid system on pushbutton.
6. Adjust calibration fluid pressure regulator knob clockwise until 4.5-5.5 psi (31-38 kPa) is indicated on the Cathode Ray Tube (CRT) monitor. Apply 17.6 volts by adjusting the variable DC voltage supply knob to energize the electric shutoff solenoid.

CAUTION

Be aware that under no circumstances should the transfer pump pressure exceed 130 psi (896 kPa). Excess pressure will damage pump.

7. Turn on main drive system by pressing main drive on pushbutton and holding button depressed for a minimum of three seconds. Increase pump speed to approximately 150 rpm to prime transfer pump and bleed air from the high-pressure fuel line nuts at each test nozzle.
8. Increase pump speed to 1000 rpm. Ensure throttle lever is secured in WOT position. Run pump approximately 10 minutes to warm the calibration fluid to 110°-115°F (43.3°-46.1°C) and bleed all air from test stand and fuel pump. Observe transfer pump pressure reading on CRT monitor. It should be 70-76 psi (483-524 kPa).

2-35. (6.2L or 6.5L) FUEL INJECTION PUMPS (DB2829-4523, DB2829-4879, or DB2831-5149) CALIBRATION USING FUEL INJECTION PUMP TEST STAND (FTIS) MODEL A8022 (Cont'd)

9. If transfer pump pressure does not read 70-76 psi (483-524 kPa), decrease pump speed to 0 rpm and turn off main drive and calibration fluid systems. Disconnect hose at hydraulic head fuel inlet port.
10. Adjust transfer pump pressure with 74-1137 5/32-in. Allen screw-driver (74-8032 adapter kit, slot 3, or equivalent) and reconnect hose. This pressure setting is critical. Repeat test and adjust until transfer pump pressure is within limits.
11. Check pump return flow. With pump speed at 1000 rpm WOT, pump return flow meter should read 225-375 cc/minute. Perform adjustment if necessary.
12. Reduce pump speed to 325 rpm. Move throttle lever to LOW IDLE position and secure. The average flow reading on CRT monitor should be 10-18 mm^3/stroke. If flow is not correct, adjust low-idle screw or MIN-MAX governor setting to obtain proper flow.
13. With pump speed still at 325 rpm, and throttle lever secured in low-idle position, check/adjust variable DC voltage to 17.6 volts as follows:
    (a) Connect positive red test lead to cold-advance solenoid terminal. Housing pressure reading on CRT monitor should be 0-1 psi (0-7 kPa).
    (b) Disconnect positive red test lead from cold-advance solenoid terminal. Reduce variable DC voltage to 0 volts. Reconnect red test lead to shutoff solenoid terminal. Reduce pump speed to 200 rpm.
    (c) Slowly increase variable DC voltage and note voltage at which time the solenoid pulls in by listening for a clicking sound from the shutoff solenoid. (That indicates proper operation.) The shutoff solenoid should pull in with a maximum of 17.6 volts applied.
14. Reduce pump speed to 75 rpm. Move throttle lever to WOT position and secure. Average flow reading on CRT monitor must be a minimum of 28 mm^3/stroke.
15. Increase pump speed to 200 rpm with throttle lever secured in WOT position. Average flow reading on CRT monitor must be a minimum of 47 mm^3/stroke. Disconnect positive red test lead from solenoid shutoff terminal. The average flow reading on CRT monitor must be 4 mm^3/stroke maximum. Reconnect positive red test lead to shutoff solenoid terminal.
16. Increase pump speed to 325 rpm. Move throttle lever to LOW IDLE position and secure. Average flow reading on CRT monitor should be 10-18 mm^3/stroke with a minimum of 1.5 degree cam movement on advance indicator. Connect positive red test lead to cold-advance solenoid terminal. Housing pressure reading on CRT monitor should be 0-1 psi (0-7 kPa) with a minimum of 2.75 degree cam movement on advance indicator. Reconnect positive red test lead to shutoff solenoid terminal.
17. Increase pump speed to 750 rpm. Adjust throttle lever to obtain average flow reading on CRT monitor of 21.5-23.5 mm^3/stroke and secure throttle lever. Advance indicator should have 2.25-4.75 degree cam movement.
18. Increase pump speed to 1000 rpm. Move and secure throttle lever in WOT position. Average flow reading on CRT monitor should be 51.0-55.0 mm^3/stroke with 0.5-2.5 degree cam movement on advance indicator.
19. Increase pump speed to 1600 rpm and ensure throttle lever is secured in WOT position. There should be 4.25-7.25 degree cam movement on advance indicator. Move throttle lever to LOW IDLE position. There must be a maximum of 10.0 degree cam movement on advance indicator. Move and secure throttle lever in WOT position.
20. Increase pump speed to 1800 rpm and ensure throttle lever is secured in WOT position. Average flow reading on CRT monitor must be a minimum of 44.5 mm^3/stroke.
21. Increase pump speed to 1950 rpm and ensure throttle lever is secured in WOT position. Average flow reading on CRT monitor must be a minimum of 42.5 mm^3/stroke.
22. Increase pump speed to 2100 rpm and ensure throttle lever is secured in WOT position. Average flow reading on CRT monitor must be a maximum of 8 mm^3/stroke. Transfer pump pressure reading on CRT monitor must be a maximum of 135 psi (931 kPa).

2-35. (6.2L or 6.5L) FUEL INJECTION PUMPS (DB2829-4523, DB2829-4879, or DB2831-5149) CALIBRATION USING FUEL INJECTION PUMP TEST STAND (FTIS) MODEL A8022 (Cont'd)

23. Reduce pump speed to 0 rpm and secure test stand. If pump met specifications, remove from test stand and mount in holding fixture. If pump did not meet specifications, repair as necessary and proceed to calibration instructions.
24. Using air timing gauge, check pump timing. Set timing to -0.75 to +1.25 degree.
25. Shut down test stand as follows:
    (a) Main drive pushbutton OFF.
    (b) Lube pushbutton OFF.
    (c) Calibration fluid pushbutton OFF.
    (d) DC selector OFF.
    (e) Master power switch OFF.

c. Removal of Fuel Injection Pump from Test Stand

1. Disconnect black test lead from DC black socket marked "P" on test stand (5), and remove alligator clip end of black test lead from ground terminal (4) on fuel injection pump (3).
2. Disconnect positive red test lead from DC power supply socket (1) marked (+) on test stand (5), and remove alligator clip end of red test lead from electric shutoff solenoid terminal (2) on fuel injection pump (3).

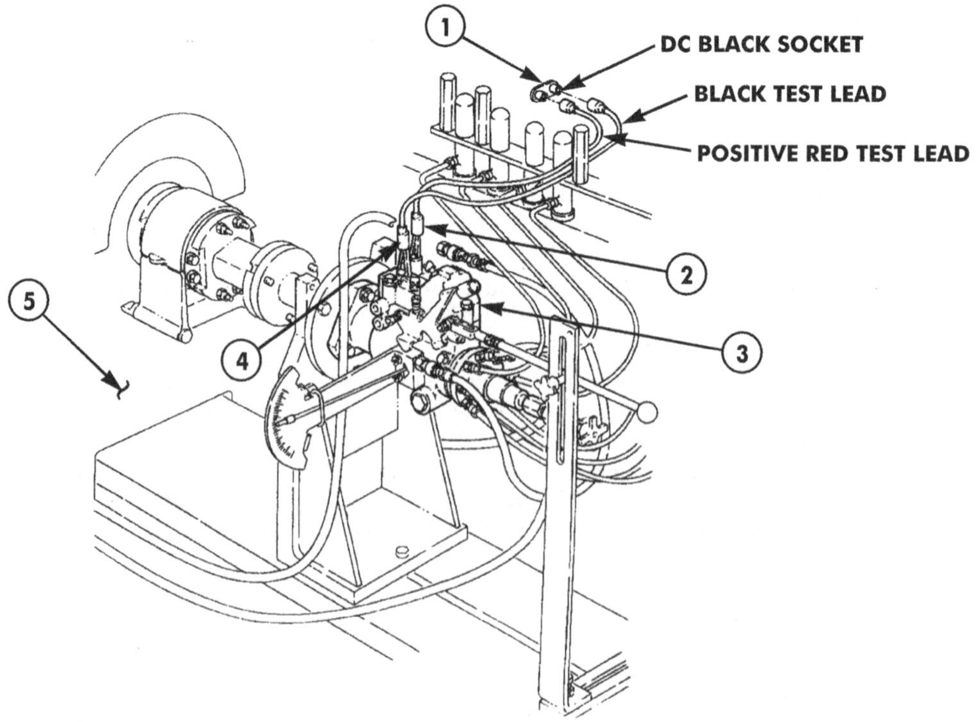

2-35. (6.2L or 6.5L) FUEL INJECTION PUMPS (DB2829-4523, DB2829-4879, or DB2831-5149) CALIBRATION USING FUEL INJECTION PUMP TEST STAND (FTIS) MODEL A8022 (Cont'd)

3. Remove rubber hose assembly from quick-connect on advance indicator block and auxiliary pressure quick-connect on front panel of test stand (2).
4. Remove quick-connect from advance indicator block and install screw plug (1) into advance indicator block.
5. Remove two screws (7) and advance indicator from fuel injection pump (4), and install gasket (5) and timing cover (6) on fuel injection pump (4) with two screws (7).
6. Remove throttle arm positioner assembly from bedplate (8) and disconnect from throttle lever (3) on fuel injection pump (4).

CAUTION

When removing high-pressure delivery lines from fuel injection pump, use two wrenches to protect discharge fittings.

7. Remove eight high-pressure delivery lines from discharge fittings (9) on fuel injection pump (4).

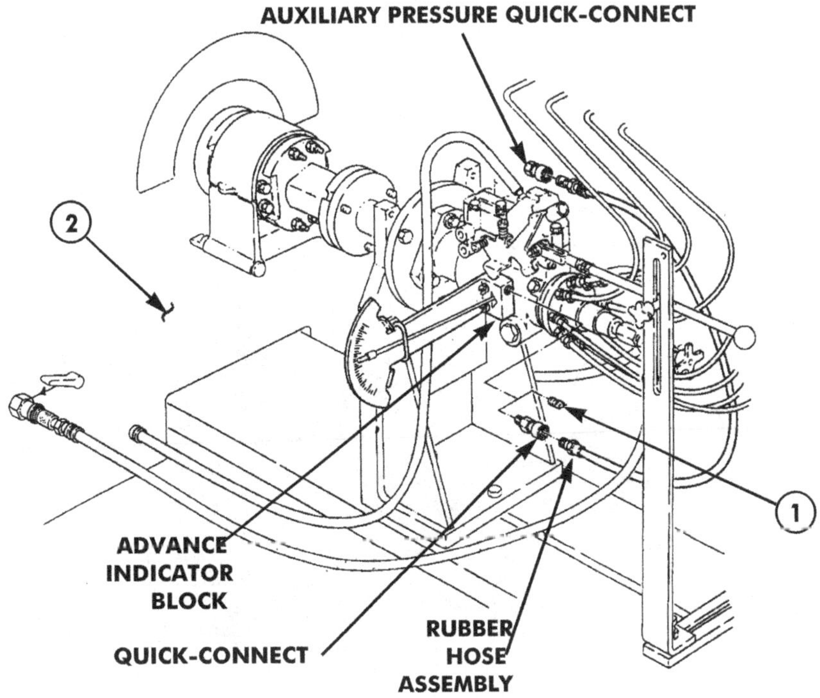

2-35. (6.2L or 6.5L) FUEL INJECTION PUMPS (DB2829-4523, DB2829-4879, or DB2831-5149) CALIBRATION USING FUEL INJECTION PUMP TEST STAND (FTIS) MODEL A8022 (Cont'd)

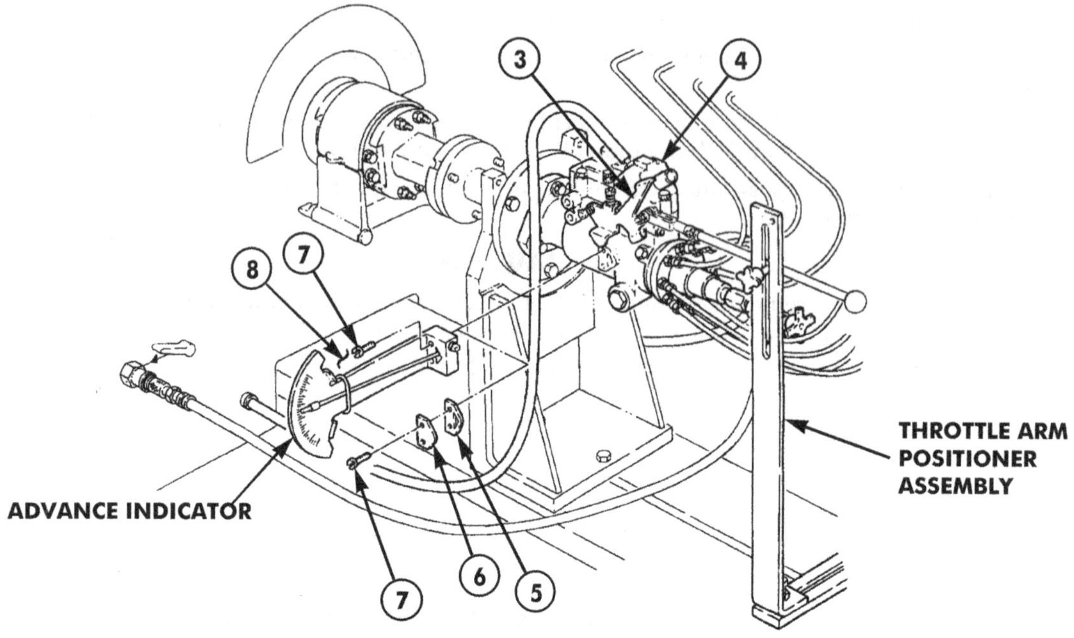

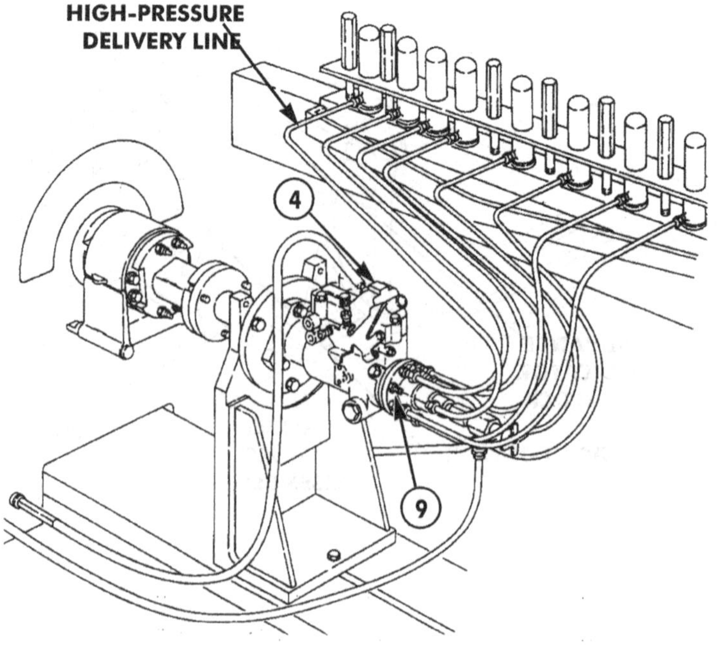

2-35. (6.2L or 6.5L) FUEL INJECTION PUMPS (DB2829-4523, DB2829-4879, or DB2831-5149) CALIBRATION USING FUEL INJECTION PUMP TEST STAND (FTIS) MODEL A8022 (Cont'd)

8. Disconnect tubing from male pipe fitting on fuel injection pump (1) and from calibration fluid push-on connector (3) on front panel of test stand (5).
9. Remove male pipe fitting from fuel injection pump (1).
10. Disconnect rubber hose assembly from quick-connect plug on test adapter and from auxiliary pressure quick-connect (11) on front panel of test stand (5).
11. Remove quick-connect plug from test adapter.
12. Remove test adapter, tab (9), washer (8), and O-ring (7) from transfer test port (6) of fuel injection pump (1).
13. Install O-ring (7), washer (8), tab (9), and screw (10) in transfer test port (6) of fuel injection pump (1).
14. Disconnect hose assembly from connector and calibration fluid supply quick-connect (4) on front panel of test stand (5).
15. Remove test adapter and connector from fuel inlet port (2) of fuel injection pump (1).
16. Remove three 3/8-16x1 hex-head capscrews (20), 3/8-in. lockwashers (19), and 3/8-in. washers (18) from slots and mounting holes on test stand drive coupling (12) and pump-mounted drive hub (17).
17. Loosen two 3/8-16x1 hex-head capscrews (15) from flange mounting bracket (16) and bedplate (21).
18. Remove flange mounting bracket (16) and fuel injection pump (1) from test stand drive coupling (12) and remove alignment gauge and barring tool.
19. Remove three 5/16-18x1 hex-head capscrews (13), 5/16-in. washers (14), and fuel injection pump (1) from adapter plate.

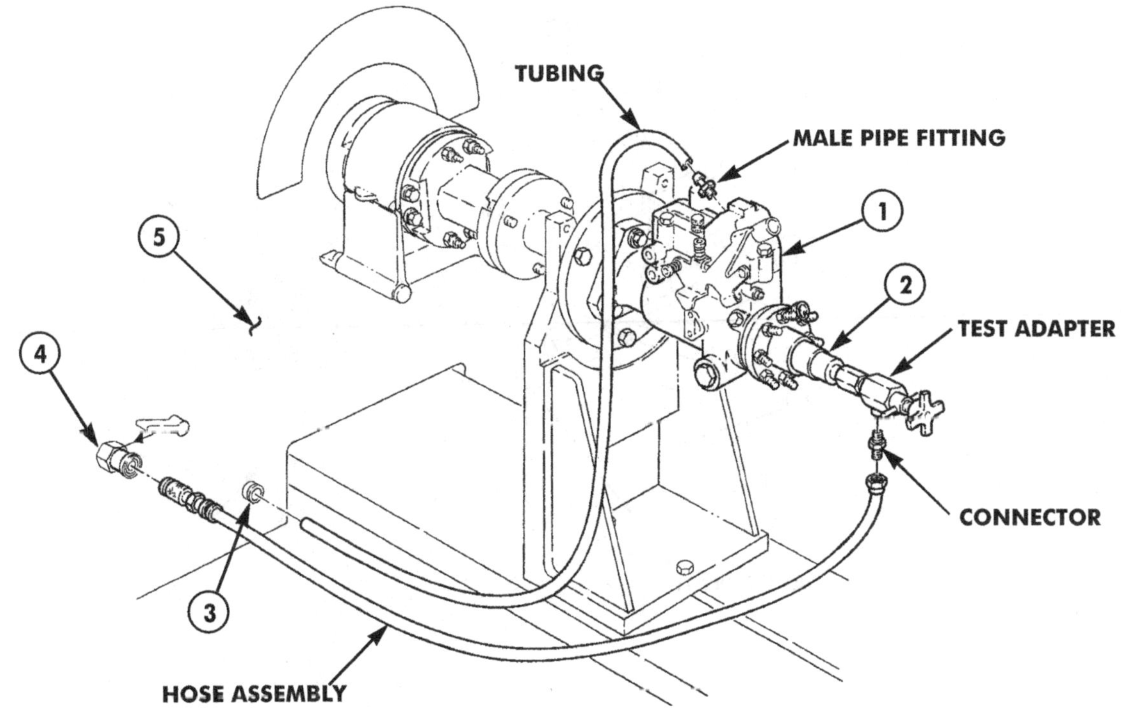

2-35. (6.2L or 6.5L) FUEL INJECTION PUMPS (DB2829-4523, DB2829-4879, or DB2831-5149) CALIBRATION USING FUEL INJECTION PUMP TEST STAND (FTIS) MODEL A8022 (Cont'd)

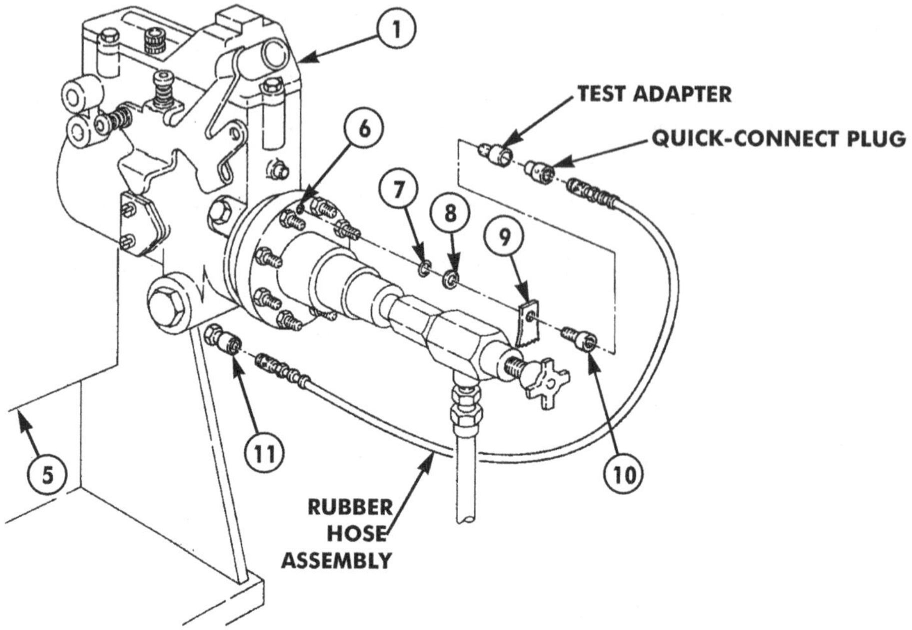

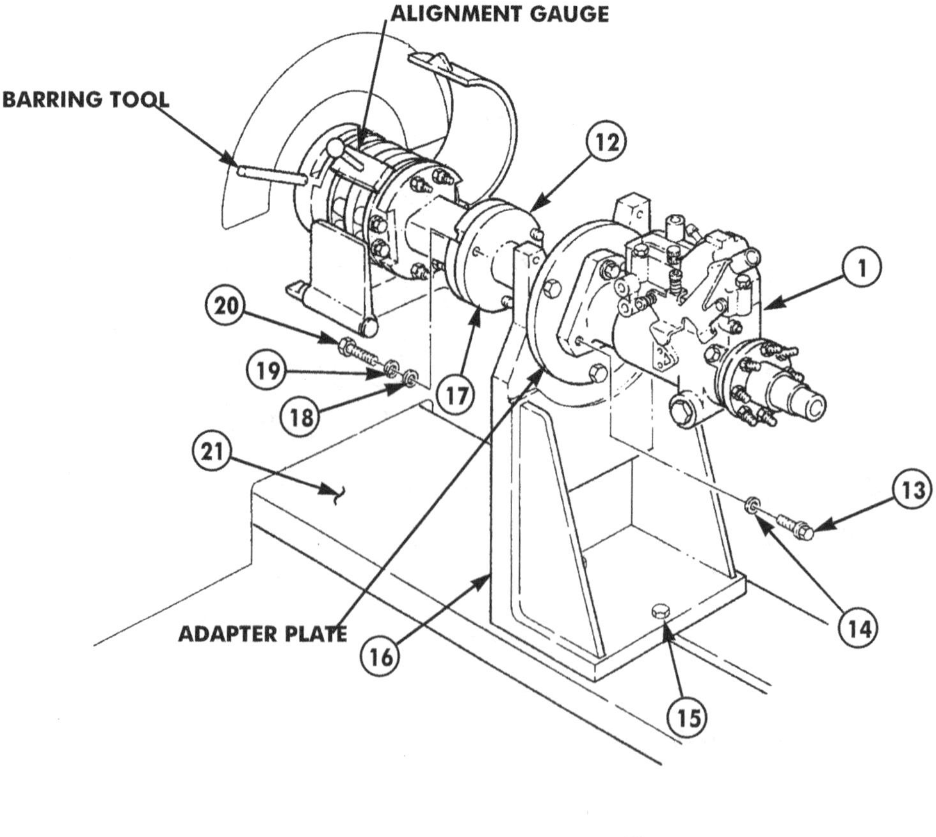

2-36. GLOW PLUG REPAIR

This task covers:

a. Cleaning
b. Inspection
c. Testing

INITIAL SETUP

Tools
General mechanic's tool kit:
automotive (Appendix E, Item 1)

Test Equipment
Multimeter (Appendix E, Item 32)

Mamual References
TM 9-2815-237-34P

Equipment Condition
Engine disassembled into subassemblies (para. 2-15).

a. Cleaning

Clean glow plug (1) in accordance with para. 2-9.

b. Inspection

NOTE

For general inspection instructions, refer to para. 2-10.

1. Inspect glow plug (1) for thread damage. Replace if threads are damaged.
2. Inspect glow plug (1) for damage. Replace if damaged.

c. Testing

1. Using multimeter, check glow plug (1) internal resistance. Connect one test lead to terminal (2) and other test lead to threaded area (3).
2. Glow plug (1) internal resistance should have continuity. If not, replace glow plug (1).

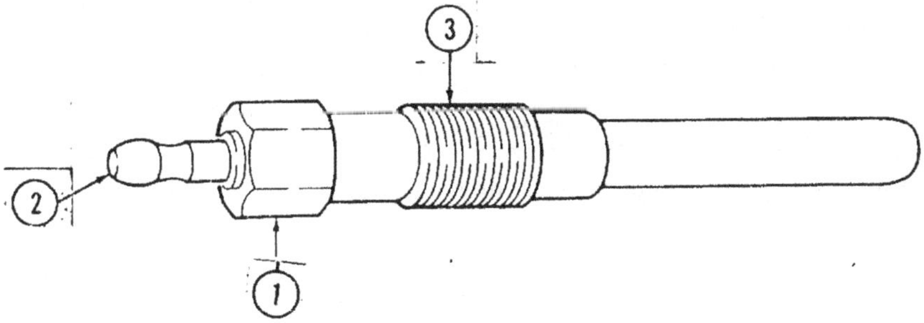

2-37. FUEL INJECTION NOZZLE REPAIR

This task covers:

a. Cleaning
b. Inspection
c. Testing

INITIAL SETUP:

Tools
General mechanic's tool kit: automotive (Appendix E, Item 1)

Test Equipment
Nozzle tester (Appendix E, Item 31)

Manual References
TM 9-2815-237-34P

Equipment Condition
Engine disassembled into subassemblies (para. 2-15).

General Safety Instructions
- Do not place hands or arms near nozzle during testing.
- Compressed air for cleaning purposes will not exceed 30 psi (207 kPa).
- Wear eyeshields when using compressed air.

a. Cleaning

Clean fuel injection nozzle (2) in accordance with para. 2-9.

b. Inspection

NOTE

For general inspection instructions, refer to para. 2-10.

1. Inspect fuel injection nozzle (2) for thread damage. Replace if damaged.
2. Inspect fuel injection nozzle return nipples (1) for damage. Replace fuel injection nozzle (2) if damaged.
3. Inspect fuel injection nozzle (2) for cracks and evidence of heat damage. Replace if cracked or damaged.

c. Testing

1. Preparation.
   (a) Connect fuel injection nozzle (2) to tester.
   (b) Connect lines to return line nipples (1).

WARNING

Do not place hands or arms near nozzle during testing.
Penetrating force of oil may cause serious injury or death.

   (c) Close shutoff valve to pressure gauge on tester.
   (d) Operate tester to prime nozzle (2).
2. Obtaining pressure check.
   (a) Open shutoff valve to pressure gauge 1/4 turn.
   (b) Depress lever on tester slowly. Note at what pressure the needle on pressure gauge stops. Opening pressure must be between 1,500 psi (10,343 kPa) and 1,960 psi (13,514 kPa). Replace nozzle (2) if opening pressure does not meet specifications.

2-37. FUEL INJECTION NOZZLE REPAIR (Cont'd)

3. Leakage test.

   (a) Open shutoff valve to pressure gauge one additional turn.

WARNING

Compressed air used for cleaning purposes will not exceed 30 psi (207 kPa). Use only with effective chip guarding and personal protective equipment (goggles/shield, gloves, etc.).

   (b) Blow-dry end of nozzle (2).

NOTE

A pressure of 1,400 psi (9,653 kPa) must be maintained for 10 seconds while checking for nozzle leakage.

   (c) Depress lever on tester until pressure gauge reads 1,400 psi (9,653 kPa) and observe tip of nozzle (2). If a droplet forms and drops off the nozzle (2) in 10 seconds or less, replace nozzle (2).

4. Chatter test.

   (a) Close shutoff valve to pressure gauge.

   (b) Depress lever on tester slowly, noting whether a chattering noise can be heard.

NOTE

Faster lever movement may cause nozzle to hiss or squeal rather than chatter; this is acceptable.

   (c) If no chatter is heard, increase speed of lever movement on tester until nozzle (2) chatters. If no chatter is heard, replace nozzle (2).

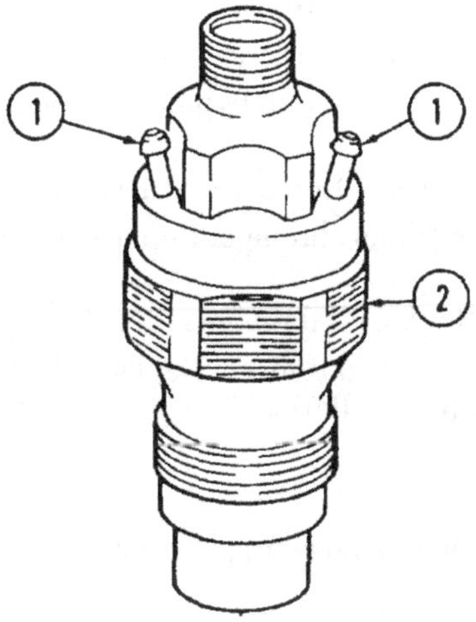

2-38. WATER PUMP REPAIR

This task covers:

a. Disassembly
b. Cleaning
c. Inspection
d. Assembly

INITIAL SETUP

Tools
General mechanic's tool kit: automotive (Appendix E, Item 1)

Materials/Parts
Pipe sealing compound (Appendix B, Item 17)
Sealing compound (Appendix B, Item 18)

Manual References
TM 9-2815-237-34P

Equipment Conndition
Engine disassembled into subassemblies (para. 2-15).

a. Disassembly

NOTE

Perform steps 1 through 3 if replacing water pump or damaged parts.

1. Loosen hose clamp (3) and remove bypass hose (4) from bypass hose adapter (5).
2. Remove 450 adapter (1) and heater hose nipple (2) from water pump (6).
3. Remove bypass hose adapter (5) from water pump (6).

b. Cleaning

Clean all components in accordance with para. 2-9.

c. Inspection

NOTE

For general inspection instructions, refer to para. 2-10.

1. Inspect heater hose nipple (2), 45° adapter (l), and bypass hose adapter (5) for thread damage. Replace if threads are damaged.
2. Inspect water pump (6) for cracks and wear. Replace if cracked or worn.
3. Inspect threaded holes in water pump (6) for damage. Replace water pump (6) if threads are damaged.
4. Inspect bypass hose (4) for damage. Replace if damaged.
5. Inspect water pump rivet (7) for damage or looseness. Replace if damaged or loose. Apply sealing compound to replacement rivet prior to installation.

d. Assembly

1. Coat threads on 45° adapter (1) and heater hose nipple (2) with pipe sealing compound and install in pump (6).
2. Coat threads on bypass hose adapter (5) with pipe sealing compound and install in water pump (6).
3. Install bypass hose (4) on bypass hose adapter (5) and secure with hose clamp (3).

NOTE

Perform step 4 if water pump was replaced.

4. Apply sealing compound to water pump rivet (7) and install in water pump (6).

2-38. WATER PUMP REPAIR (Cont'd)

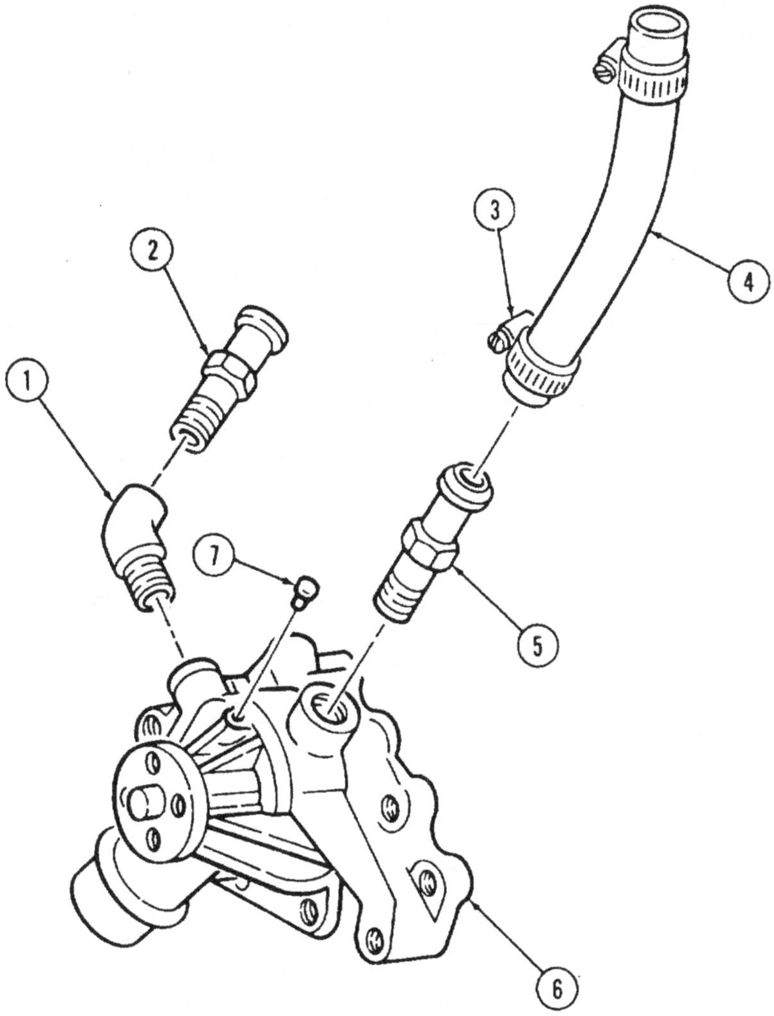

2-39. WATER CROSSOVER REPAIR

This task covers:

a. Disassembly
b. Cleaning
c. Inspection
d. Assembly

INITIAL SETUP:

Tools
General mechanic's tool kit:
automotive (Appendix E, Item 1)

Materials/Parts
Water crossover gasket (Appendix D, Item 19)
Pipe sealing compound (Appendix B, Item 17)

Manual References
TM 9-2815-237-34P

Equipment Condition
Engine disassembled into subassemblies (para. 2-15).

NOTE

In some cases, flanged head fasteners maybe present instead of standard fasteners and washers. In all cases, washers should be used when replacing a flanged head fastener with a standard fastener.

a. Disassembly

1. Remove capscrew (6), stud (5), two washers (4), water outlet (7), thermostat (9), and gasket (8) from water crossover (11). Discard gasket (8).

NOTE

Only perform steps 2 through 4 if replacing water crossover or damaged parts.

2. Remove glow plug controller (1).
3. Remove temperature switch (3).
4. Remove de-airation nipple (2), bypass hose nipple (10), and heater hose nipple (12) from water crossover (11).

b. Cleaning

Clean all components in accordance with para. 2-9.

c. Inspection

NOTE

For general inspection instructions, refer to para. 2-10.

1. Inspect water crossover (11) and water outlet (7) for damage. Replace if damaged.
2. Inspect all threaded holes in water crossover (11) for thread damage. Replace if threads are damaged.
3. Inspect heater hose nipple (12), de-airation nipple (2), and bypass hose nipple (10) for damage. Replace if damaged.
4. Inspect glow plug controller (1), thermostat (9), and temperature switch (3) for damage. Replace if damaged.

2-39. WATER CROSSOVER REPAIR (Cont'd)

d. Assembly

CAUTION

Thermostat must be installed correctly or engine damage may result.

NOTE

For general assembly instructions, refer to para. 2-12.

1. Install thermostat (9), gasket (8), and water outlet (7) on water crossover (11) with two washers (4), capscrew (6), and stud (5). Tighten capscrew (6) and stud (5) to 25 lb-ft (34 N•m).
2. Coat threads on heater hose nipple (12), bypass hose nipple (10), and de-airation nipple (2) with pipe sealing compound and install in water crossover (11).
3. Coat threads on glow plug controller (1) and temperature switch (3) with pipe sealing compound and install in water crossover (11).

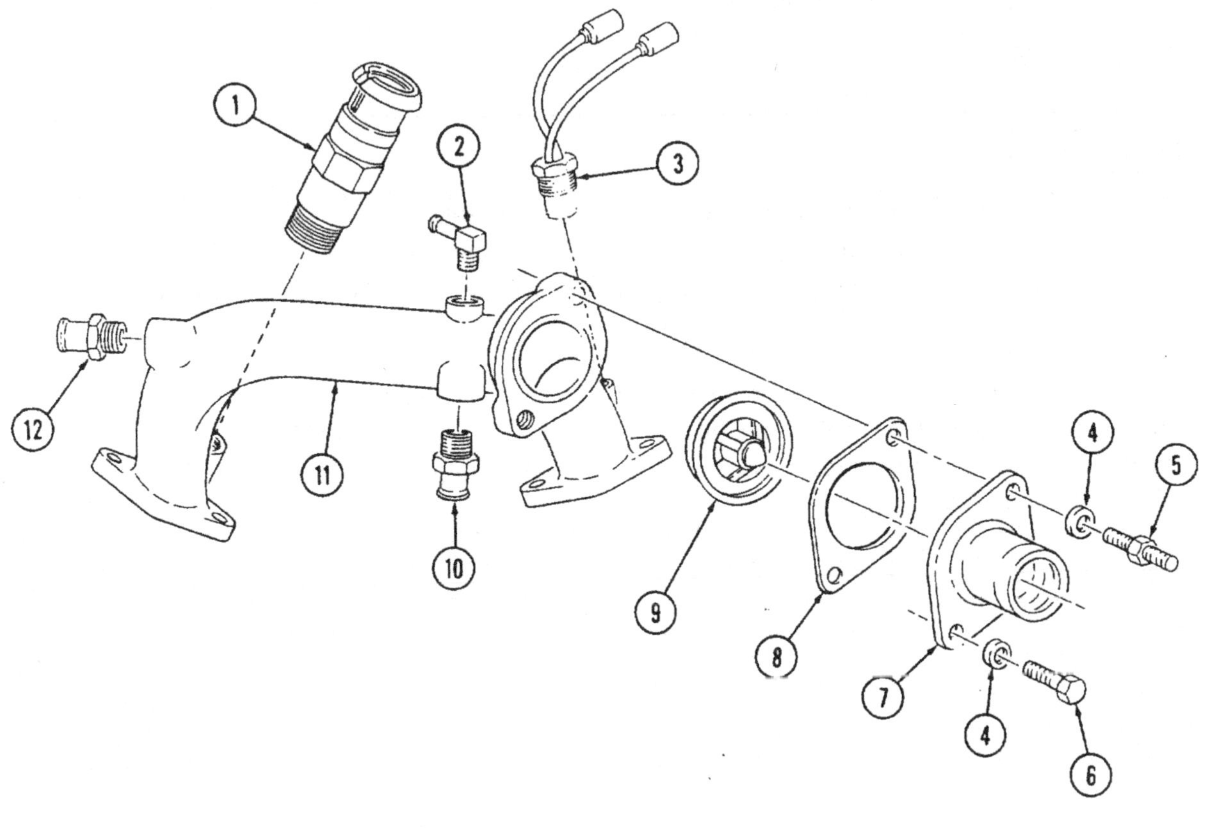

2-40. FAN DRIVE REPAIR

This task covers:

a. Disassembly
b. Cleaning
c. Inspection
d. Assembly

INITIAL SETUP

Tools

General mechanic's tool kit:
automotive (Appendix E, Item 1)

Materials/Parts

Locknut (Appendix D, Item 24)
Spring (Appendix D, Item 50)
Two shaft O-rings (Appendix D, Items 32 or 33)
Friction lining (Appendix D, Item 23)
Cylinder O-ring (Appendix D, Item 30)
Dust seal (Appendix D, Item 42)
Lithium grease (Appendix B, Item 11)

Manual References

TM 9-2815-237-34P

Personnel Required

One mechanic
One assistant

Equipment Condition

Engine disassembled into subassemblies (para. 2-15).

General Safety Instructions

- Use eyeshields when cleaning with compressed air.
- Do not use compressed air to spin bearings.

a. Disassembly

1. Remove six screws (8) and three retaining plates (9) from housing (5).
2. Remove friction lining (10) from housing (5). Discard friction lining (10).
3. Remove locknut (1) and tabwasher (2) securing shaft assembly (11) and cylinder assembly (3) to housing (5). Discard locknut (1).
4. Remove cylinder assembly (3) from front of housing (5).
5. Remove O-ring (4) from inside cylinder assembly (3). Discard O-ring (4).
6. Remove shaft assembly (11) and spring (6) from housing (5). Discard spring (6).
7. Remove two O-rings (7) from shaft assembly (11). Discard O-rings (7)-

2-40. FAN DRIVE REPAIR (Cont'd)

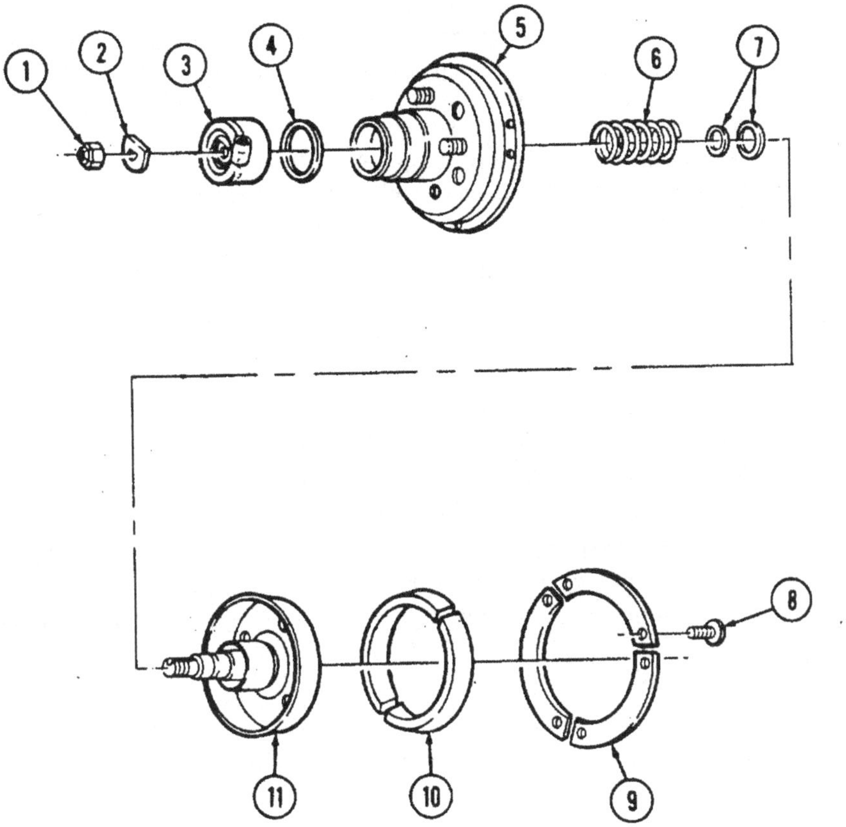

2-40. FAN DRIVE REPAIR (Cont'd)

b. Cleaning

WARNING

- Compressed air used for cleaning purposes will not exceed 30 psi (207 kPa). Use only with effective chip guarding and personal protective equipment (goggles/shield, gloves, etc.).
- Do not use compressed air to spin bearings. Spinning a dry bearing with compressed air may cause injury to personnel or damage to equipment.

1. Clean shaft assembly (7) and housing (2) with compressed air and wipe with clean cloth.
2. Clean all other parts in accordance with para. 2-9.

c. Inspection

NOTE

For general inspection instructions, refer to para. 2-10.

1. Inspect cylinder assembly (9) for damage. Replace if damaged.
2. Inspect dust seal (10) for cracks in the seal or pieces missing from it. Replace dust seal (10) if cracked or pieces missing from it. Ensure lip of seal is facing out.
3. Inspect roller bearings (3) and (4) in housing assembly (2) for damage. Replace fan drive assembly if damaged.
4. Inspect lip seal (5) in housing assembly (2) for damage. Replace fan drive assembly if damaged.
5. Inspect studs (1) on housing assembly (2) for stripped threads. Replace if damaged.
6. Inspect seal surface (6) on shaft assembly (7) for damage. Replace fan drive assembly if damaged.
7. Inspect bearing (8) on shaft assembly (7) for damage. Replace fan drive assembly if damaged.

2-40. FAN DRIVE REPAIR (Cont'd)

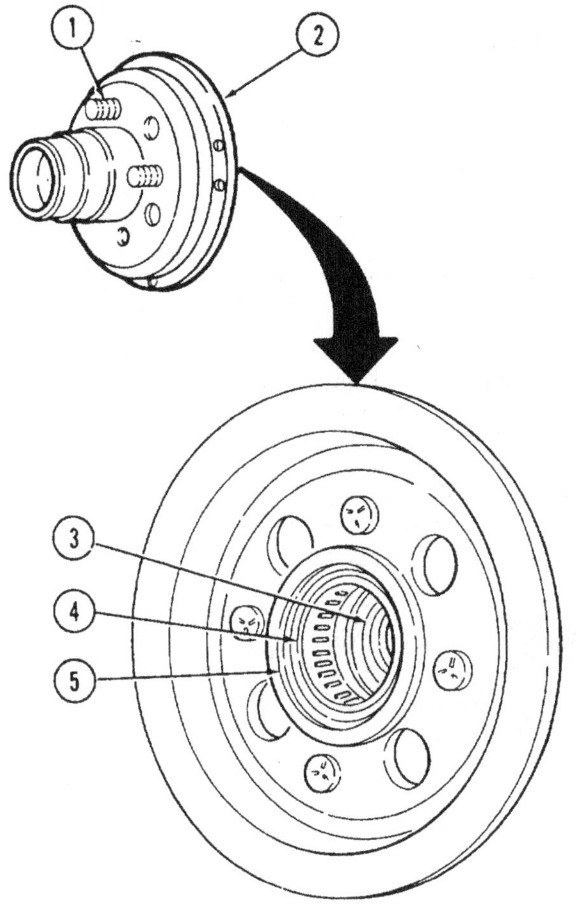

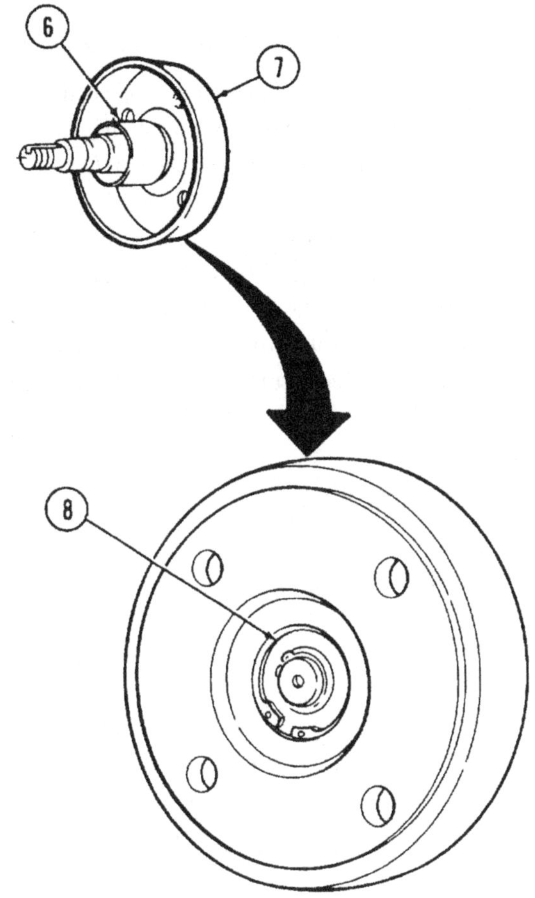

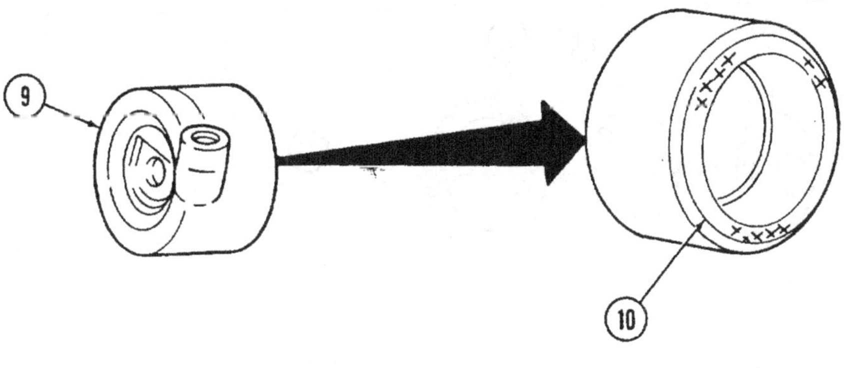

2-40. FAN DRIVE REPAIR (Cont'd)

d. Assembly

NOTE

For general assembly instructions, refer to para. 2-12.

1. Install O-ring (4) in groove inside cylinder assembly (3).
2. Install cylinder assembly (3) on housing (5).
3. Install large O-ring (9) and small O-ring (8) on shaft assembly (13).
4. Lubricate roller bearing (6) with lithium grease,
5. Install shaft assembly (13) and spring (7) in housing (5).
6. Install shaft assembly (13) on cylinder assembly (3) with tabwasher (2) and locknut (1). Tighten locknut (1) to 96 lb-in. (11 N•m).
7. Install friction lining (12) in housing (5).
8. Install three retaining plates (11) on housing (5) with six screws (10). Tighten screws (10) to 22 lb-in. (3 N•m),

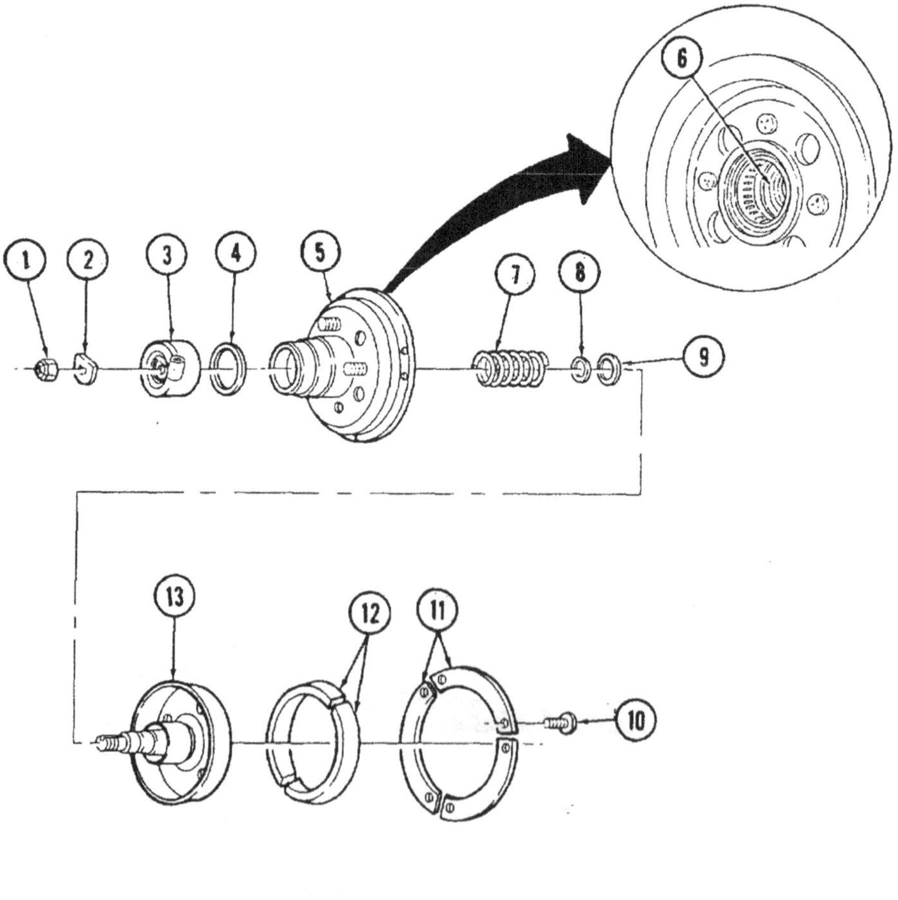

2 - 4 1ENGINE ASSEMBLY FROM SUBASSEMBLIES

This task covers:

- a. **Crankshaft and Main Bearings**
- b. **Flywheel**
- c. **Pistons and Connecting Rods**
- d. **Oil Pump**
- e. **Camshaft**
- f. **Timing Chain and Drive Sprockets**
- g. **Timing Gear Cover**
- h. **Oil Pan**
- i. **Oil Filter and Adapter**
- j. **Fuel Pump**
- k. **Oil Pump Drive**
- l. **Fuel Injection Pump**
- m. **Water Pump and Adapter Plate**
- n. **Valve Lifters**
- o. **Cylinder Heads**
- p. **Glow Plugs**
- q. **Fuel Injection Nozzles**
- r. **Rocker Arm Shafts and Pushrods**
- s. **Rocker Arm Covers**
- t. **Modulator Link**
- u. **Fuel Supply and Return Lines**
- v. **Fuel Injection Lines**
- w. **Water Crossover**
- x. **Intake Manifold**
- y. **Exhaust Manifold**
- y.1. **Turbocharger**
- y.2. **Manifold-to-Turbocharger Exhaust Pipe**
- y.3. **Wastegate Housing**
- y.4. **Wastegate Actuator**
- z. **Torsional Damper**
- aa. **Crankshaft Pulley**
- bb. **Fan Drive and Water Pump Pulley**

INITIAL SETUP:

Tools

General mechanic's tool kit:
automotive (Appendix E, Item 1)
Feeler gauge (Appendix E, Item 13)
Piston ring compressor (Appendix E, Item 35)

Special Tools

Rear main seal installer
(Appendix E, Items 33 or 34)
Injector nozzle socket (Appendix E, Item 5)
Glow plug socket (Appendix E, Item 6)
Hex-head driver, 6 mm (Appendix E, Item 7)
Hex-head driver, 8 mm (Appendix E, Item 8)
Hex-head driver, 5/16-in. (Appendix E, Item 9)
Crowfoot, 16 mm (Appendix E, Item 36)
Crowfoot, 19 mm (Appendix E, Item 37)

Personnel Required

One mechanic
One assistant

Manual References

TM 9-2815-237-34P

Materials/Parts

Turbocharger oil gasket (Appendix D, Item 19)
Intake manifold gasket kit (Appendix D, Item 8)
Two center intake manifold gaskets,
turbocharged engine (Appendix D, Item 8.1)
Two O-rings (Appendix D, Item 31)
Fuel pump gasket (Appendix D, Item 13)
Adapter plate gasket (Appendix D, Item 10)
Oil pump drive gasket (Appendix D, Item 17)

Materials/Parts (Cont'd)

Water pump gasket (Appendix D, Item 22)
Injection pump gasket (Appendix D, Item 15)
Two cylinder head gaskets (Appendix D, Item 11)
Two water crossover gaskets (Appendix D, Item 20)
Eight injector nozzle gaskets (Appendix D, Item 16)
Two exhaust manifold gaskets (Appendix D, Item 12)
Front cover oil seal (Appendix D, Item 43)
Rear oil pan seal (Appendix D, Item 46)
Oil filter (Appendix D, Item 6)
Eight rod bearings (Appendix D, Item 1)
Oil filter adapter bolt gasket (Appendix D, Item 7)
Three woodruff keys (Appendix D, Item 54)
Oil pan gasket set (Appendix D, Item 9)
Two rocker arm cover gaskets
(Appendix D, Item 18)
Rear main oil seal (Appendix D, Items 44 or 45)
O-ring (Appendix D, Item 28)
Adhesive (Appendix B, Item 1)
Bearing clearance gauge (Appendix B, Item 9)
Sealer, anaerobic gasket (Appendix B, Item 10)
Lubricating oil OE/HDO (Appendix B, Item 13)
Silicone sealant (Appendix B, Item 16)
Pipe sealing compound (Appendix B, Item 17)
Sealing compound (Appendix B, Item 18)

General Safety Instructions

- Crankshaft must be supported during removal and installation.
- Do not perform this procedure near fire, flame, or sparks.

2-41. ENGINE ASSEMBLY FROM SUBASSEMBLIES (Cont'd)

NOTE

- During assembly operations, it is important to closely inspect each unit to make sure nothing has been overlooked during inspection and repair. Plugs should be checked for tightness, parts kept clean, openings covered, and machined surfaces protected. Application of lubricant should be performed from covered containers.
- Work area should be clean, well-ventilated, and free of blowing dirt and dust.
- In some cases, flanged head fasteners may be present instead of standard fasteners and washers. In all cases, washers should be used when replacing a flanged head fastener with a standard fastener.
- For general assembly instructions, refer to para. 2-12.
- Assembly procedures for pre-1990, 1990 and above 6.2L and 6.5L engines are basically the same. Any differences in procedures are noted. Some replacement parts are different. Refer to para. 1-10 to determine engine model before ordering replacement parts.

a. Crankshaft and Main Bearings

NOTE

- Measure outside diameter of crankshaft main bearing journals to determine what thickness main bearings to install.
- Main bearings are of the precision insert type and do not utilize shims for adjustment. If clearances are found to be excessive, a new bearing, both upper and lower halves, will be required. Service bearings are available in standard size and undersize.
- Selective fitting of both rod and main bearing inserts is necessary in production in order to obtain close tolerances. For this reason you may find one-half of a standard insert with one half of a 0.001 in. (0.025 mm) undersize insert, which will decrease the clearance 0.0005 in. (0.013 mm) from using a full standard bearing.

1. Remove all capscrews (1) and (2) and bearing caps (3) and install upper halves of main bearings (7) in cylinder block (6).
2. Install lower halves of main bearings (4) in bearing caps (3).

WARNING

Crankshaft must be supported during removal and installation. Failure to support crankshaft may cause injury to personnel or damage to equipment.

3. Install crankshaft (8) in cylinder block (6).

NOTE

The gaging plastic shall be positioned the full width of the bearing and centered on the bearing journal when the bearing cap is installed.

4. Position gauging plastic on bearing journal (5) the full width of the bearing (4) and parallel to the center line of the crankshaft (8).

2 - 4 1ENGINE ASSEMBLY FROM SUBASSEMBLIES (Cont'd)

CAUTION

Bearing caps are numbered 1 through 5 and must be installed in correct order, starting with number 1 at front of cylinder block. Arrow on bearing cap must point forward or engine damage will result.

5. Install bearing cap (3) on cylinder block (6) with two long capscrews (1) and short capscrews (2). Tighten capscrews (1) and (2) finger tight.
6. Tighten all long capscrews (1) to 110 lb-ft (149 N·m). Tighten all short capscrews (2) to 100 lb-ft (136 N·m).
7. Remove two long capscrews (1), short capscrews (2), and bearing cap (3) from cylinder block (6).
8. On edge of envelope (9) there is a scale which is graduated in thousandths of an inch. Without removing gauging plastic, measure its thickness at the widest point with scale on envelope (9).
9. If gauging plastic tapers toward the middle or ends, there is a difference in clearance indicating taper, low spot, or irregularity of bearing or crankshaft journal. If gauging plastic indicates more than 0.001 in. (0.025 mm) taper, measure crankshaft journal with micrometer (para. 2-17).

NOTE

If unable to obtain the proper crankshaft journal clearance, it is acceptable to use a standard bearing with a 0.0005 in. (0.013 mm) or 0.001 in. (0.025 mm) undersize bearing to produce clearance. If unable to produce proper clearance, replace crankshaft.

10. If the measurement is within the specifications of 0.002-0.004 in. (0.051-0.102 mm), go to step 11. If measurement is not within specifications, a 0.0005 in. (0.013 mm) or 0.001 in. (0.025 mm) undersize bearing may produce proper bearing clearance. Replace bearings and repeat steps 4 through 10.
11. Repeat steps 4 through 10 for remaining main bearings.

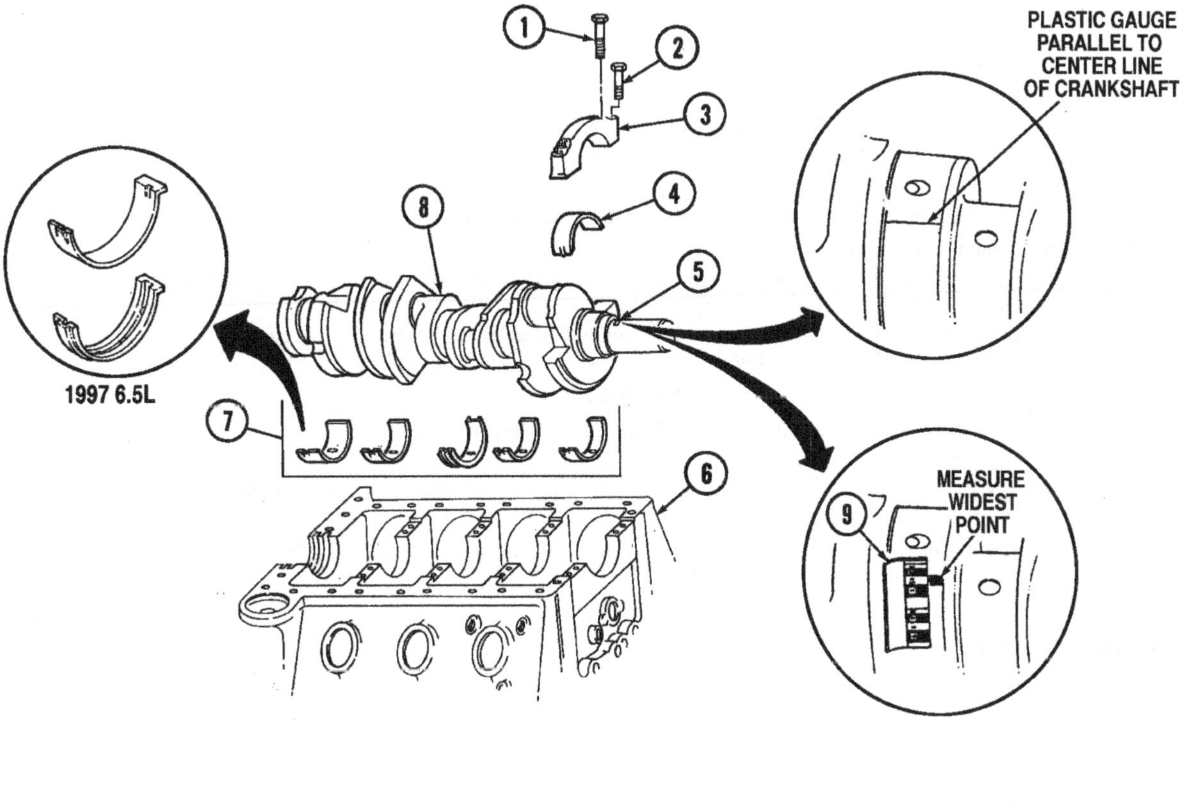

2-41. ENGINE ASSEMBLY FROM SUBASSEMBLIES (Cont'd)

WARNING

Crankshaft must be supported during removal and installation. Failure to support crankshaft may cause injury to personnel or damage to equipment.

CAUTION

Bearing caps are numbered one through five and must be installed in correct order, starting with number one at front of cylinder block. Arrow on bearing cap must point forward or engine damage will result.

12. After all main bearing clearances have been set, remove crankshaft (7) and gauging plastic and apply OE/HDO to all main bearings. Install crankshaft (7), all bearing caps (1), (2), and (3), long cap-screws (4) and short capscrews (5). Tighten long capscrews (4) and short capscrews (5) to 12 lb-ft (16 N·m).
13. Tap crankshaft (7) rearward, then frontward; this will align main bearing thrust surfaces (6). Tighten long capscrews (4) and short capscrews (5) to 40 lb-ft (54 N ·m).
14. Tighten all long capscrews (4) to 110 lb-ft (149 N ·m). Tighten all short capscrews (5) to 100 lb-ft (136 N·m).
15. Rotate crankshaft (7) to ensure crankshaft (7) rotates freely.
16. Force the crankshaft (7) to the extreme front position and hold in place. Measure end play at the front end of the number three bearing with a feeler gauge. End play should be 0.004-0.010 in. (0.101-0.250 mm).

NOTE

- Perform steps 17 through 24 for 6.2L engines ONLY when using a 2-piece nylon rope type rear main seal.
- Perform steps 17 and 21 through 24 when using a 2-piece rubber rear main seal.

17. Remove long capscrews (4), short capscrews (5), bearing caps (1), (2), and (3), and crankshaft (7) from cylinder block (8).
18. Apply adhesive to seal groove in rear main bearing cap (1) and cylinder block (8).
19. Install rear main oil seal (9) into cylinder block (8) and rear main bearing cap (1) using seal installer.
20. Cut excess oil seal (9) off flush with bearing cap (1) and cylinder block (8) mounting surface.
21. Repeat steps 4 through 10 to ensure rear oil seal (9) did not affect rear main bearing clearance. If bearing clearance changed, remove bearing cap (1) and inspect rear oil seal (9).
22. Remove gauging plastic.
23. Install number three bearing cap (3), two long capscrews (4), and short capscrews (5). Tighten capscrews (4) and (5) to 12 lb-ft (16 N ·m).
24. Tap crankshaft (7) rearward, then frontward; this will align main bearing thrust surfaces (6). Tighten capscrews (4) and (5) on number three bearing cap (3) to 40 lb-ft (54 N ·m).

NOTE

Immediately install rear main bearing cap after application of sealant.

25. Apply thin film of sealer to rear main bearing cap (1) mating surface. Keep sealer off bearing and oil seal (9) (installed on 6.2L engines).
26. Install rear main bearing cap (1) on cylinder block (8) with long capscrews (4), and short cap-screws (5). Tighten capscrews (4) and (5) to 40 lb-ft (54 N ·m).

2-41. ENGINE ASSEMBLY FROM SUBASSEMBLIES (Cont'd)

27. Install remaining bearing caps (2) on cylinder block (8) with long capscrews (4) and short capscrews (5). Tighten capscrews (4) and (5) to 40 lb-ft (54 N•m).
28. Evenly tighten all long capscrews (4) to 110 lb-ft (149 N•m).
29. Evenly tighten all short capscrews (5) to 100 lb-ft (136 N•m).

NOTE

Perform steps 30 through 32 for 6.5L engines.

30. Coat rear crankshaft (7) surface with engine oil, and lightly coat lip of rear main oil seal (9) with engine oil.
31. Insttall rear main oil seal (9), with spring Cavity towards cylinder block (8), on crankshaft (7).
32. Using seal installer, drive rear main oil seal (9) in until tool bottoms against cylinder block (8) and

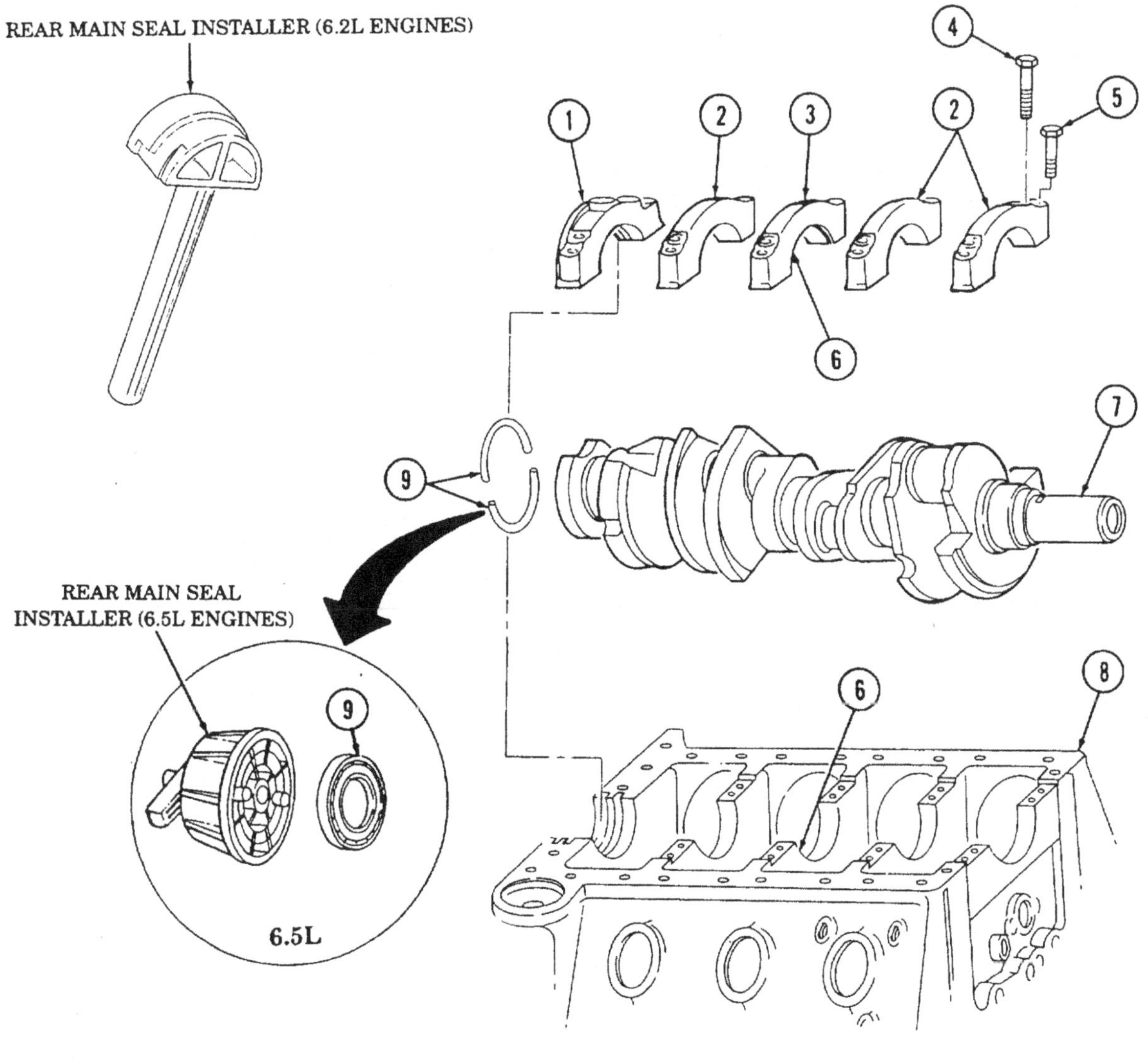

2-41. ENGINE ASSEMBLY FROM SUBASSEMBLIES (Cont'd)

b. Flywheel

1. Install flywheel (3) on crankshaft (4) so that converter mounting surface (2) faces outward.
2. Secure flywheel (3) to crankshaft (4) with six capscrews (1). Tighten capscrews (1) to 65 lb-ft (88 N•m).

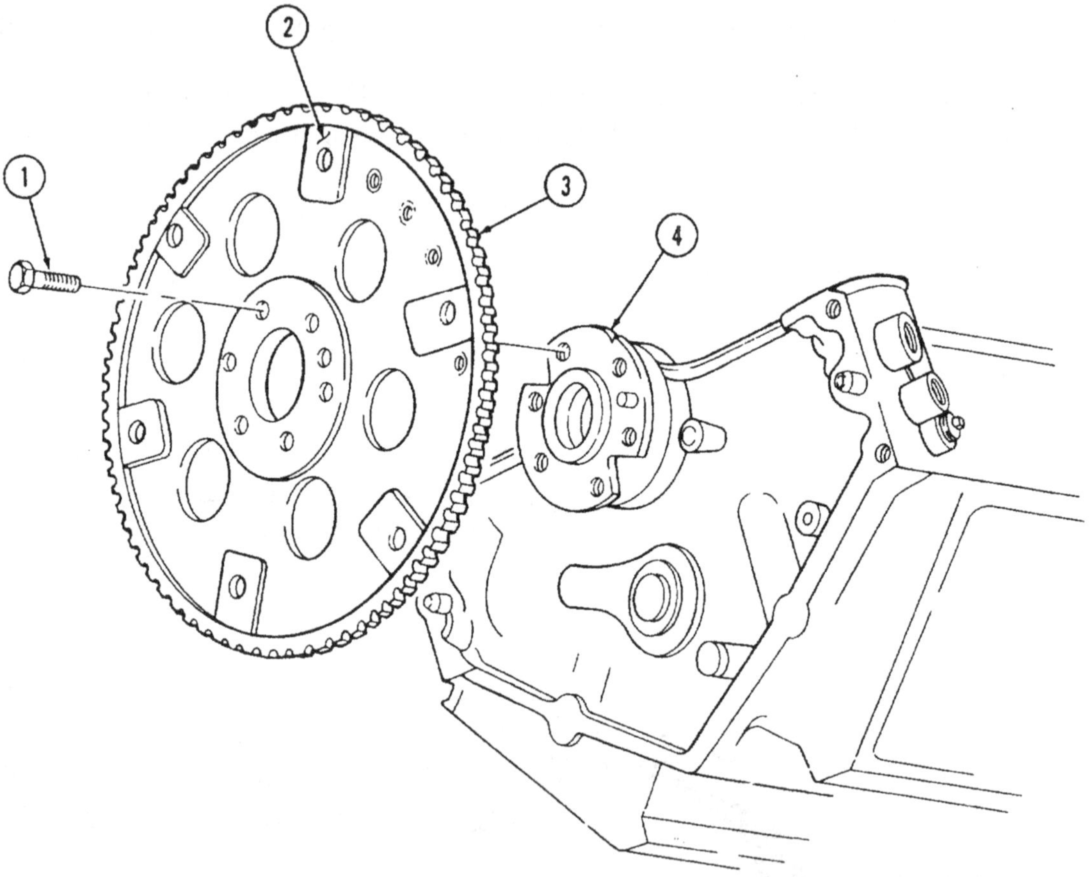

2-41. ENGINE ASSEMBLY FROM SUBASSEMBLIES (Cont'd)

c. Pistons and Connecting Rods

1. Rotate cylinder block (6) approximately 90°.

NOTE

Measure outside diameter of crankshaft rod bearing journals to determine what thickness connecting rod bearings to install.

2. Check identification number on side of connecting rod (9) and install upper rod bearing half (7) in connecting rod (9) and lower rod bearing half (12) in corresponding rod cap (5).
3. Rotate crankshaft (4) so rod journal (13) faces outward.
4. Cover connecting rod bolts (8) with 3/8-in. inside diameter rubber hose.
5. Apply OE/HDO to piston rings (11) and piston (10).

NOTE

Make sure all ring gaps are staggered so they are not in line with each other or piston pin.

6. Compress piston rings (11) with ring compressor.

NOTE

- Depression on top of piston must be assembled toward outside of engine.
- Starting from front of cylinder block, cylinders 1, 3, 5, and 7 are on left side and cylinders 2, 4, 6, and 8 are on right side.
- For remaining pistons, rotate crankshaft so rod journal faces outward.

7. Check identification number on side of connecting rod (9) and install piston (10) and connecting rod (9) into corresponding cylinder.
8. Tap piston (10) into cylinder until connecting rod (9) seats on crankshaft (4) and remove rubber hoses.

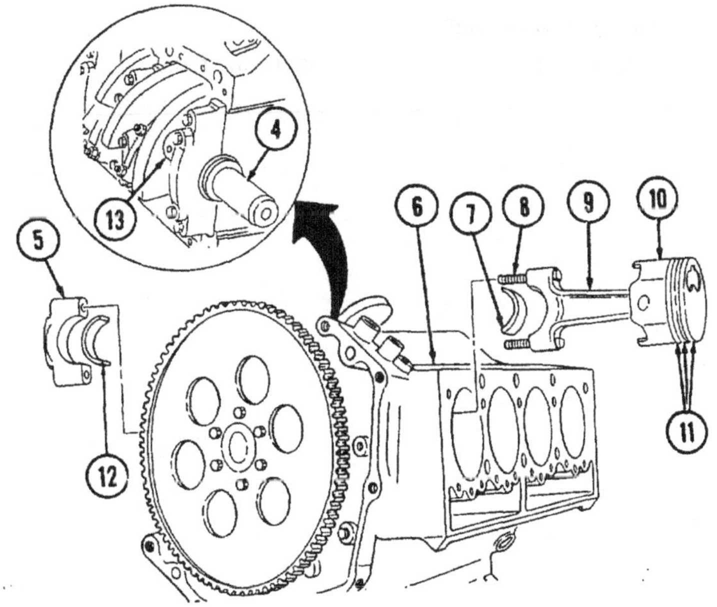

2-41. ENGINE ASSEMBLY FROM SUBASSEMBLIES (Cont'd)

NOTE

The gauging plastic shall be positioned the full width of the bearing and centered on the bearing journal when the bearing cap is installed.

9. Position gauging plastic on bearing journal (3) the fill width of the bearing (4) and parallel to the center line of the crankshaft.

NOTE

Identification numbers on connecting rods and rod caps must be installed on the same side.

10. Check identification number on side of connecting rod (6) and install corresponding rod cap (2).
11. Secure rod cap (2) to connecting rod (6) with two nuts (1) and tighten to 44-52 lb-ft (60-71 N•m).
12. Remove two nuts (1) and rod cap (2) from connecting rod (6).
13. Measure gauging plastic at its widest point.
14. If the gauging plastic tapers toward the middle or ends, there is a difference in clearance indicating taper, low spot, or irregularity of bearing or crankshaft journal. If gauging plastic indicates more than 0.001 in. (0.025 mm) taper, measure crankshaft journal with micrometer (para. 2-17).

NOTE

It is acceptable to use a standard bearing with a 0.001 in. (0.025 mm) undersize bearing to obtain the proper rod bearing clearance. If unable to produce proper clearance, replace crankshaft.

15. If the measurement is within specification, 0.002-0.004 in. (0.051-0.102 mm) bearing is satisfactory. If measurement is not within specifications, a 0.001 in. (0.025 mm) undersize bearing may produce proper bearing clearance. Replace bearings and repeat steps 9 through 15.
16. Remove gauging plastic.
17. Apply OE/HDO to rod bearings. Check identification number on side of connecting rod (6) and install corresponding rod cap (2).
18. Secure rod cap (2) to connecting rod (6) with two nuts (1) and tighten to 44-52 lb-ft (60-71 N•m).
19. Rotate crankshaft (5) to ensure crankshaft (5) rotates.
20. Repeat steps 2 through 19 for remaining pistons and connecting rods.

2-41. ENGINE ASSEMBLY FROM SUBASSEMBLIES (Cont'd)

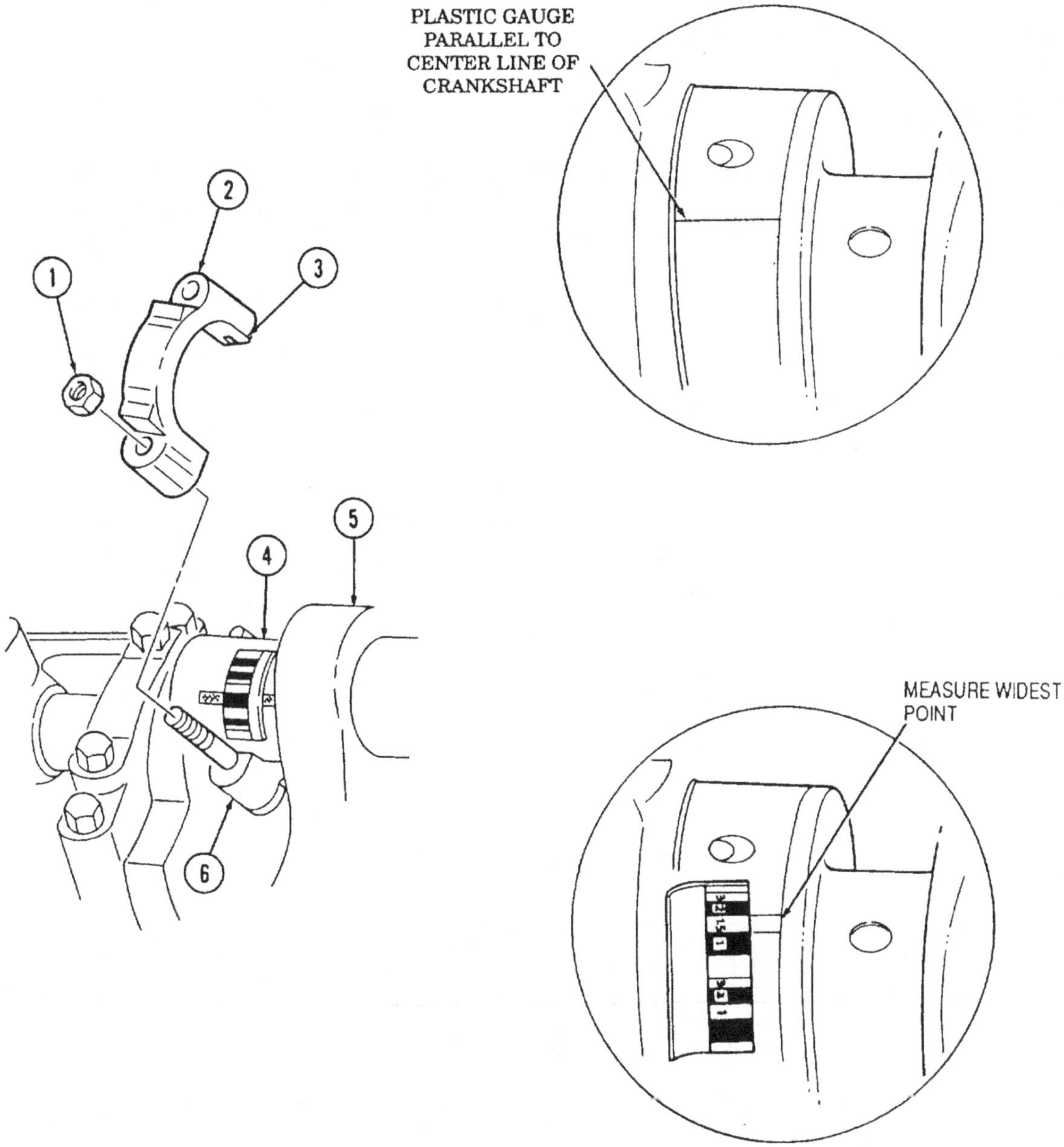

2-41. ENGINE ASSEMBLY FROM SUBASSEMBLIES (Cont'd)

d. Oil Pump

1. Rotate cylinder block (7) so crankshaft faces upward.
2. Install oil pump (8) on rear main bearing cap (6).
3. Install stud (1) and oil pump (8) on rear main bearing cap (6) and tighten stud (1) to 59-74 lb-ft (80-100 N•m).
4. Install pickup tube bracket (4) on stud (1) with nut (2). Tighten nut (2) to 35 lb-ft (48 N•m).
5. Tighten capscrew (5) securing pickup tube bracket (4) to pickup tube (3) to 12 lb-ft (16 N•m).

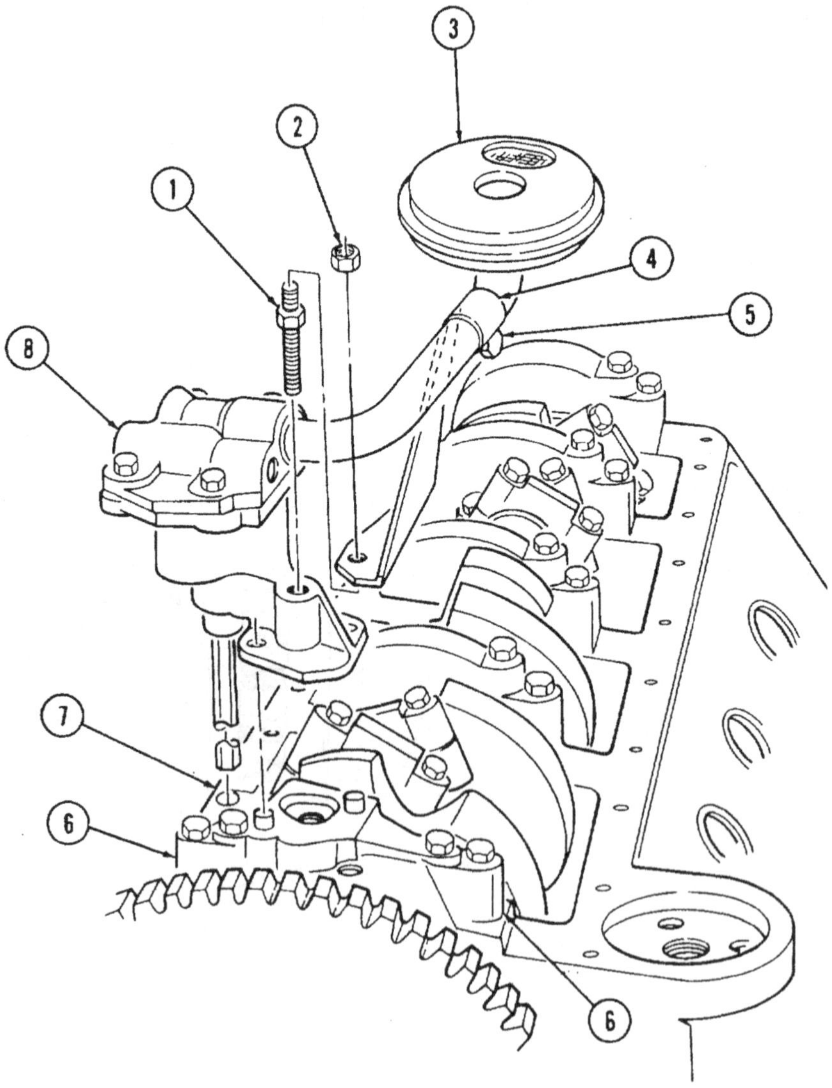

2-41. ENGINE ASSEMBLY FROM SUBASSEMBLIES (Cont'd)

e. Camshaft

1. Rotate cylinder block (7) 180°.
2. Apply OE/HDO to all bearing journals (12) and cam lobes (11) on camshaft (13).

CAUTION

Support camshaft during installation to prevent damage to camshaft bearings.

3. Install camshaft (13) into cylinder block (7).

NOTE

Beveled edge of spacer must seat against front bearing journal on camshaft.

4. Install spacer (10) and camshaft thrust plate (9) with two capscrews (14). Tighten capscrews (14) to 18 lb-ft (24 N•m).

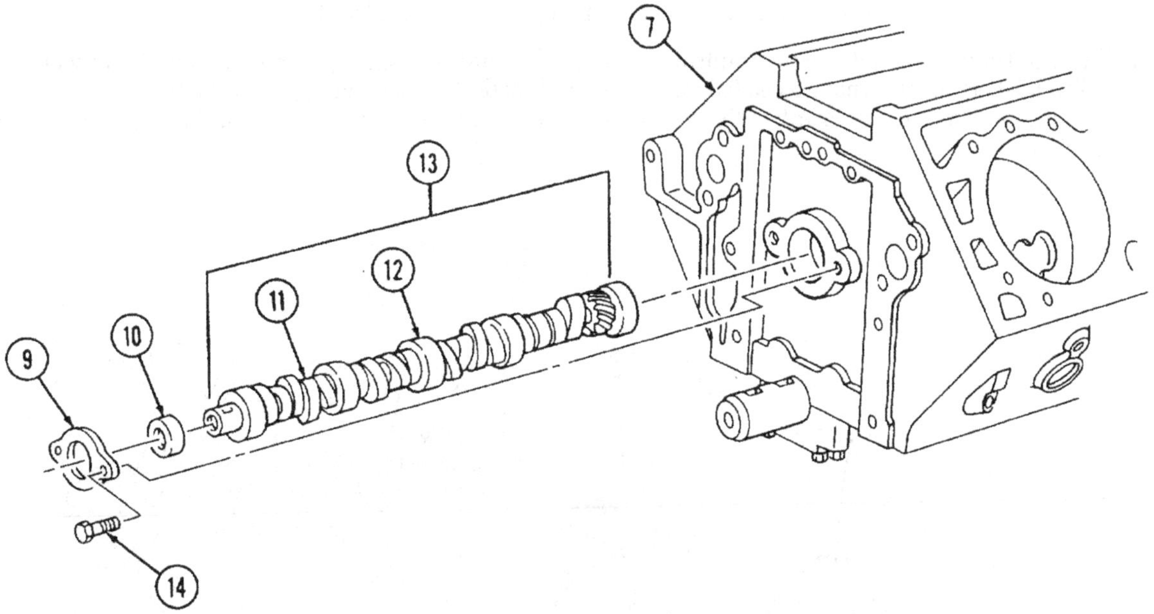

2-41. ENGINE ASSEMBLY FROM SUBASSEMBLIES (Cont'd)

f. Timing Chain and Drive Sprockets

1. Install woodruff key (7) in camshaft (6) and woodruff key (8) in crankshaft (9).
2. Install crankshaft sprocket (10) and camshaft sprocket (4) in timing chain (5) and align timing marks (11).
3. Rotate crankshaft (9) and camshaft (6) so woodruff keys (7) and (8) align with keyway slots in crankshaft sprocket (10) and camshaft sprocket (4).

CAUTION

Timing marks on crankshaft sprocket and camshaft sprocket must remain aligned after installation or engine damage may result.

4. Install crankshaft sprocket (10), camshaft sprocket (4), and timing chain (5) as an assembly onto camshaft (6) and crankshaft (9).
5. Install pump drive gear (3) on camshaft (6) with washer (2) and capscrew (1). Tighten capscrew (1) to 55-66 lb-ft (75-90 N•m).
6. Apply OE/HDO to timing chain (5).

NOTE

Perform step 7 only if timing chain was replaced.

7. Check timing chain deflection midway between camshaft sprocket (4) and crankshaft sprocket (10). Total deflection must not exceed 0.5 in. (12.7 mm). If deflection exceeds specification, inspect camshaft sprocket (4) and crankshaft sprocket (10) for damage. Replace eitherif damaged and recheck deflection.

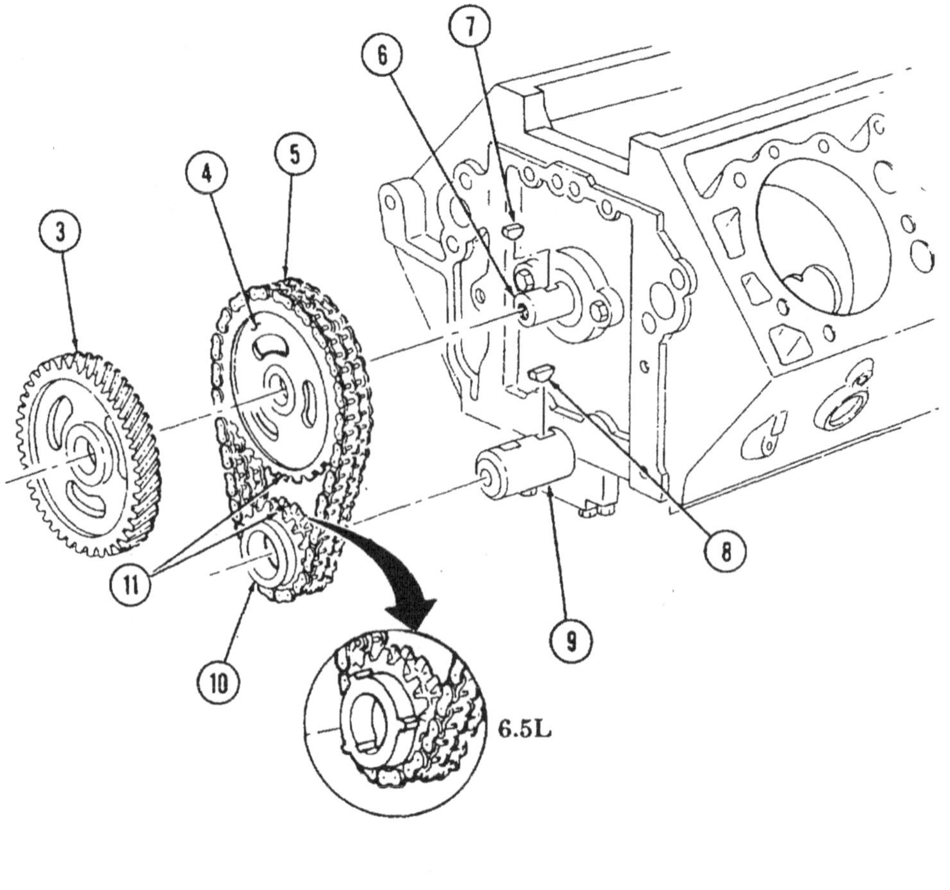

2-41. ENGINE ASSEMBLY FROM SUBASSEMBLIES (Cent'd)

g. Timing Gear Cover

1. Install front cover seal (19) in timing gear cover (15), ensuring lip of seal (19) faces inward.
2. Apply a 3/32 in. (2 mm) bead of sealer to sealing surface on timing gear cover (15), following diagram shown.
3. Install timing gear cover (15) on cylinder block (16) with four washers (17), capscrews (18), and short capscrews (20). Tighten four capscrews (18) and short capscrews (20) to 33 lb-ft (45 N•m).

CAUTION

Maintain 0.040 in. (1.02 mm) minimum clearance between baffle plate and pump drive gear to avoid noise.

4. Install baffle plate (14) in timing gear cover (15) with two capscrews (12) and nut (13). Tighten capscrews (12) and nut (13) to 33 lb-ft (45 N•m).

NOTE

Perform step 5 for 6.5L engines.

5. Install rpm sensor (22) in timing gear cover (15) with capscrew (21).

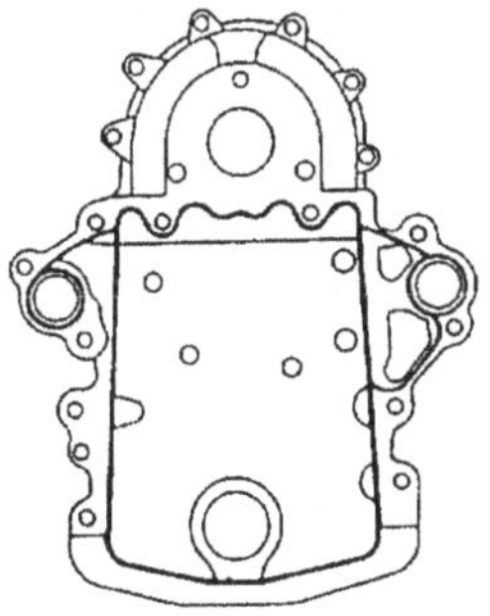

SEALER DIAGRAM

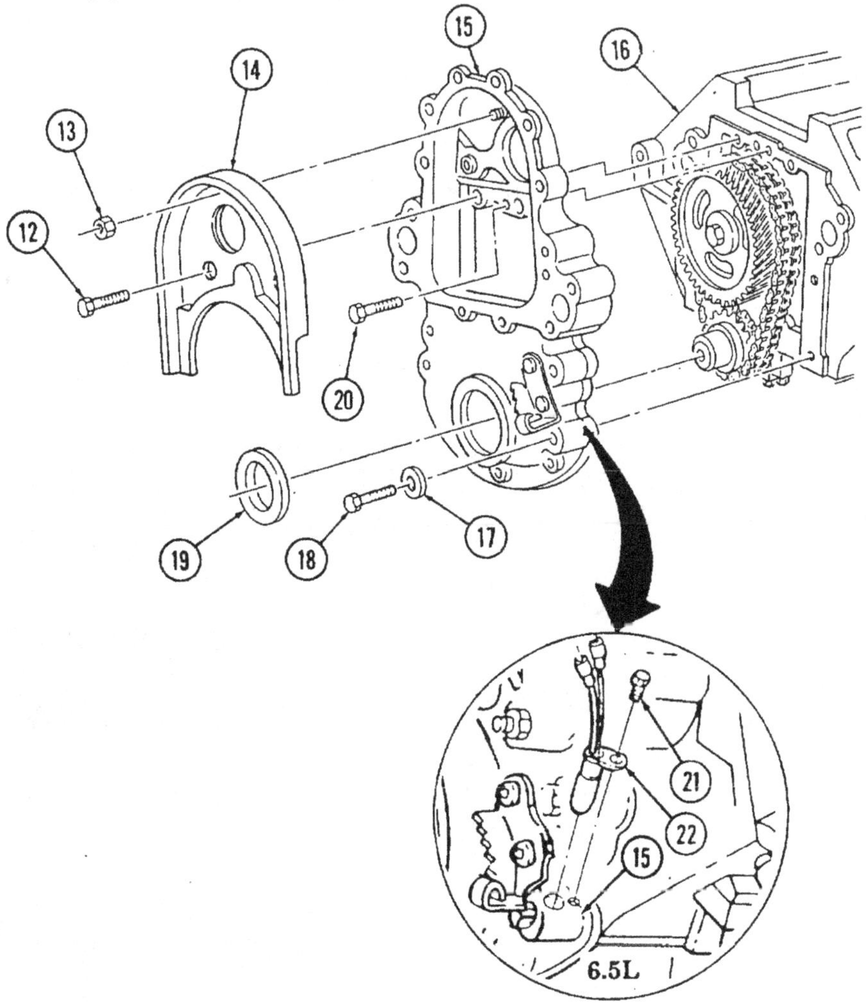

2-41. ENGINE ASSEMBLY FROM SUBASSEMBLIES (Cont'd)

h. Oil Pan

1. Rotate cylinder block (9) 180°.
2. Install oil pan rear seal (7) in rear main cap (8) and apply a drop of silicone sealant at each end of seal (7).

NOTE

- Install oil pan immediately after application of sealant.
- If applying silicone sealant, perform step 4.
- If installing new engine oil pan gasket, perform step 3.
- Silicone sealant can be applied with the gasket, but is not required.

3. Install engine oil pan gasket (11) on oil pan sealing surface (12) and align with capscrew holes.
4. Apply a 3/16 in. (5 mm) bead of silicone sealant around two large holes (10) on cylinder block (9). Apply a 3/16 in. (5 mm) bead of silicone sealant around oil pan sealing surface (12), following diagram shown.
5. Install oil pan (5) on cylinder block (9) with twenty capscrews (1), two studs (4), and two large capscrews (6). Tighten studs (4) and capscrews (1) to 48-120 lb-in. (5-14 N•m) and capscrews (6) to 156-240 lb-in. (18-27 N•m).
6. Install drainplug (2) and washer (3) in oil pan (5). Tighten drainplug (2) to 20 lb-ft (27 N•m).

SEALANT DIAGRAM

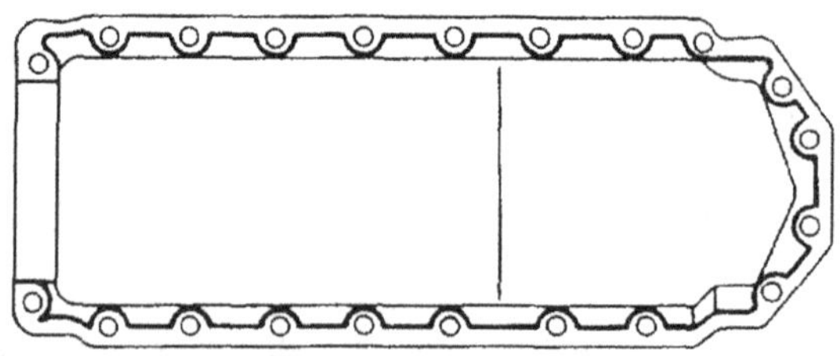

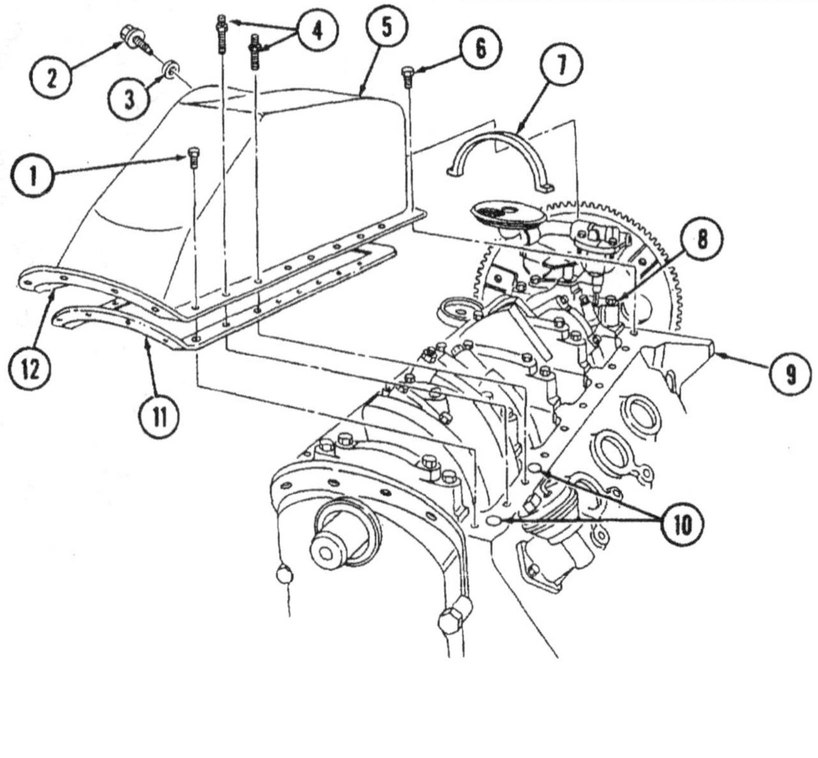

2-41. ENGINE ASSEMBLY FROM SUBASSEMBLIES (Cont'd)

i. Oil Filter and Adapter

1. Install two O-rings (15) on adapter bolt (13).
2. Install O-ring (18) on adapter (16).
3. Install adapter (16) on cylinder block (9) with adapter bolt (13) and gasket (14). Tighten adapter bolt (13) to 40 lb-ft (54 N•m).
4. Install oil filter (17) on adapter (16). Tighten oil filter (17) until gasket contacts adapter (16) and tighten an additional 1/2 to 3/4 turn.
5. Coat threads on fitting (22) with pipe sealing compound and install in cylinder block (9).
6. Coat threads on oil pressure sending unit (21) with pipe sealing compound and install in fitting (22).
7. Coat threads on cooler line adapters (19) and (20) with pipe sealing compound ad insall in cylinder. block (9).

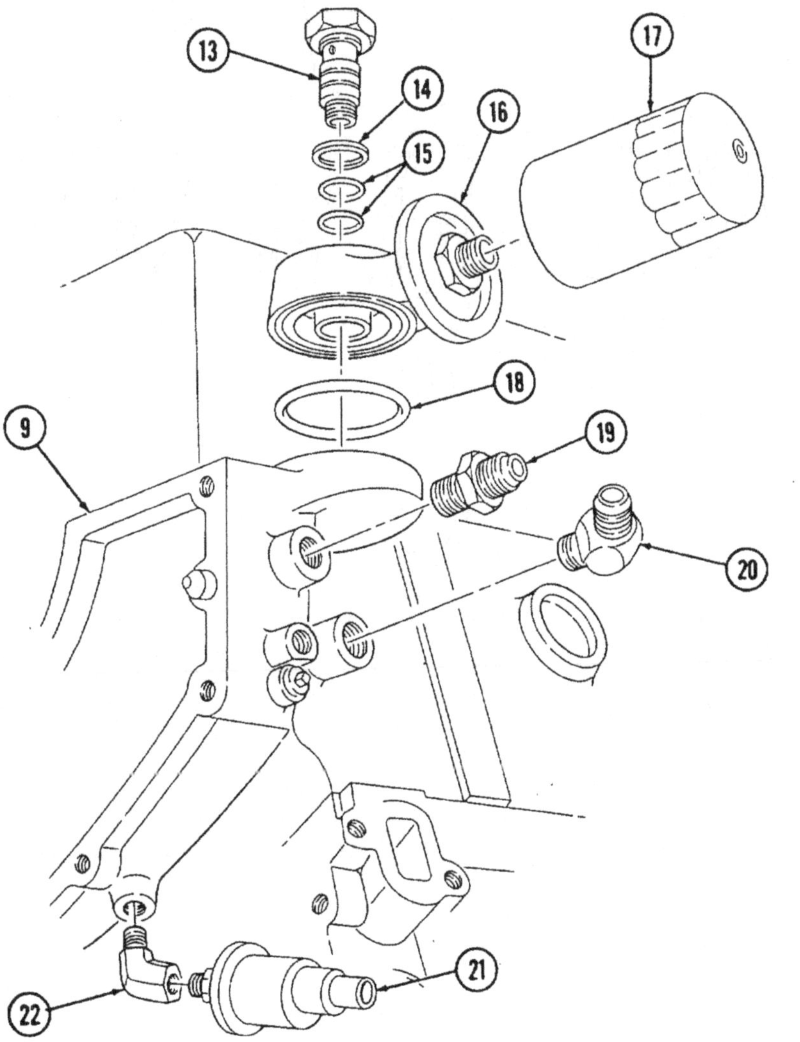

2-41. ENGINE ASSEMBLY FROM SUBASSEMBLIES (Cont'd)

j. Fuel Pump

1. Apply OE/HDO to pushrod (8) and install into cylinder block (1).
2. Install gasket (2) and adapter plate (3) on cylinder block (1) with two capscrews (4). Tighten capscrews (4) to 48-84 lb-in. (5-10 N•m).
3. Install gasket (5) and fuel pump (6) on adapter plate (3) with two capscrews (7). Tighten capscrews (7) to 20-30 lb-ft (27-41 N•m).

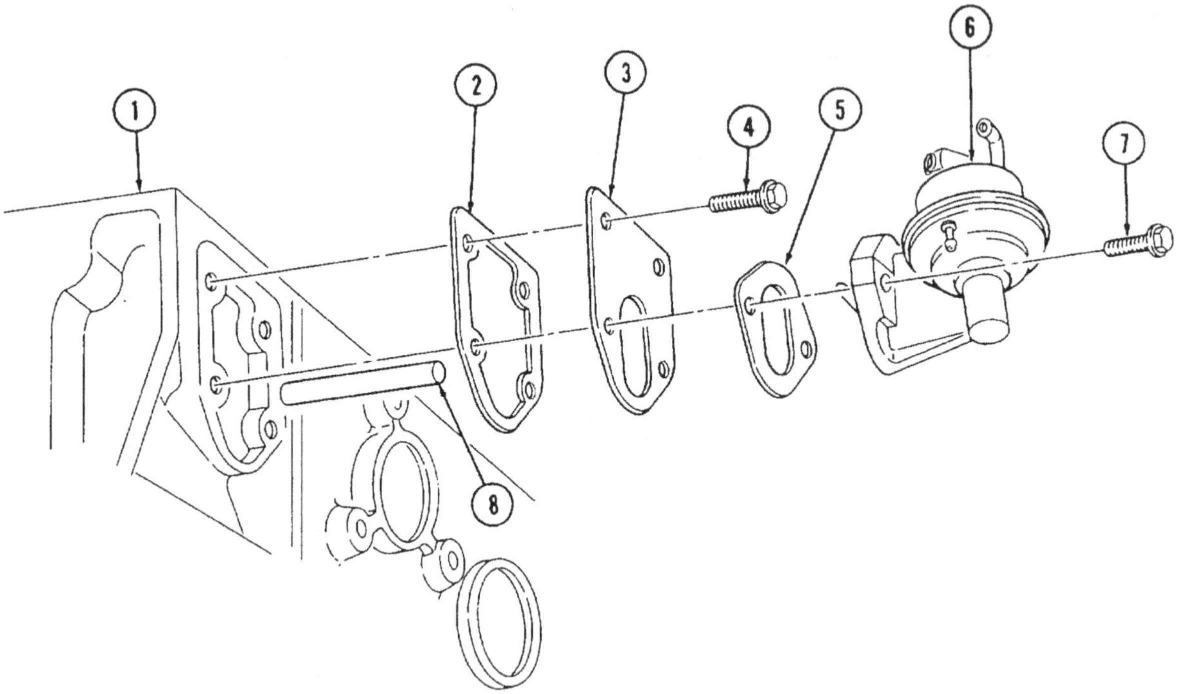

2 - 4 1ENGINE ASSEMBLY FROM SUBASSEMBLIES (Cont'd)

k. Oil Pump Drive

1. Rotate cylinder block (1) 180 degrees.
2. Install gasket (11) and oil pump drive (10) into cylinder block (1).
3. Secure oil pump drive (10) with clamp (12) and capscrew (9). Tighten capscrew (9) to 25-37 lb-ft (34-50 N·m).

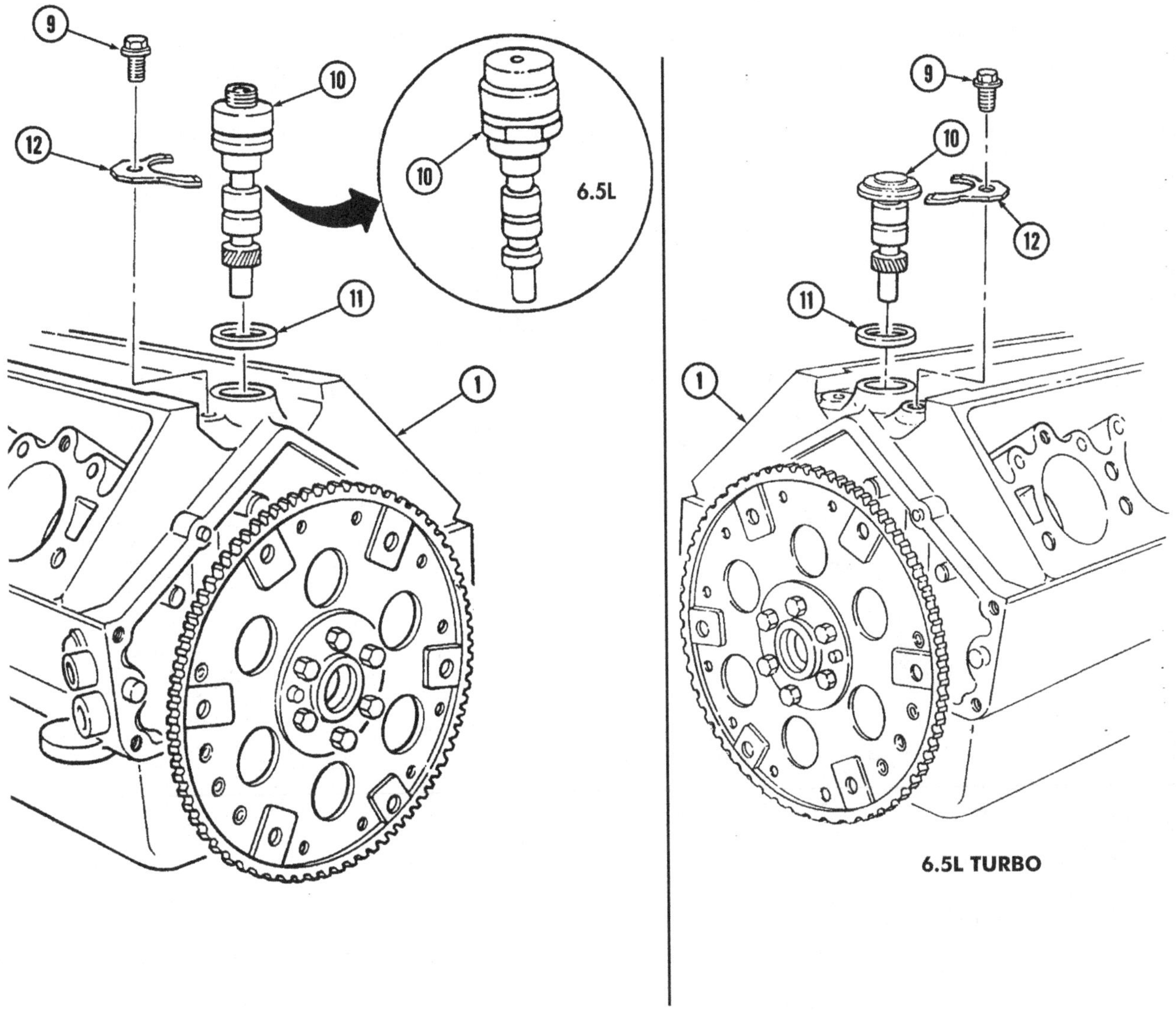

2-41. ENGINE ASSEMBLY FROM SUBASSEMBLIES (Cont'd)

l. Fuel Injection Pump

1. Install gasket (5) and injection pump (6) on timing gear cover (3).
2. Install accelerator cable bracket (12) on fuel injection pump (6) with two capscrews (11). Tighten capscrews (11) to 13 lb-ft (18 N•m).
3. Install idle return spring (8) on throttle lever (7) and accelerator cable bracket (12).
4. Align timing marks (4) on injection pump (6) and timing gear cover (3).
5. Install injection pump (6) to timing gear cover (3) with three washers (10) and nuts (9).
6. Rotate crankshaft until timing mark (14) on pump drive gear (13) is straight up.
7. Align timing marks (14) on pump drive gear (13) and pump driven gear (2) and install driven gear (2) into timing gear cover (3).
8. Install three capscrews (1) on driven gear (2) and tighten capscrews (1) to 13-20 lb-ft (18-27 N•m).

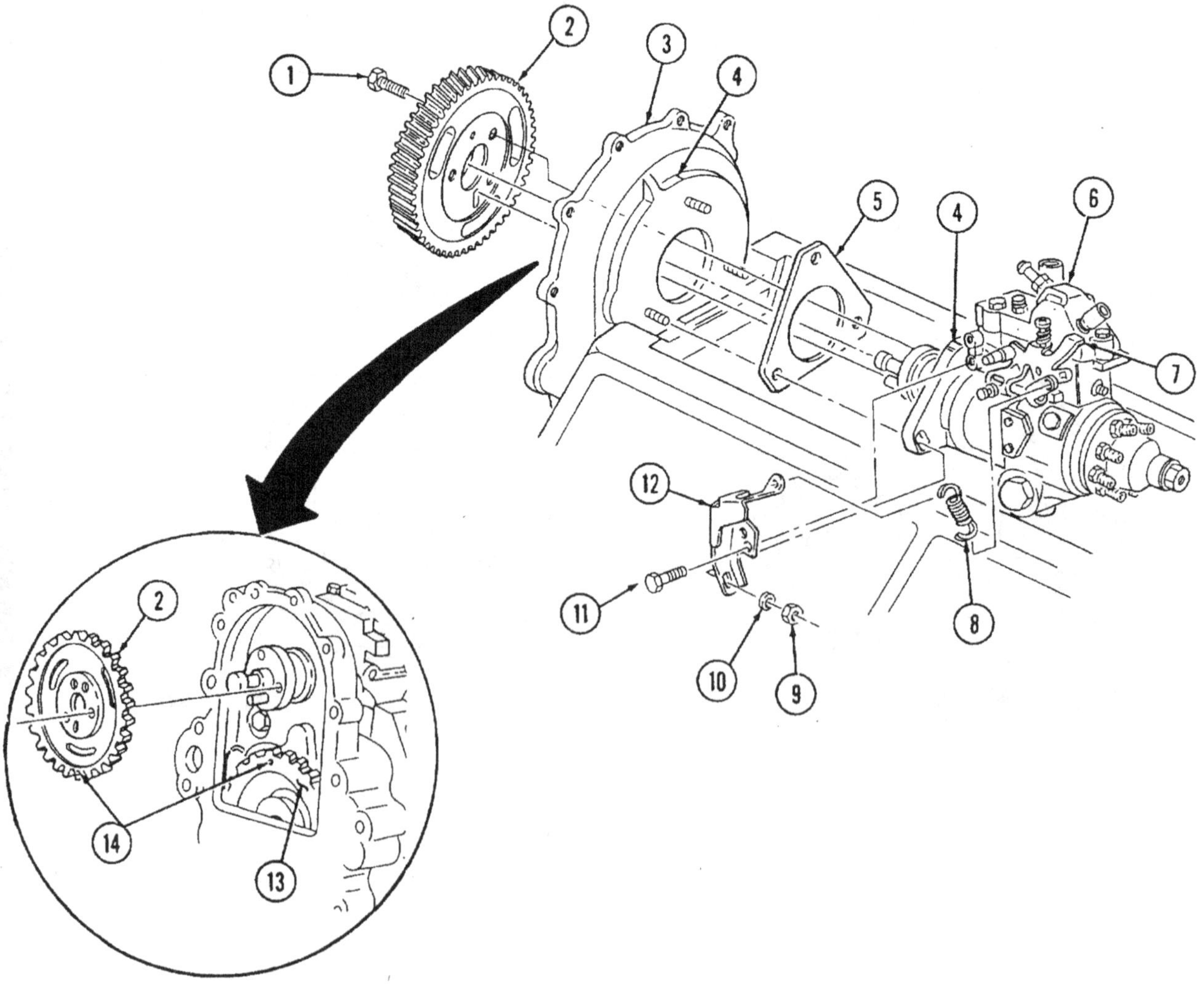

2-41. ENGINE ASSEMBLY FROM SUBASSEMBLIES (Cont'd)

m. Water Pump and Adapter Plate

CAUTION

Ensure water pump P/N 23500085 is used on 6.5 L engines or damage to equipment will result.

1. Install gasket (16) and water pump (15) on adapter plate (17) with seven capscrews (18). Tighten capscrews (18) to 13-20 lb-ft (18-27 N•m).
2. Apply sealer to sealing surfaces on adapter plate (17), following diagram shown.
3. Apply pipe sealing compound to capscrews (24) and (23).
4. Install adapter plate (17) and water pump (15) on timing gear cover (3) with two long studs (25), short stud with thick hex (21), stud (22), and capscrew (24).
5. Install two long capscrews (23), studs (19), four washers (20), and capscrews (26). Tighten studs (19) and capscrews (26) and (23) to 13-20 lb-ft (18-27 N•m). Tighten studs (21), (22), and (25), and capscrew (24) to 24-37 lb-ft (34-50 N•m).
6. Install oil fill tube (29) on adapter plate (17) with two washers (28) and nuts (27). Tighten nuts (27) to 13-20 lb-ft (18-27 N•m).

SEALER DIAGRAM

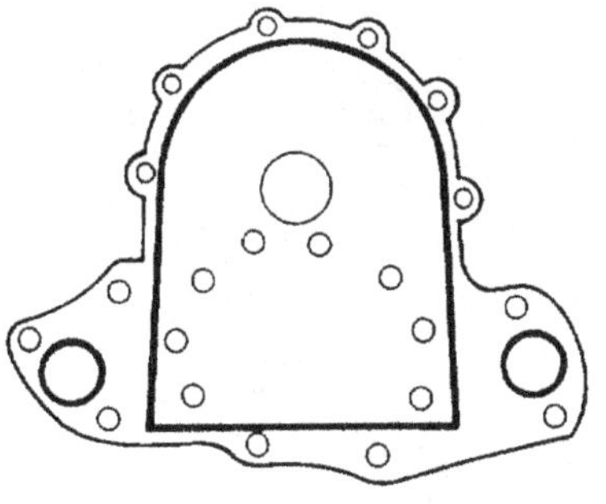

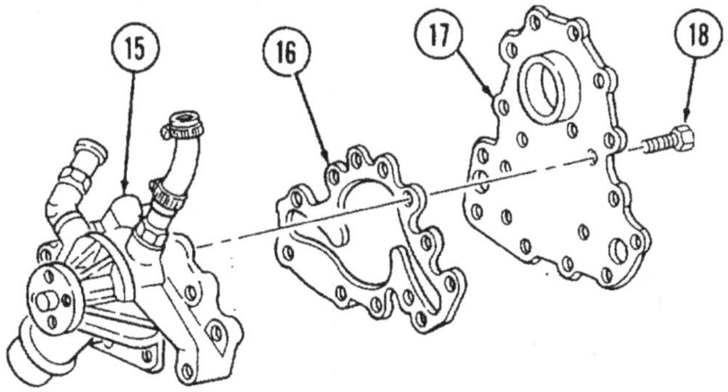

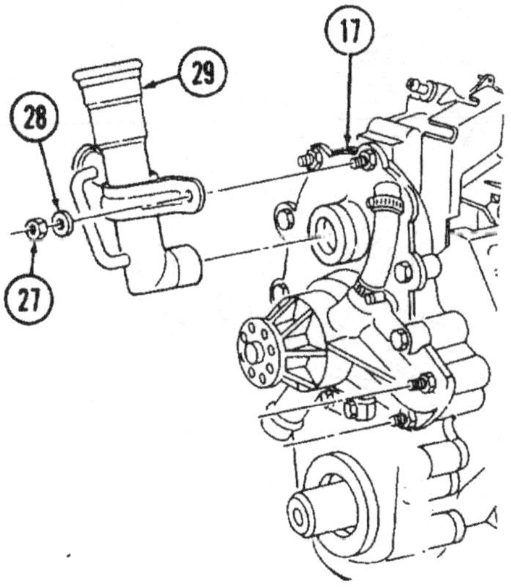

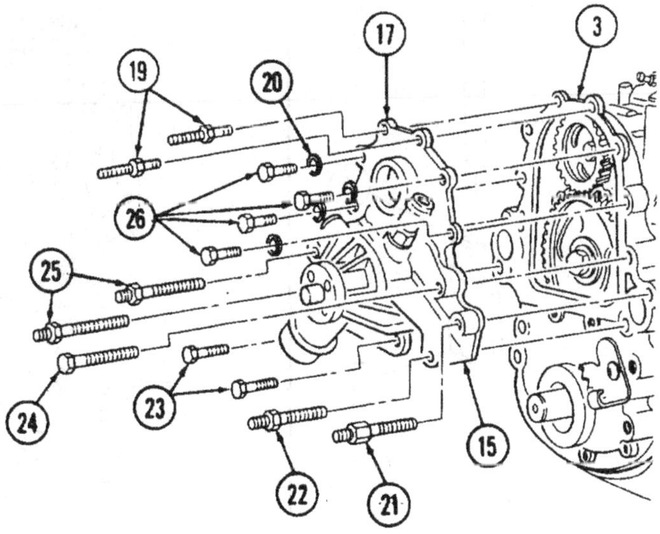

2-41. ENGINE ASSEMBLY FROM SUBASSEMBLIES (Cont'd)

n. Valve Lifters

WARNING

Diesel fuel is highly flammable. Do not perform this procedure near fire, flame, or sparks. Severe injury or death may result.

1. Submerge eight lifters (3) in clean diesel fuel or kerosene and work plunger (2) up and down with pushrod (1) to prime lifters (3).
2. Apply OE/HDO to lifters (3).
3. Install lifters (3) into cylinder block (7).
4. Install four guide plates (6) on lifters (3).
5. Install two guide plate clamps (4) with two capscrews (5). Tighten capscrews (5) to 15-20 lb-ft (20-27 N•m).
6. Repeat steps 1 through 5 for opposite side.
7. Manually rotate crankshaft (8) two complete revolutions to ensure free movement of lifters (3).

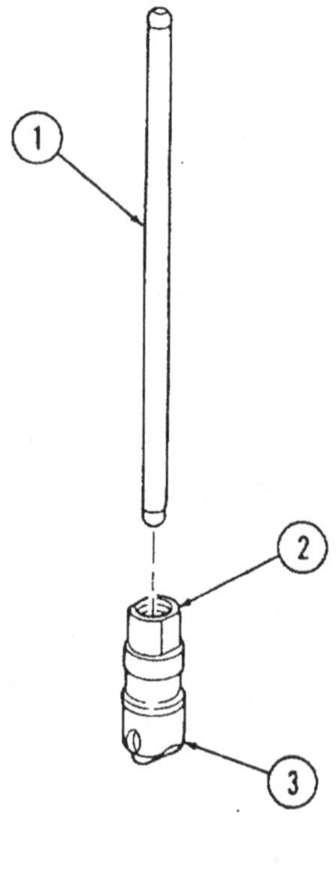

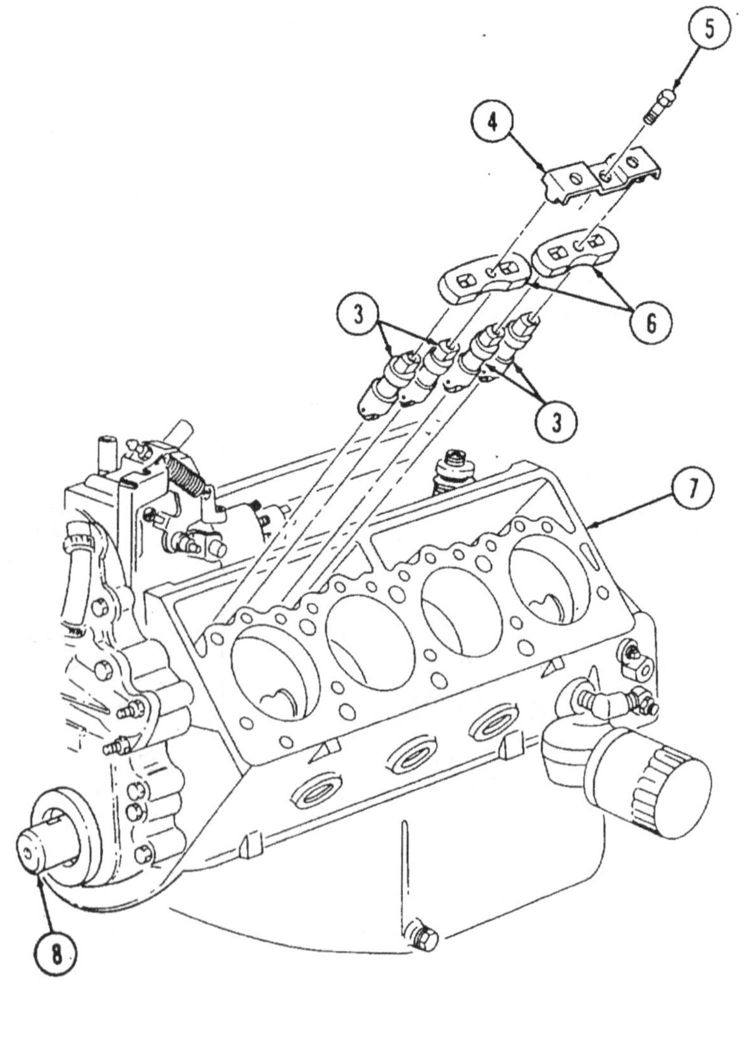

2-41. ENGINE ASSEMBLY FROM SUBASSEMBLIES (Cont'd)

o. Cylinder Heads

CAUTION

Head gasket must be used without sealer. Additional sealant may cause leaks or damage to engine.

1. Install head gasket (11) over dowel pins (12) on cylinder block (7).

NOTE

Use of cylinder head lifting device is optional.

1.1. Install cylinder head lifting device on center of cylinder head (9) with two washers (10.2) and capscrews (10.1).

1.2. Install cylinder head (9) on cylinder block (7) and remove two capscrews (10.1), washers (10.2), and cylinder head lifting device.

2. Clean threads of capscrews (10) with a wire brush to remove old pipe sealing compound. Apply pipe sealing compound to threads and under heads of capscrews (10).

CAUTION

Failure to tighten cylinder head capscrews in proper sequence may result in leaks or damage to cylinder head.

3. Install cylinder head (9) on cylinder block (7) with seventeen capscrews (10). Tighten capscrews (10) to 20 lb-ft (27 N·m), following torque sequence shown.

4. For 6.2 L engine, tighten seventeen capscrews (10) to 50 lb-ft (68 N·m), following torque sequence shown.

4.1 For 6.5 L engine, tighten seventeen capscrews (10) to 55 lb-ft (75 N·m), following torque sequence shown.

4.2 For 6.5 L engine, retighten seventeen capscrews (10) to 55 lb-ft (75 N·m), following torque sequence shown.

5. Tighten seventeen capscrews (10) an additional 90°, following torque sequence shown and mark capscrews (10).

6. Repeat steps 1 through 5 for opposite side.

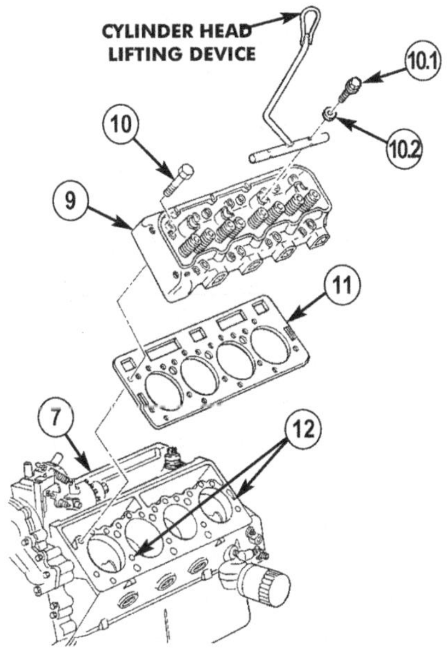

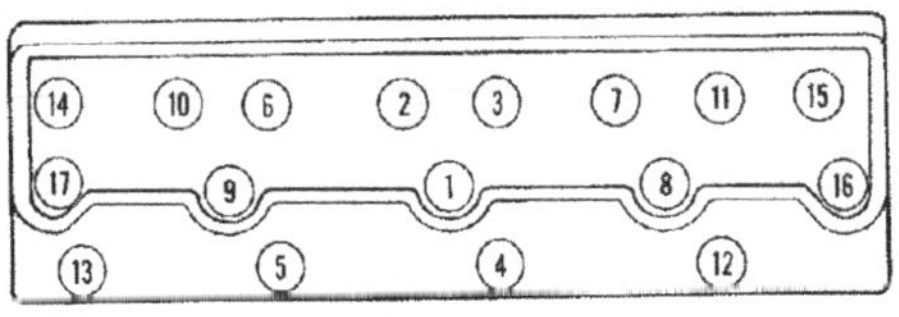

TORQUE SEQUENCE

2-41. ENGINE ASSEMBLY FROM SUBASSEMBLIES (Cont'd)

p. Glow Plugs

1. Install four glow plugs (2) into cylinder head (l). Using glow plug socket, tighten glow plugs (2) to 8-12 lb-ft (11-16 N•m).
2. Repeat step 1 for opposite side.

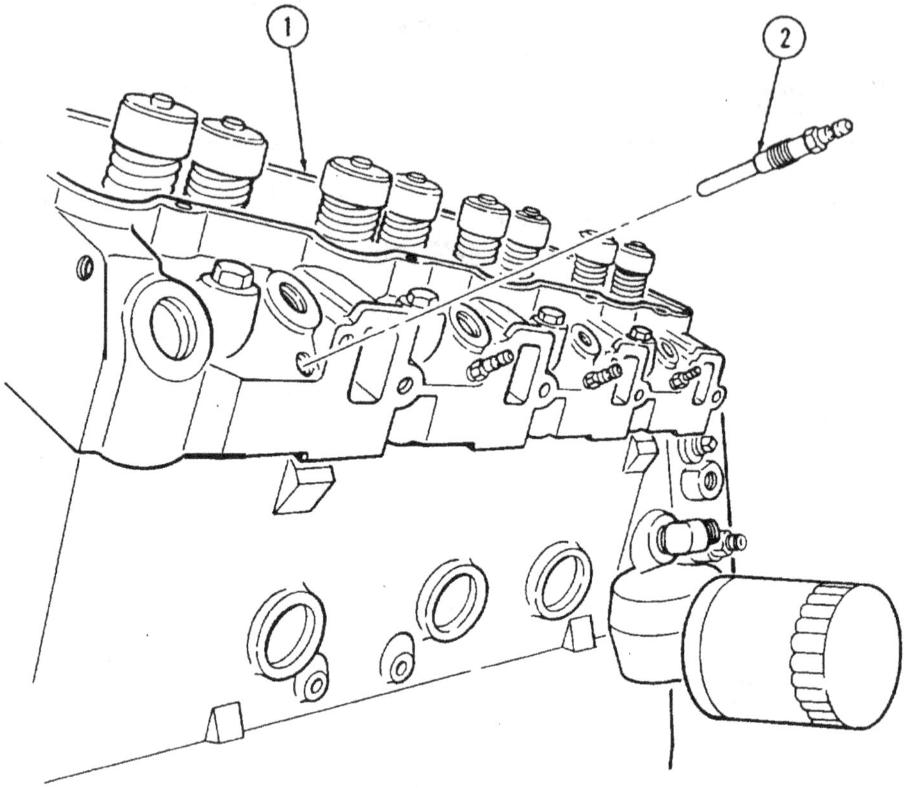

2-41. ENGINE ASSEMBLY FROM SUBASSEMBLIES (Cont'd)

q. Fuel Injection Nozzles

1. Install four gaskets (3) and fuel injection nozzles (4) into cylinder head (1). Using injector nozzle socket, tighten fuel injection nozzles (4) to 44-60 lb-ft (60-81 N•m).
2. Install three hoses (6) on fuel injection nozzles (4) with six clamps (5).
3. Install cap (7) on rear fuel injection nozzle (4) with clamp (5).
4. Repeat steps 1 through 3 for opposite side.

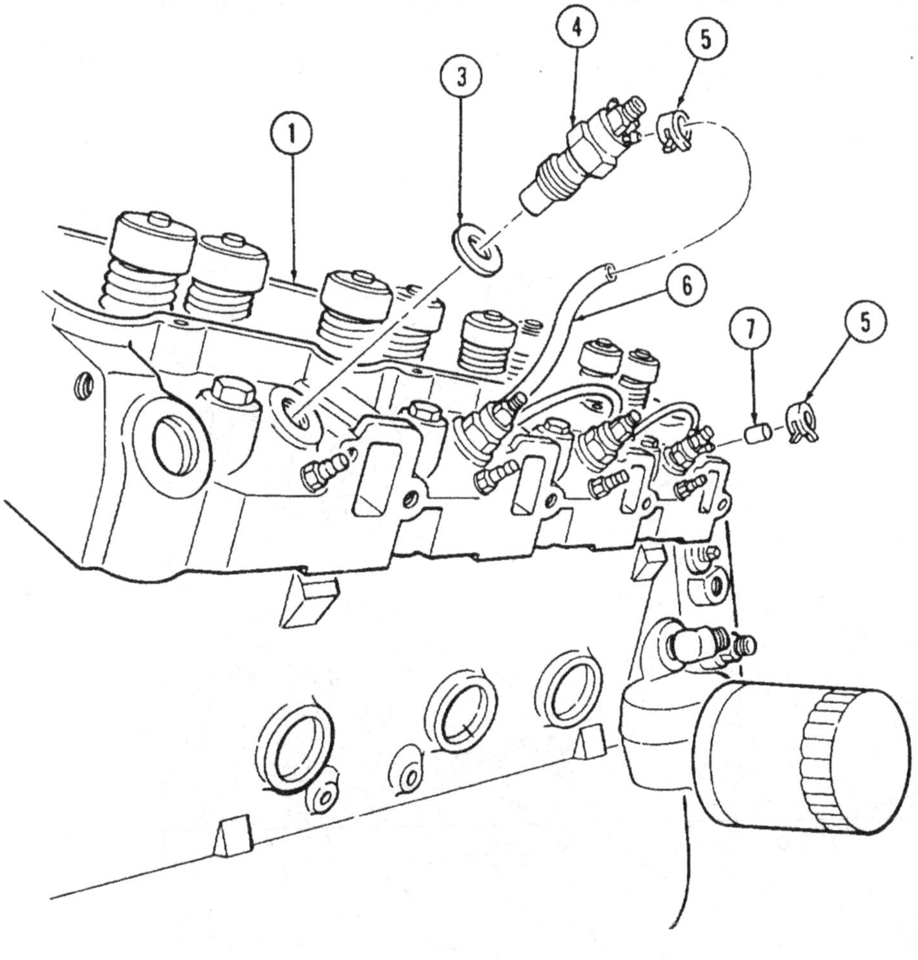

2-41. ENGINE ASSEMBLY FROM SUBASSEMBLIES (Cont'd)

r. Rocker Arm Shaft and Pushrods

CAUTION

Marked ends of pushrods must point up when installed or engine damage may result.

1. Install eight pushrods (1) in proper location with marked end of pushrods (1) up. Make sure pushrods (1) are properly seated.
2. Install two rocker arm and shaft assemblies (2) on cylinder head (6). Make sure pushrods (1) properly seat in rocker arms (3).
3. Secure rocker arm and shaft assemblies (2) with four retainers (4) and capscrews (5). Tighten capscrews (5) to 41 lb-ft (56 N•m).
4. Repeat steps 1 through 3 for opposite side.

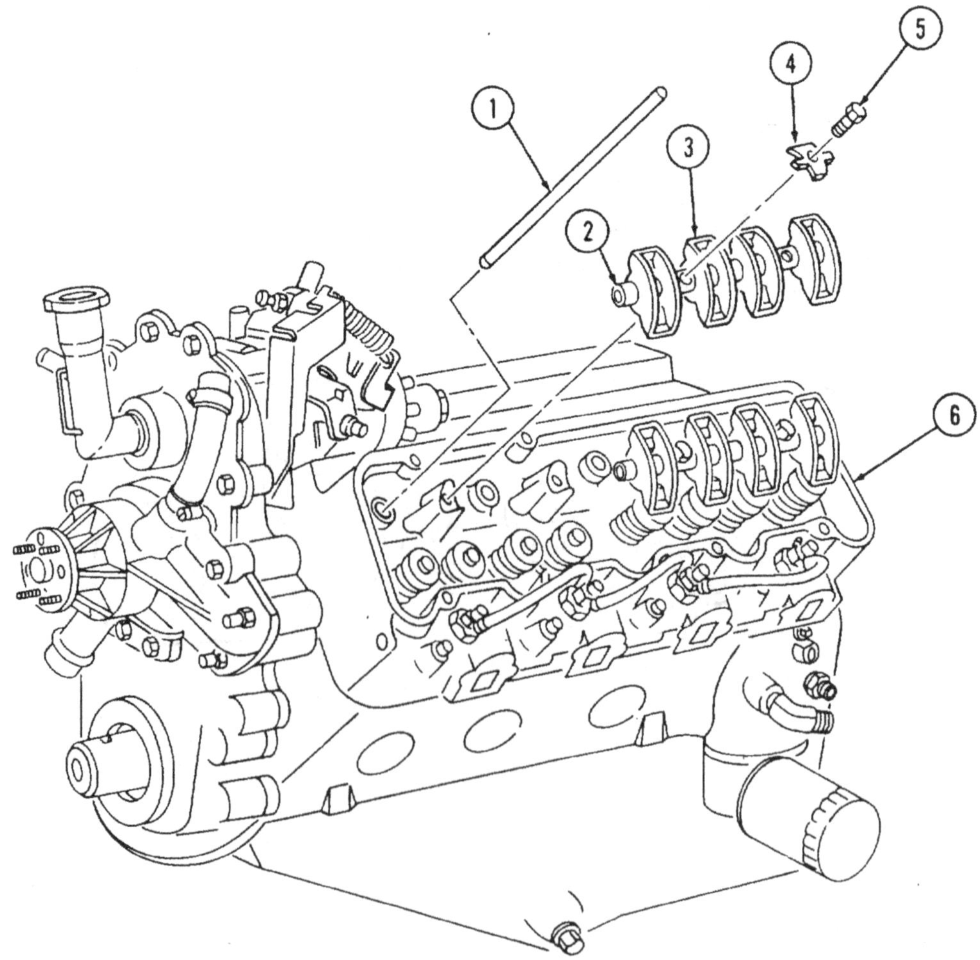

2-41. ENGINE ASSEMBLY FROM SUBASSEMBLIES (Cont'd)

s. Rocker Arm Covers

NOTE

- If applying sealant, perform step 2. Keep sealant out of capscrew holes.
- If installing new rocker arm cover gaskets, perform step 1, then step 3.
- Silicone sealant can be used with the gasket, but is not required.

1. Install rocker arm cover gasket (13) on lip of cover (12) and align with capscrew holes.
2. Apply a 1/16-in. (4 mm) bead of silicone sealant around sealing surface of rocker arm covers (11).
3. Install rocker arm covers (11) on cylinder head (6) with sixteen washers (10), seven capscrews (9), and nine studs (14), following assembly diagram.
4. Tighten capscrews (9) and studs (14) to 13-25 lb-ft (18-34 N•m).
5. Install two injection line support brackets (8) on studs (12) with four nuts (7). Tighten nuts (7) to 13-20 lb-ft (18-27 N•m).

ASSEMBLY DIAGRAM

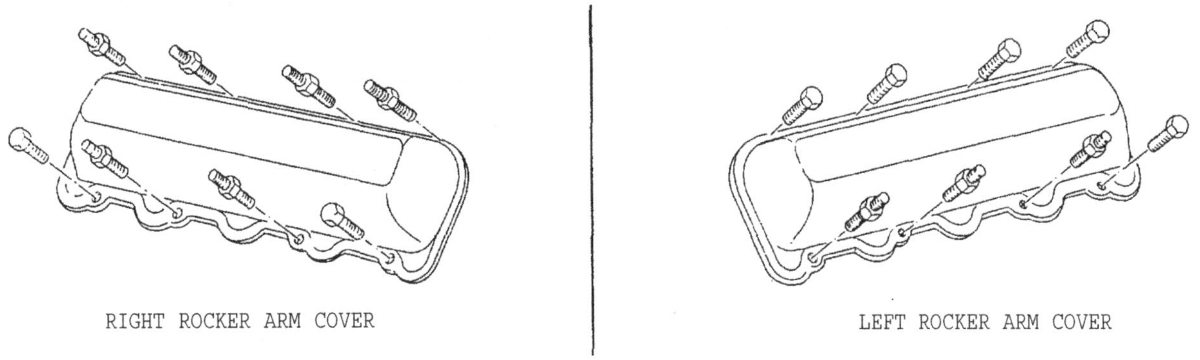

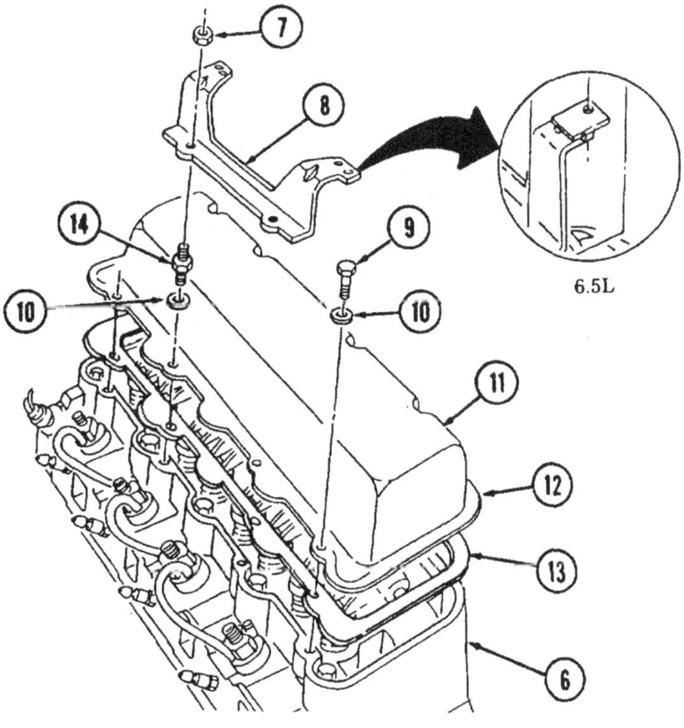

2-41. ENGINE ASSEMBLY FROM SUBASSEMBLIES (Cont'd)

t. Modulator Link

NOTE

Installation of the modulator link is applicable only to 6.2L engines.

1. Attach modulator link (1) to throttle lever (7).
2. Install water jacket cover gasket (5), water jacket cover (4), and modulator link bracket (3) on cylinder head (6) and secure fith capscrews (2).
3. Tighten capscrews (2) to 25-37 lb-ft (3450 N•m).

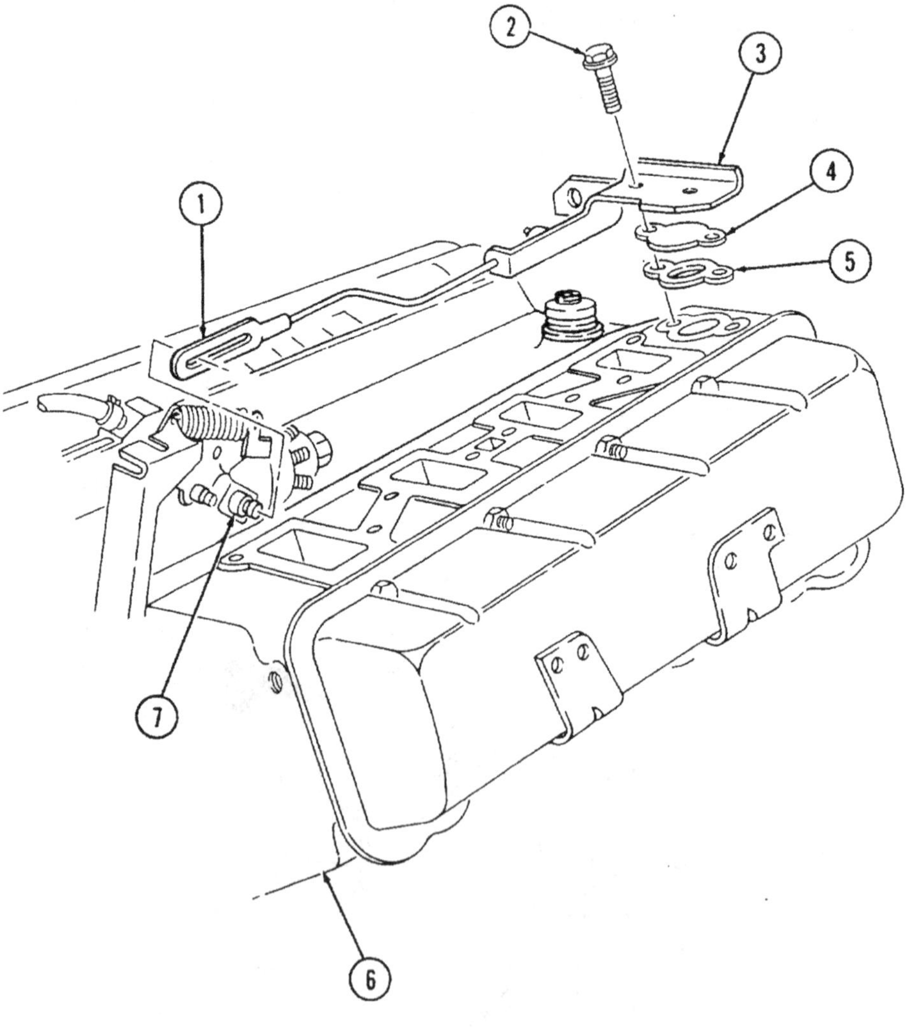

2-41. ENGINE ASSEMBLY FROM SUBASSEMBLIES (Cont'd)

u. Fuel Supply and Return Lines

1. Install fuel supply line (8) on fuel pump (27) and finger tighten tube nut (26).
2. Install fuel return line (17) on timing gear cover (25) with clamp (13), washer (12), and capscrew (11). Finger tighten capscrew (11).
3. Install return line (17) to valve cover stud (21) with clamp (20), washer (19), and nut (18). Tighten nut (18) to 13-20 lb-ft (18-27 N•m).
4, Install supply line (8) and return line (17) to cylinder block (29) with two clamps (28), washer (24), and stud (23). Tighten stud (23) and capscrew (11) to 25 lb-ft (34 N•m).
5. Tighten tube nut (26).
6. Connect three rubber hoses (16) to front fuel injection nozzles (22) and fuel injection pump (14) with, three clamps (15).
7. Connect fuel filter inlet hose (10) to fuel supply line (8) with hose clamp (9).

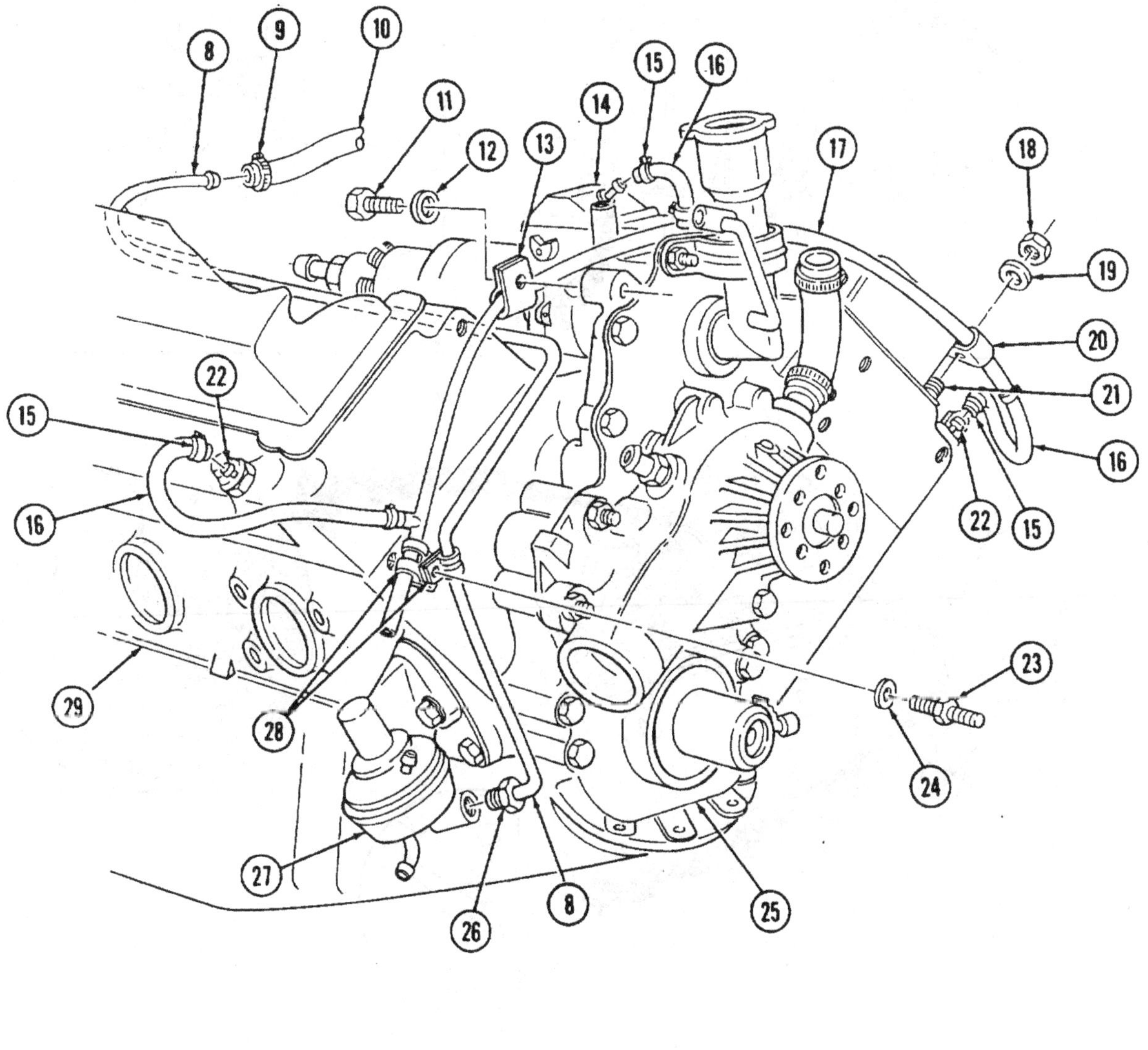

2-41. ENGINE ASSEMBLY FROM SUBASSEMBLIES (Cont'd)

v. Fuel Injection Lines

1. Install number 1 and 3 injection lines (7) and connect t o fuel injection pump (1) and fuel injection nozzles (4).
2. Using crowfoot wrenches, tighten four nuts (3) and fuel injection lines (7) on fuel injection nozzles (4) and fuel injection pump (1) to 20 lb-ft (27 N•m).
3. Repeat steps 1 and 2 for each set of remaining fuel injection lines (7), number 5 and 7, number 6 and 8, and number 2 and 4.
4. Install four clamps (5) securing fuel injection lines (7) to support brackets (2) with four screw-aesembled washers (6).
5. Install boot (8) onto fuel injection pump (1) and tighten clamp (9).

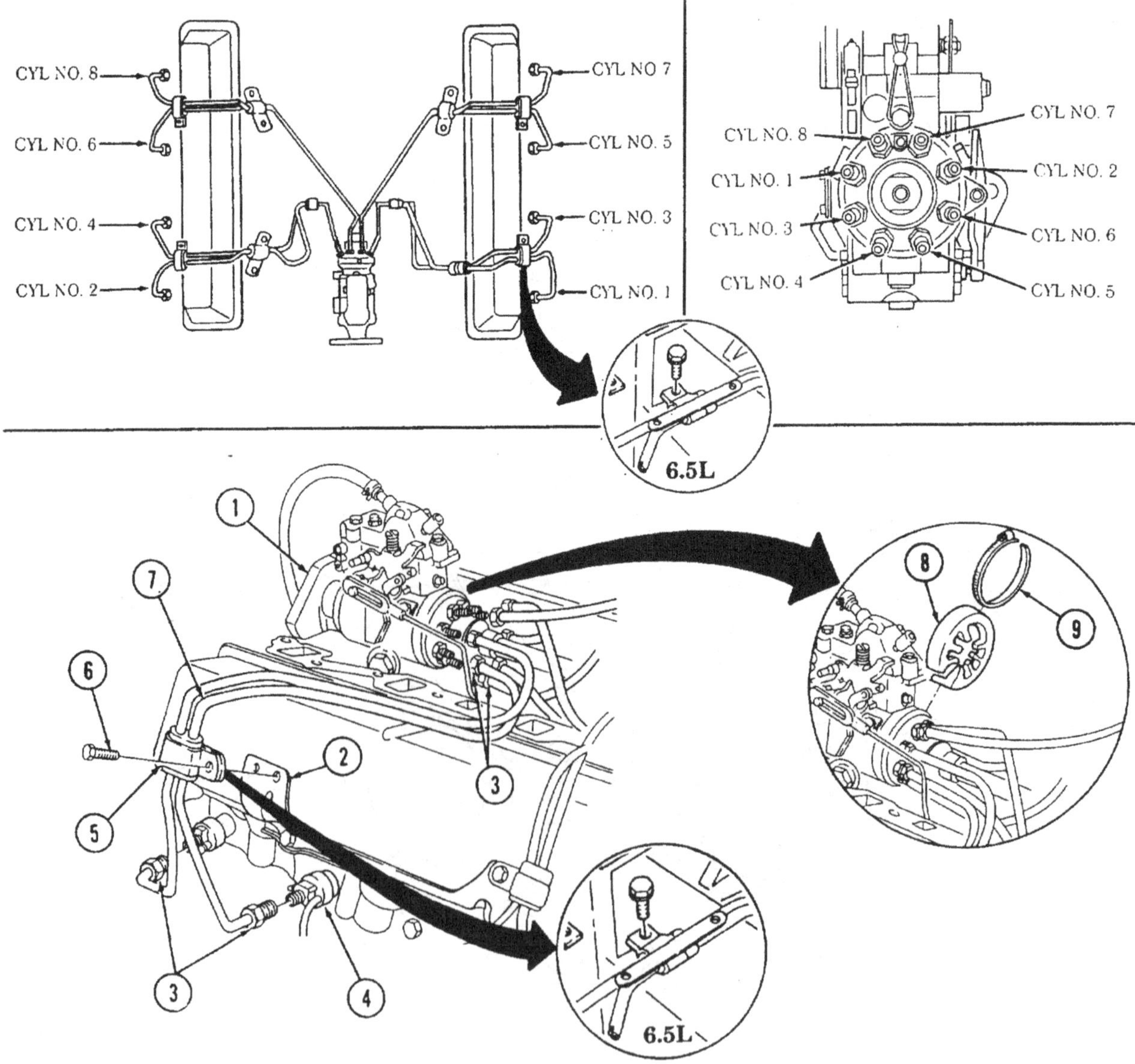

2-41. ENGINE ASSEMBLY FROM SUBASSEMBLIES (Cont'd)

w. Water Crossover

1. Install two gaskets (11) and water crossover (16) on cylinder heads (15).
2. Install four capscrews (10) on water crossover (16) and cylinder heads (15). Tighten capscrews (10) t o 25-37 lb-ft (34-50 N•m).
3. Connect thermostat bypass hose (12) to nipple (14) and tighten hose clamp (13).

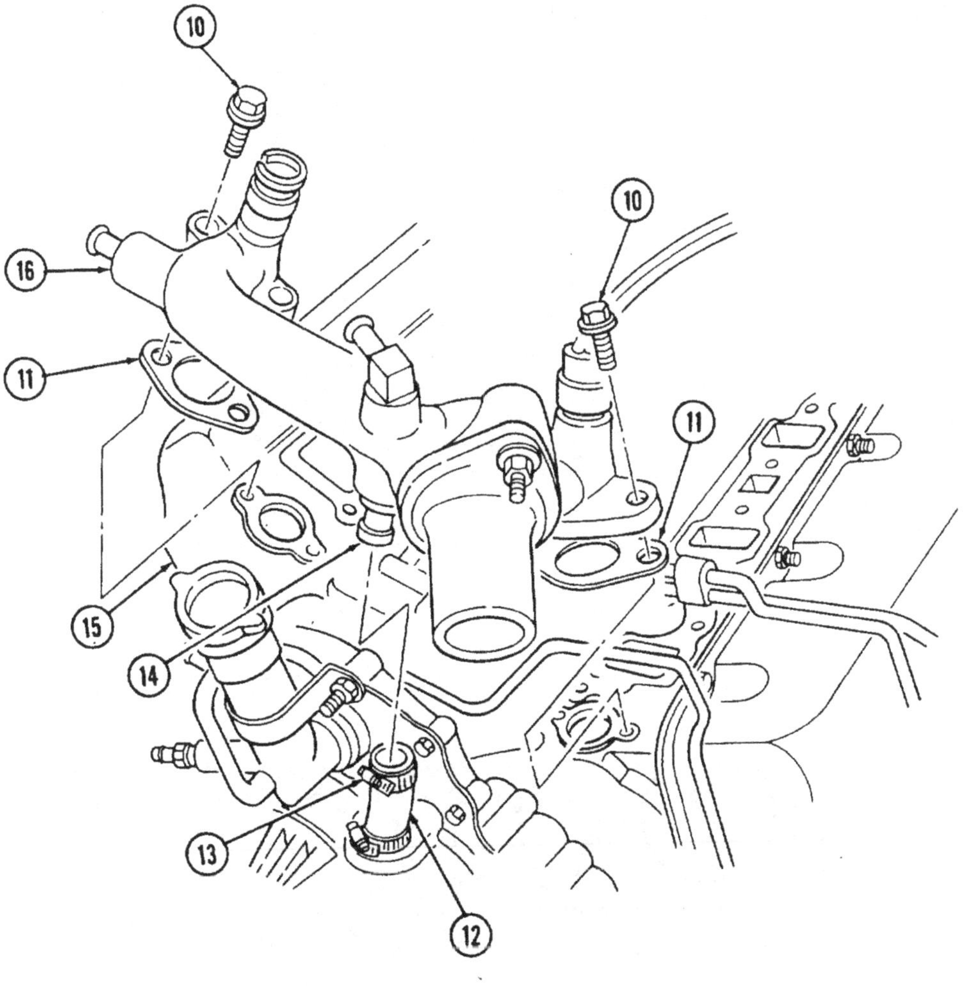

2-41. ENGINE ASSEMBLY FROM SUBASSEMBLIES (Cont'd)

x. Intake Manifold

1. Install two gaskets (2) and intake manifold (1) on cylinder heads (3).
2. Install injection line clip (10), three injection line clips (5), and fuel supply line clamp (11) on intake manifold (1), following assembly diagram shown.

NOTE

Perform step 3 for 6.2L engines. Perform step 4 for 6.5L engines.

3. Install four long studs (6), long capscrew (9), six studs (7), five capscrews (8), and sixteen washers (4), following assembly diagram shown.
4. Install four long studs (6), long capscrew (9), seven studs (7), four capscrews (8), and sixteen washers (4), following assembly diagram.

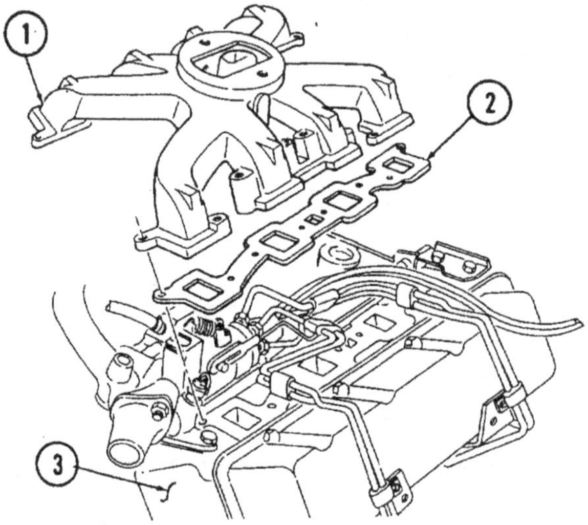

ASSEMBLY DIAGRAM

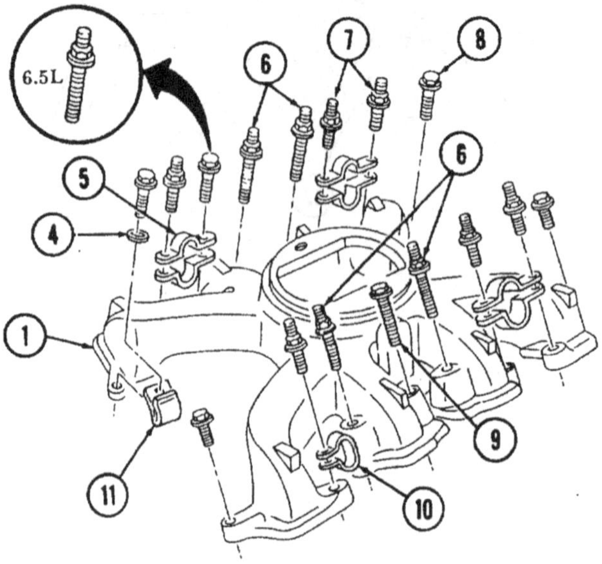

2 - 4 1ENGINE ASSEMBLY FROM SUBASSEMBLIES (Cont'd)

5. Tighten all studs (6) and (7) and capscrews (8) and (9) to 30 lb-ft (41 N·m), following torque sequence shown.

NOTE

Perform steps 6 through 9 for turbocharged engines.

6. Install gasket (17) and intake manifold (16) on cylinder head (18) with eight studs (15).
7. Beginning at center of intake manifold (16) and working toward ends, tighten studs (15) to 30 lb-ft (41 N·m).
8. Repeat steps 6 and 7 for opposite side.

NOTE

Install turbocharger before installing center intake manifold.

9. Install two gaskets (12) and center intake manifold (14) on intake manifolds (16) with six capscrews (13). Tighten capscrews (13) to 17 lb-ft (23 N·m).
10. Secure fuel supply line (21) and hose (20) to stud (7) with clamp (19).

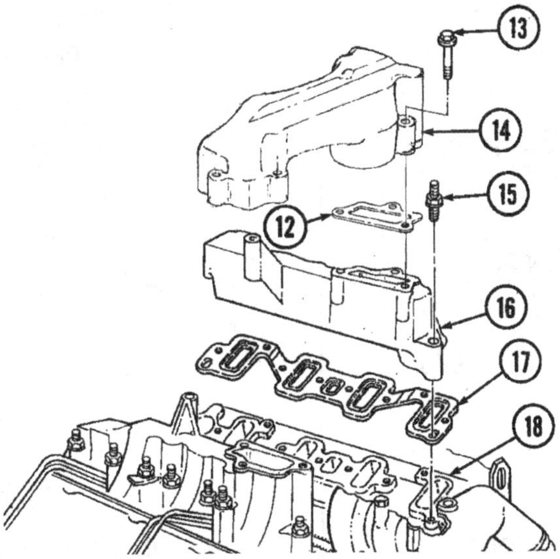

TURBOCHARGED ENGINE

TORQUE SEQUENCE

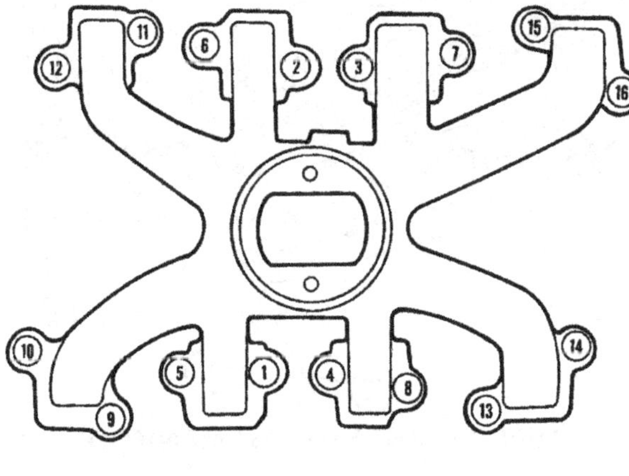

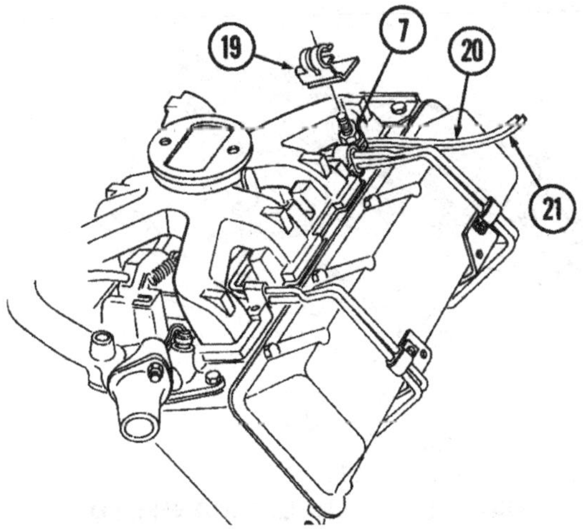

2-41 ENGINE ASSEMBLY FROM SUBASSEMBLIES (Cont'd)

y. Exhaust Manifolds

NOTE

- Early production vehicles may have a socket-head screw in place of stud.
- Turbo exhaust manifolds use hex-head capscrews and do not have gaskets or stud as indicated in steps 1 through 3.

1. Install gasket (5) and left exhaust manifold (3) on cylinder head (4) with seven washers (6) and socket-head screws (7). Using 8-mm hex-head driver, tighten socket-head screws (7) to 18-25 lb-ft (24-34 N·m).
2. Install washer (2) and stud (1) on left exhaust manifold (3) and cylinder head (4). Tighten stud (1) to 18-25 lb-ft (24-34 N·m).
3. Install gasket (8) and right exhaust manifold (9) on cylinder head (4) with eight washers (10) and socket-head screws (11). Using 8-mm hex-head driver, tighten socket-head screws (11) to 18-25 lb-ft (24-34 N·m).

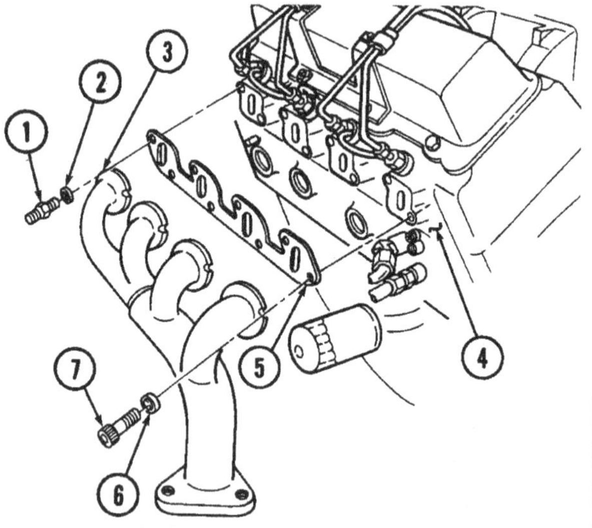

LEFT EXHAUST MANIFOLD

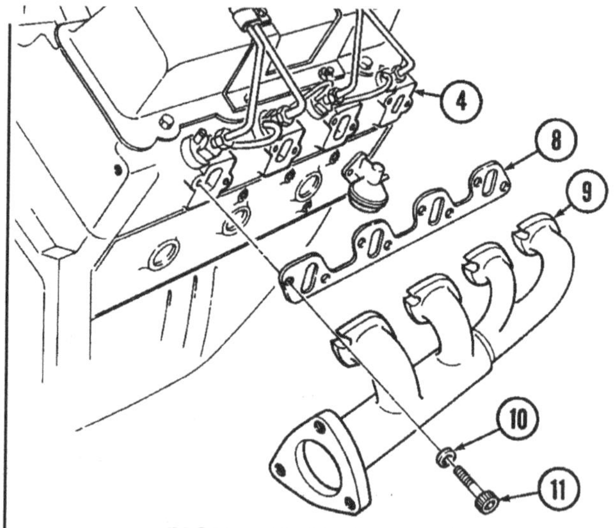

RIGHT EXHAUST MANIFOLD

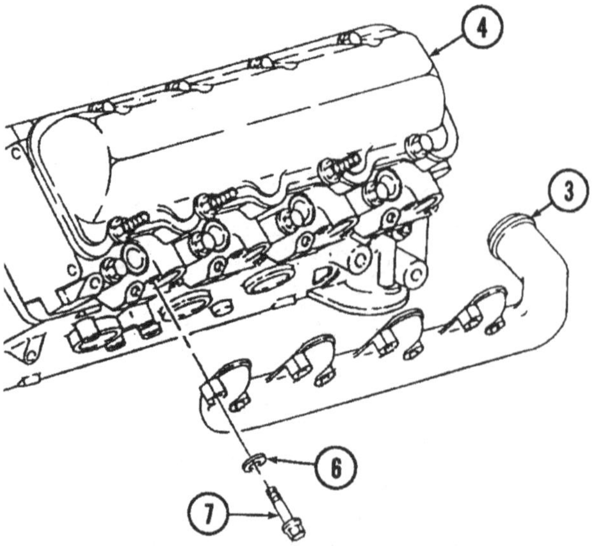

TURBO LEFT EXHAUST MANIFOLD

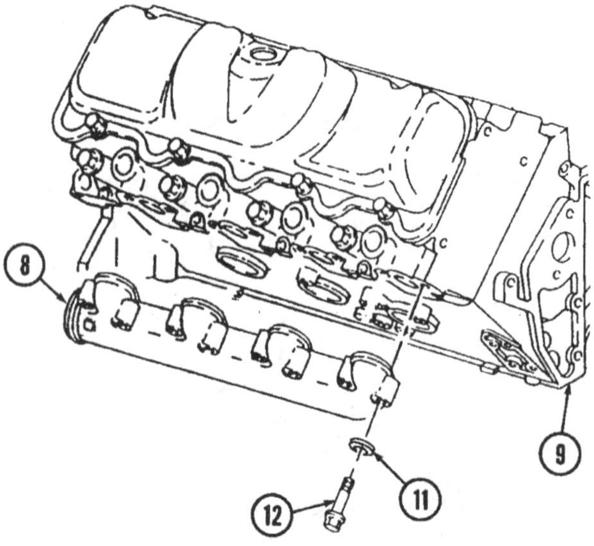

TURBO RIGHT EXHAUST MANIFOLD

2 - 4 1ENGINE ASSEMBLY FROM SUBASSEMBLIES (Cont'd)

NOTE

Perform tasks y.1 through y.4 for 6.5L turbo engines only.

y.1. Turbocharger

1. Install turbocharger oil gasket (12) and turbocharger (15) on engine block (16) with two capscrews (13).
2. Install O-ring (14) on turbocharger (15) outlet.

y.2. Manifold-to-Turbocharger Exhaust Pipe

1. Install two clamps (18) on exhaust pipe (17).
2. Install exhaust pipe (17) on turbocharger (15) and tighten clamp (18).
3. Install exhaust pipe (17) on exhaust manifold (19) and tighten clamp (18).
4. Repeat steps 1 through 3 for opposite side.

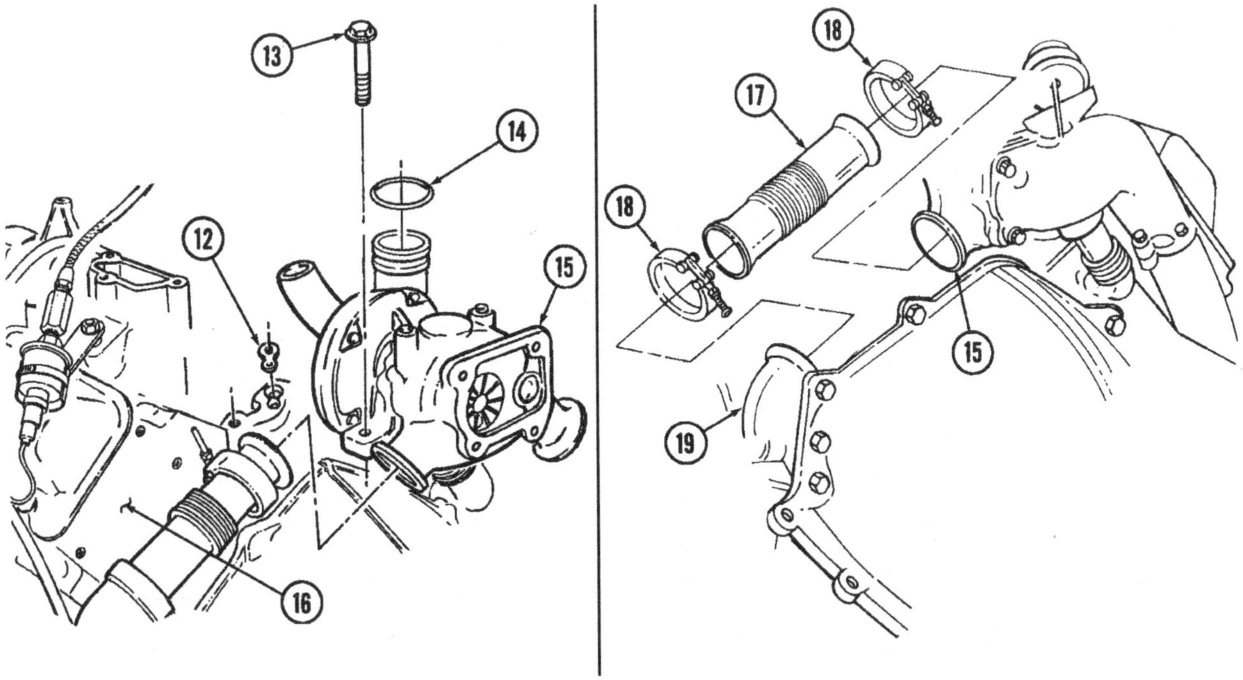

2 - 4 1ENGINE ASSEMBLY FROM SUBASSEMBLIES (Cont'd)

y.3. Wastegate Housing

Install wastegate housing (5) on turbocharger (5.5) with four capscrews (5.4).

y.4. Wastegate Actuator

1. Install wastegate actuator (2) on cylinder head (4) with capscrew (5.3).
2. Install wastegate actuator (2) on wastegate housing (5) with two capscrews (3).
3. Install actuator rod (1) on wastegate bellcrank (5.2) with clip (5.1).

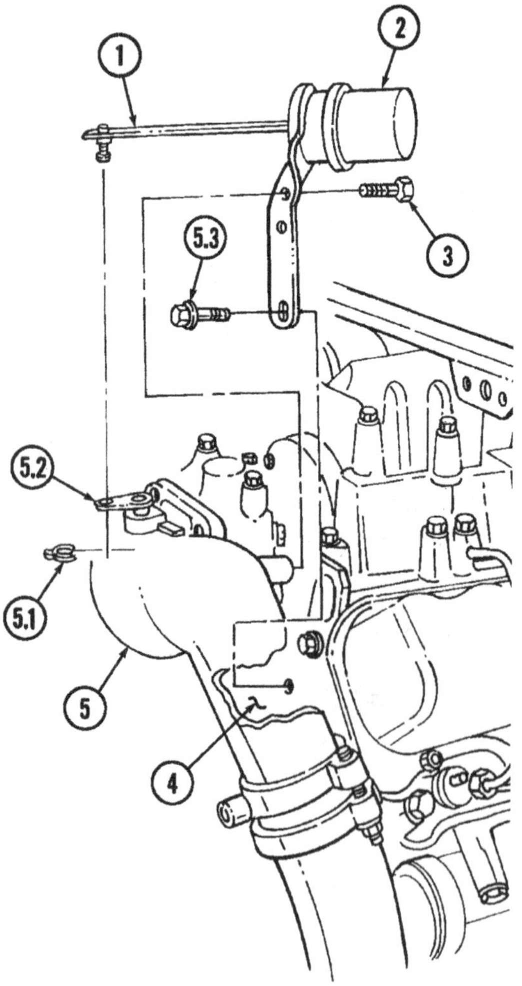

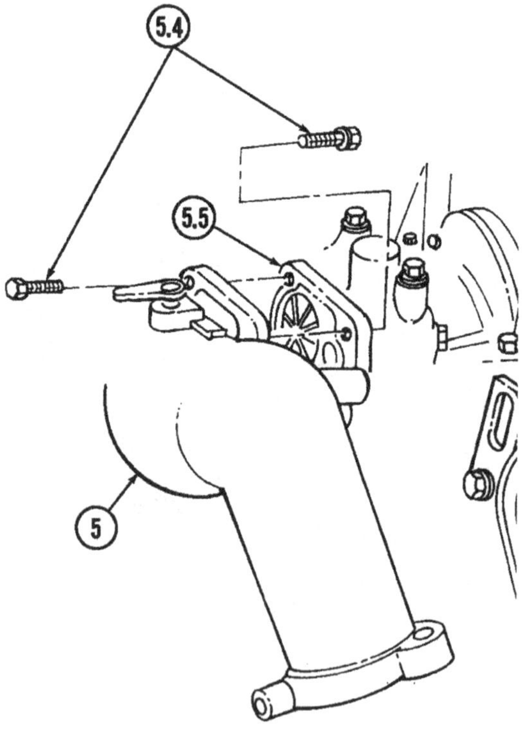

2-41. ENGINE ASSEMBLY FROM SUBASSEMBLIES (Cont'd)

z. Torsional Damper

1. Install woodruff key (9) in crankshaft (10).
2. Apply lubricating oil to seal surface (11) on torsional damper (8).
3. Install torsional damper (8) on crankshaft (10) far enough to install washer (7) and capscrew (6).
4, Hold flywheel stationary with pry bar and tighten capscrew (6) to 140-162 lb-ft (190-220 N•m).

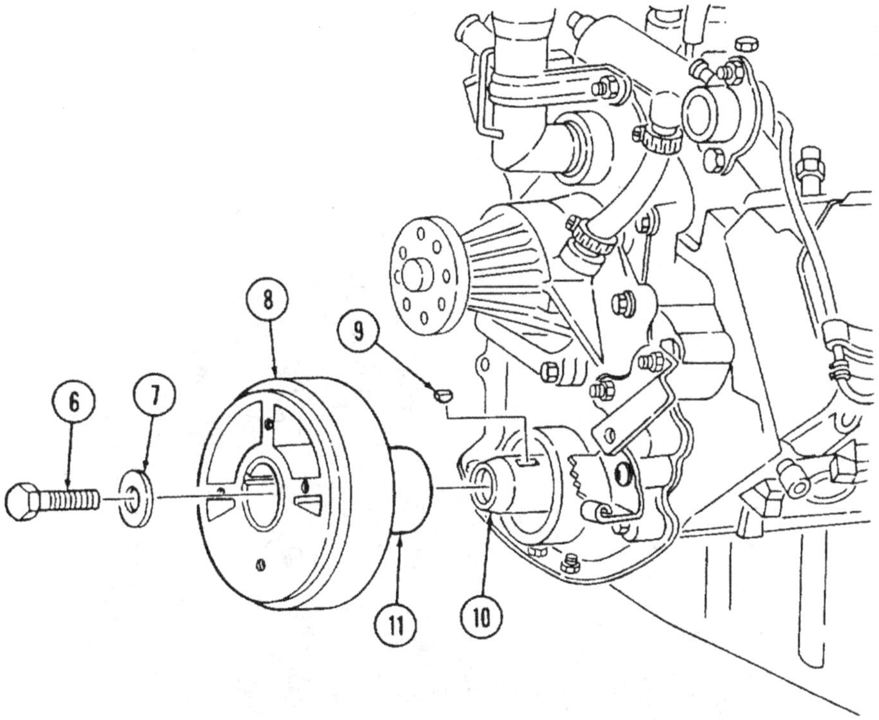

2-41 ENGINE ASSEMBLY FROM SUBASSEMBLIES (Cont'd)

aa. Crankshaft Pulley

1. Apply sealing compound to four capscrews (1).
2. Install crankshaft pulley (2) and spacer (2.1) (if removed) on torsional damper (3) with four capscrews (1).
3. Hold flywheel stationary with pry bar and tighten capscrews (1) to 48 lb-ft (65 N·m).

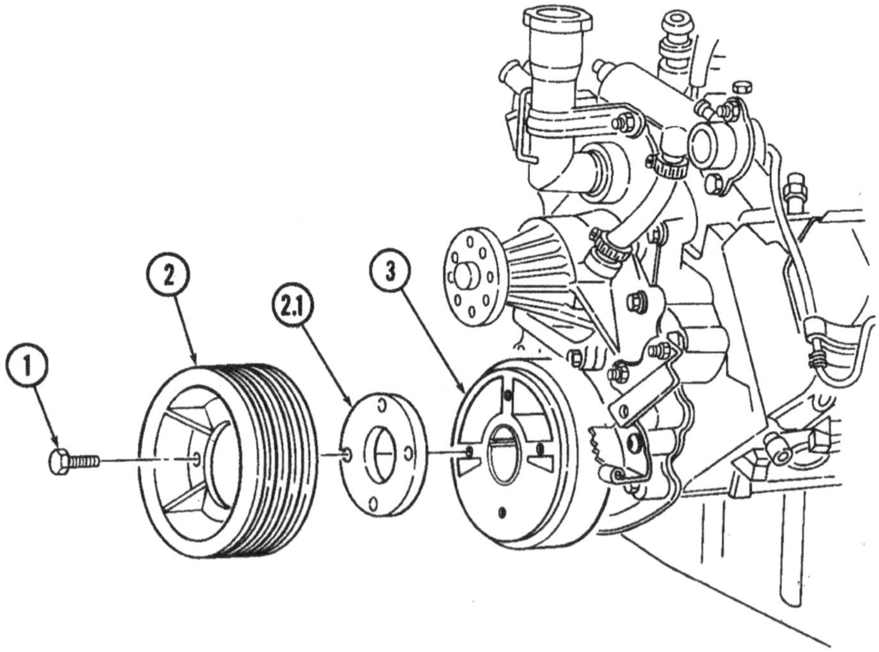

2-41. ENGINE ASSEMBLY FROM SUBASSEMBLIES (Cont'd)

bb. Fan Drive and Water Pump Pulley

1. Apply sealing compound to four socket head screws (8).
2. Install water pump pulley (6) on water pump (7) with four socket head screws (8).
3. position pry bar between water pump pulley (6) and crankshaft pulley (2) and apply pressure to prevent rotation. Using 6 mm hex head driver, tighten socket head screws (8) to 20 lb-ft (27 N•m).
4. Apply sealing compound to four socket head screws (4).
5. Install fan drive (5) on water pump pulley (6) with four socket head screws (4). Repeat step 3. Using 5/16-in. hex head driver, tighten socket head screws (4) to 45 lb-ft (61 N•m).

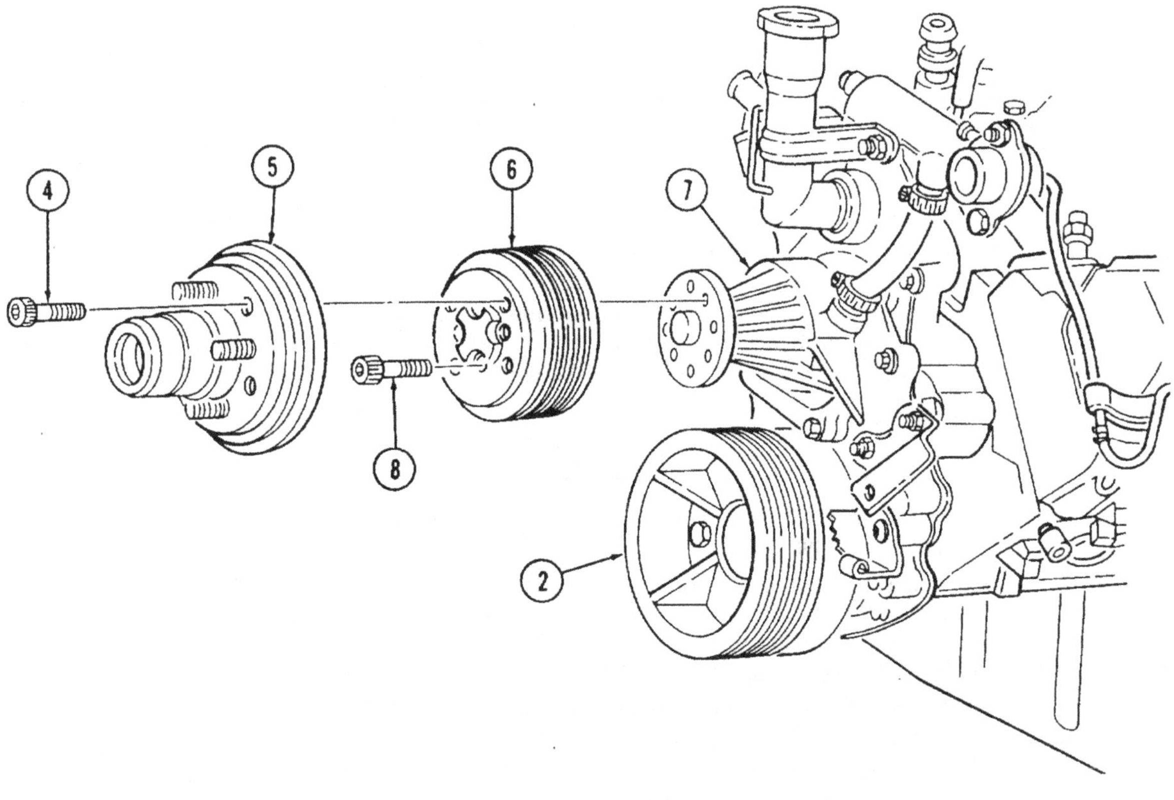

FOLLOW-ON TASKS: • Remove engine from repair stand (para. 2-14).
• Perform engine run-in (para. 2-43).

Section VI. FINAL INSPECTION

2-42. GENERAL

Following any major repair to the engine, the engine must be run-in, and tests made prior to release for issue to user. Paragraph 2-43 contains instructions for engine run-in.

2-43. ENGINE RUN-IN

This task covers:

Out of Chassis Run-In

INITIAL SETUP

Test Equipment	Manual References
Dynamometer	TM 9-2320-280-34

Out of Chassis Run-In

NOTE

- This procedure should be performed in a well-ventilated work area.
- If engine dynamometer is available, follow manufacturer's instructions for its use. If dynamometer is not available, refer to TM 9-2320-280-34 for In Chassis Run-In.

1. Connect engine to dynamometer,

CAUTION

- Operating an engine without sufficient lubricating oil will cause damage to engine.
- If any leaks or abnormal noise is noted, stop engine immediately and correct as necessary. Any abnormalities must be corrected before proceeding to the next test interval.
- Oil pressure should be 40-50 psi (276-345 kPa) under load and 10 psi (69 kPa) minimum at idle. If oil pressure does not meet specifications, stop engine immediately and correct as necessary.
- Oil temperature must not exceed 260°F (127°C) and coolant temperature must not exceed 230°F (110°C). If coolant temperature or oil temperature exceed limits, stop engine immediately and correct as necessary.

2. Start engine and run-in according to schedule in table 2-1.

2 - 43 ENGINE RUN-IN (Cont'd)

Table 2-1. 6.2 L Engine Run-In Schedule.

Speed (rpm)	Time (Minutes)	Torque	
		Lb-Ft	Nm
650	5	0	0
1200	30	80	109
2000	30	94	128
3000	40	140	190
3600	15	WOT*	
4000	0.5	WOT*	

* WOT – Wide OpenThrottle. Final test interval at 3600 rpm will verify the power level of the engine. Power output should be approximately 150 hp (112 kW) and 219 lb-ft (297 N·m) of torque.

NOTE

Power output should be 135 hp (100.8 kW) and 197.1 lb-ft (267.3 N·m) of torque when using JP-8 fuel.

Table 2-1.1. 6.5 L (NA) and 6.5 L (T) Engine Run-In Schedule.

Speed (rpm)	Time (Minutes)	Torque	
		Lb-Ft	Nm
725	5	0	0
2000	30	103	140
3000	30	133	180
3400	30	146	198
Max	30	No Load	
3400	30	WOT*	

* WOT – Wide OpenThrottle. Final test interval at 3400 rpm will verify the power level of the engine. Power output should be approximately 160 hp (119 kw) at 247 lb-ft (335 N·m) of torque for naturally aspirated engine. For turbocharged engine, power output should be 190 hp (142 kw) at 293.5 lb-ft (398 N·m) of torque.

3. Remove engine load and idle until temperatures drop.

3.1. Test governor speed with a WOT no-load test for 30 seconds.

4. Stop engine and inspect for and correct any leaks.

5. Disconnect engine from dynamometer.

Section VII. REPAIR AND REPLACEMENT STANDARDS

2-44. GENERAL

All parts, components or assemblies which do not meet the standards specified in para. 2-45 will be replaced with new material. Parts must be closely inspected to ensure there is no damage in areas not listed, which would make them unfit for further use.

2-45. REPAIR AND REPLACEMENT STANDARDS

Tables 2-2 and 2-3 list general specifications to be utilized while repairing the engine.

Table 2-2. General Repair Standards (6.2L).

	STANDARD	METRIC
Cylinder Bore		
Diameter	3.976-3.979 in.	100.990-101.066 mm
Out of Round	0.0008 in.	0.0203 mm max.
Taper - Thrust Side	0.0008 in.	0.0203 mm max.
Piston		
Piston Clearance – # 1,2, 3,4,5,6	0.0035-0.0045 in.	0.0889-0.1143 mm
Piston Clearance - #7, 8	0.004-0.005 in.	0.102-0.127 mm
Compression Rings		
Groove Clearance:		
Top	0.003-0.007 in.	0.076-0.178 mm
Second	0.002-0.003 in.	0.051-0.076 mm
Gap:		
Top	0.012-0.022 in.	0.30-0.55 mm
Second	0.030-0.039 in.	0.76-1.0 mm
Oil Ring		
Groove Clearance	0.002-0.004 in.	0.051-0.102 mm
Gap	0.010-0.020 in.	0.25-0.51 mm
Piston Pin		
Diameter	1.2203-1.2206 in.	30.9956-31.0032 mm
Clearance	0.0003-0.0006 in.	0.0076-0.0153 mm
Fit in Rod	0.0003-0.0010 in.	0.0076-0.025 mm
Camshaft		
Lift:		
Intake	0.281±.002 in.	7.137±0.05 mm
Exhaust	0.281±.002 in.	7.137±0.05 mm
Journal Diameter – # 1,2,3,4	2.164-2.166 in.	54.965-55.016 mm
Journal Diameter - # 5	2.007-2.009 in.	50.978-51.029 mm
Journal Clearance	0.001-0.004 in.	0.025-0.102 mm
Crankshaft		
Main Journal Diameter - # 1,2,3,4	2.949-2.950 in.	74.905-74.930 mm
Main Journal Diameter – # 5	2.949-2.950 in.	74.905-74.930 mm
Taper	0.0002 in. max.	0.0051 mm max.
Out of Round	0.0002 in. max.	0.0051 mm max.
Main Bearing Clearance - # 1,2,3,4,5	0.002-0.004 in.	0.051-0.102 mm
End Play	0.004-0.010 in.	0.102-0.254 mm

2-45. REPAIR AND REPLACEMENT STANDARDS (Cont'd)

Table 2-2. General Repair Standards (6.2L) (Cont'd).

	STANDARD	METRIC
Crankshaft (Cent'd)		
Crankpin:		
Diameter	2.398-2.399 in.	60.909-60.935 mm
Taper	0.0002 in. max.	0.0051 mm max.
Out of Round	0.0002 in. max.	0.0051 mm max.
Rod Bearing Clearance	0.002-0.004 in.	0.051-0.102 mm
Rod Side Clearance	0.007-0.025 in.	0.178-0.635 mm
Valve Train		
Rocker Arm Ratio	1.5:1	1.5:1
Face Angle	45°	45°
Seat Angle	46°	46°
Seat Runout	0.002 in.	0.05 mm
Seat Width:		
Intake	0.035-0.060 in.	0.89-1.52 mm
Exhaust	0.062-0.093 in.	1.57-2.36 mm
Stem Clearance:		
Intake	0.001-0.003 in.	0.025-0.076 mm
Exhaust	0.001-0.003 in.	0.025-0.076 mm
Valve Spring Force		
Closed	80 lb @ 1.8 in.	356 N @ 46.0 mm
Open	230 lb@ 1.39 in.	1023 N @ 35.3 mm
Installed Height	1.8 in.	46.0 mm
Timing Chain Deflection		
New Chain	0.5 in. max.	12.7 mm max.
Old Chain	0.81 in. max.	20.6 mm max.

Table 2-3. General Repair Standards (6.5L).

	STANDARD	METRIC
Cylinder Bore		
Diameter	3.978-3.981 in.	101.041-101.12 mm
Out of Round	0.0008 in.	0.0203 mm max.
Taper - Thrust Side	0.0008 in.	0.0203 mm max.
Piston		
Piston Clearance - # 1,2,3,4, 5,6	0.0035-0.0045 in.	0.0889-0.1143 mm
Piston Clearance - # 7,8	0.004-0.005 in.	0.102-0.127 mm
Compression Rings		
Groove Clearance:		
Top	0.003-0.007 in.	0.076-0.178 mm
Second	0.002-0.003 in.	0,051-0.076 mm
Gap:		
Top	0.012-0.022 in.	0.30-0.55 mm
Second	0.030-0.039 in.	0.76-1.0 mm
Oil Ring		
Groove Clearance	0.002-0.004 in.	0.051-0.102 mm
Gap	0.010-0.020 in.	0.25-0.51 mm

2-45. REPAIR AND REPLACEMENT STANDARDS (Cont'd)

Table 2-3. General Repair Standards (6.5L) (Cont'd).

	STANDARD	METRIC
Piston Pin		
Diameter	1.2203-1.2206 in.	30.9956-31.0032 mm
Clearance	0.0003-0.0006 in.	0.0076-0.0153 mm
Fit in Rod	0.0003-0.0010 in.	0.0076-0.025 mm
Camshaft		
Lift:		
Intake	0.281±0.002 in.	7.137±0.05 mm
Exhaust	0.281±.002 in.	7.137±0.05 mm
Journal Diameter - # 1,2,3,4	2.164-2.166 in.	54.965-55.016 mm
Journal Diameter – # 5	2.007-2.009 in.	50.978-51.029 mm
Journal Clearance	0.001-0.004 in.	0.025-0.102 mm
Crankshaft		
Main Journal Diameter - # 1,2, 3,4	2,949-2.950 in.	74.905-74.930 mm
Main Journal Diameter – # 5	2.949-2.950 in.	74.905-74.930 mm
Taper	0.0002 in. max.	0.0051 mm max.
Out of Round	0.0002 in. max.	0.0051 mm max.
Main Bearing Clearance – # 1,2,3,4,5	0.002-0.004 in.	0.051-0.102 mm
End Play	0.004-0.010 in.	0.102-0.254 mm
Crankpin:		
Diameter	2.398-2.399 in.	60.909-60.935 mm
Taper	0.0002 in. max.	0.0051 mm max.
Out of Round	0.0002 in. max.	0.0051 mm max.
Rod Bearing Clearance	0.002-0.004 in.	0.051-0.102 mm
Rod Side Clearance	0.007-0.025 in.	0.178-0.635 mm
Valve Train		
Rocker Arm Ratio	1.5:1	1.5:1
Face Angle	45°	45°
Seat Angle	46°	46°
Seat Runout	0.002 in.	0.05 mm
Seat Width:		
Intake	0.035-0.060 in.	0.89-1.52 mm
Exhaust	0.062-0.093 in.	1.57-2.36 mm
Stem Clearance:		
Intake	0.001-0.003 in.	0.025-0.076 mm
Exhaust	0.001-0.003 in.	0.025-0.076 mm
Valve Spring Force		
Closed	80 lb @ 1.8 in.	356 N @ 46.0 mm
Open	230 lb@ 1.39 in.	1023 N @ 35.3 mm
Installed Height	1.8 in.	46.0 mm
Timing Chain Deflection		
New Chain	0.5 in. max.	12.7 mm max.
Old Chain	0.81 in. max.	20.6 mm max.

2-45 REPAIR AND REPLACEMENT STANDARDS (Cont'd)

Table 2-4. 6.2L Diesel Engine Piston-Cylinder Specification.

PISTON	CODE	PISTON SIZE	CYLINDER SIZE
Standard PN 23500391	S5	3.9749 in. to 3.9754 in. (100.962-100.975 mm)	3.9778 in. to 3.9784 in. (101.036-101.051 mm)
Standard Hi-Limit PN 23500392	S7	3.9759 in. to 3.9765 in. (100.988-101.003 mm)	3.9788 in. to 3.9794 in. (101.061-101.077 mm)
Oversized PN 23500393	2	4.0044 in. to 4.0050 in. (101.712-101.727 mm)	4.0074 in. to 4.0079 in. (101.788-101.801 mm)
Oversized PN 23500393	3	4.0050 in. to 4.0055 in. (101.727-101.740 mm)	4.0079 in. to 4.0085 in. (101.801-101.816 mm)

Use standard ring set, part number (11862) 15537018, NSN 2815-01-163-7838, with standard and standard Hi-Limit pistons.

Oversized piston (0.75 mm) ring set part number is (11862) 15537020, NSN 2815-01-163-9999.

Table 2-5. 6.5L Diesel Engine Piston-Cylinder Specification.

PISTON	PISTON SIZE	CYLINDER SIZE
Standard (11862) 12550059	4.0529 in. to 4.0534 in. (102.944-102.956 mm)	4.0571 in. to 4.0576 in. (103.050-103.063 mm)
0.50 mm Oversized (11862) 12550062	4.0739 in. to 4.0745 in. (103.477-103.492 mm)	4.0782 in. to 4.0787 in. (103.586-103.599 mm)

Piston ring set part numbers are:
Standard – (11862) 12510752
0.50 mm Oversized – (11862) 12510753

The 6.5L engine does not have a standard Hi-Limit piston.

APPENDIX A
REFERENCES

A-1. SCOPE

This appendix lists all forms, field manuals, and technical manuals for use with this manual.

A-2. PUBLICATIONS INDEX

The following index should be consulted frequently for latest changes or revisions and for new publications relating to materiel covered in this manual.

Consolidated Index of Army Publications and Blank Forms . . . . DA Pam 25-30

A-3. FORMS

Recommended Changes to Publications and Blank Forms . . . . DA Form 2028
Recommended Changes to Equipment Technical Publications . . . . DA Form 2028-2
Hand Receipt/Annex Number . . . . DA Form 2062
Exchange Tag . . . . DA Form 2402
Equipment Inspection and Maintenance Worksheet . . . . DA Form 2404
Maintenance Request . . . . DA Form2407
Preventive Maintenance Schedule and Record . . . . DD Form 314
Quality Deficiency Report . . . . SF 368
Processing and Reprocessing Record of Shipping, Storage, and Issue of Vehicles and Spare Engines . . . . DD Form 1397
The Army Maintenance Management System (TAMMS) . . . . DA Pam 738-750

A-4. FIELD MANUALS

Operation and Maintenance of Ordnance Material in Cold Weather (0° to -65°F) . . . . FM 9-207
First Aid for Soldiers . . . . FM 21-11
Basic Cold Weather Manual . . . . FM 31-70
Northern Operations . . . . FM 31-71
Army Motor Transport Units and Operations . . . . FM 55-30
Mountain Operations . . . . FM 90-6

APPENDIX A (Cont'd)

A-5. TECHNICAL MANUALS

Procedures for Destruction of Equipment to Prevent Enemy Use . . . . TM 750-244-6
Marking, Packaging, and Supplies and Equipment:
General Packaging Instructions for Field Use . . . . TM 746-10
Operator's Manual: Welding Theory and Application . . . . TC 9-237
Direct Support and General Support Maintenance
Repair Parts and Special Tools List Engine, Diesel, 6.2 Liter . . . . TM 9-2815-237-34P
Unit Maintenance, M998 Series Vehicles . . . . TM 9-2320-280-20
Unit Maintenance Repair Parts and Special Tools List . . . . TM 9-2320-280-24P
Direct Support and General Support Maintenance,
M998 Series Vehicles . . . . TM 9-2320-280-34
Inspection, Care, and Maintenance of Antifriction Bearings . . . . TM 9-214
Materials Used for Cleaning, Preserving, Abrading, and
Cementing Ordnance Materiel and Related Materials
(Including Chemicals) . . . . TM 9-247

A-6. TECHNICAL BULLETINS

Equipment Improvement Report and Maintenance Digest . . . . TB 43-0001-62
Use of Antifreeze Solutions, and Cleaning Compounds
in Engine Cooling Systems . . . . TB 750-651
Calibration and Repair Requirements for the Maintenance
of Army Materiel . . . . TB 43-180

A-7. ARMY REGULATIONS

Identification and Distribution of DA Publications and
Issue of Agency and Command Administration Publications . . . . AR 310-2
Dictionary of United States Army Terms . . . . AR 310-25

APPENDIX B
EXPENDABLE/DURABLE SUPPLIES AND MATERIALS LIST

Section I. INTRODUCTION

B-1. SCOPE

This appendix lists expendable/durable supplies and materials you will need to maintain pre 1990 and 1990 and above 6.2L engines and 6.5L engines. These items are authorized to you by CTA 50-970, Expendable Items (except Medical, Class V, Repair Parts, and Heraldic Items).

B-2. EXPLANATION OF COLUMNS

a. Column (1) - Item Number. This number is assigned to each entry in the listing and is referenced in the "initial Setup" of applicable tasks under the heading of Materials/Parts/"

b. Column (2)- Level. This column identifies the lowest level of maintenance that requires the listed item.

C - Operator/Crew
O - Unit Maintenance
F - Direct Support Maintenance
H - General Support Maintenance

c. Column (3) - National Stock Number. This is the National stock number assigned to the item; use it to request or requisition the item.

d. Column (4) - Description. Indicates the Federal item name and, if required, a description to identify the item. The last line for each item listing indicates the Commercial and Government Entity Code (CAGEC) in parentheses followed by the part number.

e. Column (5) - Unit of Measure (U/M). Indicates the measure used in performing the actual maintenance function. This measure is expressed by an alphabetical abbreviation (QT, GAL.). If the unit of mesdure differs from the unit of issue, requisition the lowest unit of issue that will satisfy your requirements.

Section II. EXPENDABLE/DURABLE SUPPLIES AND MATERIALS LIST

(1) ITEM NUMBER	(2) LEVEL	(3) NATIONAL STOCK NUMBER	(4) DESCRIPTION	(5) U/M
1	O		ADHESIVE: methyl 1-2-cyanoacrylate, medium viscosity (80244) MIL-A-46050, type I, class 2	
		8040-00-826-3535	16 Ounce Bottle	OZ
		8040-01-043-7537	1 Pound Container	LB
2	H		ACETONE, TECHNICAL: (81348) O-A-51	
		6810-0-223-2739	1 Pint Can	PT'
3	F		ANTISEIZE COMPOUND: mica-base (80244) MIL-A-13881	
		8030-00-753-4953	1 Pound Container	LB
4	F		CALIBRATING FLUID:	
		4910-00-779-6851	5 Gallon Drum	GAL.
5	F		CLOTH: abrasive, crocus (58536) A-A-1206	
		5350-00-221-0872	50 Sheet Package	SH
		5350-00-268-3116	50 Yard Roll	YD
6	C		DETERGENT:	
		7930-00-282-9699	1 Gallon Container	GAL.
7	C		DRYCLEANING SOLVENT: (81348) P-D-680, type II	
		6850-00-110-4498	1 Pint Can	PT
		6850-00-274-5421	5 Gallon Drum	GAL.
		6850-00-285-8011	55 Gallon Drum	GAL.
		6850-00-637-6135	Bulk	GAL.
8	C		FUEL OIL: diesel, regular, DF-2 (81348) VV-F-800	
		9140-00-286-5295	5 Gallon Drum	GAL.
		9140-00-286-5296	55 Gallon Drum	GAL.
9	H		GAUGE: bearing clearance (77220) PG-1	
		5210-00-640-6177	Box	EA
10	F		SEALER: anaerobic gasket (05972) 51831	
		8030-01-3743504	50 Milliliter Tube	ML
		8030-01-3742338	300 Milliliter Cartridge	ML
11	F		GREASE: lithium base w/molybdenum disulf (60218) LS 2267	
		9150-01-015-1542	14.5 Ounce cartridge	OZ

Section II. EXPENDABLE/DURABLE SUPPLIES AND MATERIALS LIST (Cont'd)

(1) ITEM NUMBER	(2) LEVEL	(3) NATIONAL STOCK NUMBER	(4) DESCRIPTION	(5) U/M
12	H		LUBRICANT: oil seal (87460) 22204	
		N/A	8 Ounce Bottle	OZ
13	c		LUBRICATING OIL: internal combustion engine, tactical service, OE/HDO 10 (81349) MIL-L-2104	
		9150-00-189-6727	1 Quart Can	QT
		9150-00-186-6618	5 Gallon Drum	GAL.
		9150-00-191-2772	55 Gallon Drum	GAL.
14	C		LUBRICATING OIL: internal combustion engine, tactical service, OE/HDO 30 (81349) MIL-L-2104	
		9150-00-186-6681	1 Quart Can	QT
		9150-00-188-9858	5 Gallon Drum	GAL.
		9150-00-189-6729	55 Gallon Drum	GAL.
15	C		RAG wiping, cotton and cotton-synthetic (58536) A-A-531	
		7920-00-205-1711	50 Pound Bale	LB
16	O		SEALANT: silicone, RTW (11862) 1052734	
		8030-01-159-4844	50 Milliliter Tube	ML
17	O		SEALING COMPOUND: pipe, anaerobic (80244) MIL-S-46163, Type I, Grade K	
		8030-01-054-0740	10 CC Bottle	CC
18	O		SEALING COMPOUND: thread-locking, anaerobic (80244) MIL-S-46163, Type I, Grade K	
		8030-00-148-9833	10 CC Bottle	CC
19	F		SEALING COMPOUND: thread-locking, penetrating, anaerobic (80244) MIL-S-46163, Type 111, Grade R	
		8030-00-111-2763	10 CC Bottle	CC
20	H		THINNER, LACQUER cellulose nitrate (58536) A-A-857	
		8010-00-160-5787	1 Gallon Can	GAL.
21	H		SEALING COMPOUND: high temp, cylindrical part bonding (05972) 62040	
		8030-01-268-5917	50 CC Bottle	CC

APPENDIX C
ILLUSTRATED LIST OF MANUFACTURED ITEMS

Section I. INTRODUCTION

C-1. SCOPE

This appendix includes complete instructions for making items authorized to be manufactured or fabricated at organizational, direct support, and general support maintenance.

C-2. GENERAL

a. A part number index in alphanumeric order is provided for cross-referencing the part number of the item to be manufactured to the figure which covers fabrication criteria.

b. All bulk materials needed for manufacture of an item are listed by part number or specification number in a tabular list on the illustration.

Table C-1. Manufactured Items Part Number Index.

PART NUMBER	FIGURE NUMBER	PART NUMBER	FIGURE NUMBER
14066305	C-1	23500084	C-4
23500023	C-2	9439363	C-5
23500024	C-3		

Section II. ILLUSTRATED MANUFACTURING INSTRUCTIONS

MATERIAL BLOCK		
STOCK SIZE	DESCRIPTION	SPECIFICATION
0.156-in. INSIDE DIA., 0.281-in. OUTSIDE DIA.	HOSE, NON-METALLIC	N/A

FUEL DRAIN BACK HOSE		
HOSE PART NUMBER	CUT LENGTH (INCHES)	MANUFACTURED FROM NSN
14066305	7	4720-01-184-0433

Figure C-1. Fuel Drain Back Hose.

INSTRUCTIONS:

Cut hose to length shown.

Section II. ILLUSTRATED MANUFACTURING INSTRUCTIONS (Cont'd)

MATERIAL BLOCK		
STOCK SIZE	DESCRIPTION	SPECIFICATION
0.359-0.391-in. INSIDE DIA., 0.648-0.602-in. OUTSIDE DIA.	HOSE, NON-METALLIC	GM-6165-M

FUEL SUPPLY HOSE		
HOSE PART NUMBER	CUT LENGTH (INCHES)	MANUFACTURED FROM NSN
23500023	9-13/16	4720-01-159-5796

Figure C-2. Fuel Supply Hose.

INSTRUCTIONS:

Cut hose to length shown.

MATERIAL BLOCK		
STOCK SIZE	DESCRIPTION	SPECIFICATION
0.234-0.266-in. INSIDE DIA., 0.523-0.477-in. OUTSIDE DIA.	HOSE, NON-METALLIC	GM-6165-M

FUEL HOSE		
HOSE PART NUMBER	CUT LENGTH (INCHES)	MANUFACTURED FROM NSN
23500024	22	4720-01-155-7784

Figure C-3. Fuel Hose.

INSTRUCTIONS:

Cut hose to length shown.

Section II. ILLUSTRATED MANUFACTURING INSTRUCTIONS (Cont'd)

MATERIAL BLOCK		
STOCK SIZE	DESCRIPTION	SPECIFICATION
0.750-in. INSIDE DIA., 1.00-in. OUTSIDE DIA.	HOSE, NON-METALLIC	N/A

THERMOSTAT BYPASS HOSE		
HOSE PART NUMBER	CUT LENGTH (INCHES)	MANUFACTURED FROM NSN
23500084	5	4720-00-845-9211

Figure C-4. Thermostat Bypass Hose.

INSTRUCTIONS:

Cut hose to length shown.

MATERIAL BLOCK		
STOCK SIZE	DESCRIPTION	SPECIFICATION
0.25-in. INSIDE DIA., 0.50-in. OUTSIDE DIA.	HOSE, NON-METALLIC	GM-6463-M

FUEL DRAIN BACK HOSE		
HOSE PART NUMBER	CUT I.ENGTH (INCHES)	MANUFACTURED FROM NSN
9439363	7	4720-01-163-7833

Figure C-5. Fuel Drain Back Hose.

INSTRUCTIONS

Cut hose to length shown.

APPENDIX D
MANDATORY REPLACEMENT PARTS

Section I. INTRODUCTION

D-1. SCOPE

This appendix lists mandatory replacement parts you will need to maintain the 6.2L and 6.5L diesel engines.

D-2. EXPLANATION OF COLUMNS

a. Column (1) - Item Number. This number is assigned to each entry in the listing and is referenced in the "Initial Setup" of applicable tasks under the heading of "Materials/Parts".

b. Column (2) - Nomenclature. Name or identification of the part.

c. Column (3) - Part Number. The manufacturer's part number.

d. Column (4) - National Stock Number. The national stock number of the part.

Section II. MANDATORY REPLACEMENT PARTS

(1) ITEM NO	(2) NOMENCLATURE	(3) PART NUMBER	(4) NATIONAL/NATO STOCK NUMBER
1	Bearing Half, Connecting Rod	18009094	3120-01-155-3317
2	Bearing, Roller	23101	3110-01-188-7682
3	Bushing Sleeve	15093	3128-00-993-4152
4	Cap, Plastic	26961	5340-01-415-9633
5	Dust Cap, Protective	10000462	5348-00-449-6408
6	Filter, Oil	6438148	2940-00-082-6034
7	Gasket, Oil Filter Adapter Bolt	5577933	5330-01-184-6500
8	Gasket Kit, Intake Manifold	15633464	5330-01-157-0856
8.1	Gasket, Center Intake Manifold (Turbocharged Engine)	10211661	5330-01-437-0547
9	Gasket Set, Oil Pan	OS30442A	5330-01-310-6780
10	Gasket, Adapter Plate	10108438	5330-00-830-1745
11	Gasket, Cylinder Head	10149600	5330-01-367-2308
12	Gasket, Exhaust Manifold	5574856	5330-01-194-0472
13	Gasket, Fuel Pump	10114141	5330-01-112-1533
14	Gasket, Governor Cover	27244	5330-01-234-2615
15	Gasket, Injection Pump	10137486	5330-01-150-5944
16	Gasket, Injector Nozzle	14025557	5330-01-150-1215
17	Gasket, Oil Pump Drive	14022649	5330-01-156-5147
18	Gasket, Rocker Arm Cover	91599	5330-01-372-0636
19	Gasket, Turbocharger Oil	10218787	
20	Gasket, Water Crossover	10137488	5330-01-149-0874
21	Gasket, Water Outlet Thermostat	12551591	5330-01-445-5459
22	Gasket, Water Pump	10137492	5330-01-147-9808
23	Lining, Friction	4026-38368-01	2930-01-189-8643
24	Locknut	3029-01371-01	5318-01-194-0481
25	Lockwasher	11582	5310-01-213-9964
26	Lockwasher	12049	5310-00-209-2947
26.1	Lockwasher	MS35338-46	5310-00-637-9541
27	Overhaul Kit, Injection Pump	30405	2910-01-299-0473
28	O-ring	5577933	5330-01-184-6500
29	O-ring, Cylinder	3018-01265-01	5330-01-192-8892
30	O-ring, Oil Filter Adapter	MS28775-236	5330-00-815-1712
31	O-ring, Oil Filter Adapter Bolt	MS28775-020	5330-00-585-7723
32	O-ring, Shaft, Front (Small)	9002-00181-48	5330-01-194-8966
33	O-ring, Shaft, Rear (Large)	9002-00741-58	5330-01-195-1500
34	Pilot, Tube	27106	2910-01-210-6938
35	Plug, Adjusting	15228	4930-00-459-6077
36	Retainer	23500076	3040-01-212-7616
37	Ring Set, Piston (Standard) (6.2L)	15537018	2815-01-163-7838

Section II. MANDATORY REPLACEMENT PARTS (Cont'd)

(1) ITEM NO	(2) NOMENCLATURE	(3) PART NUMBER	(4) NATIONAL/NATO STOCK NUMBER
38	Ring Set, Piston (Standard) (6.5L)	12510752	
39	Ring, Retaining	121-1	5365-00-598-1428
40	Ring, Retaining	23566	5365-01-188-8962
41	Seal, Drain Plug	27609	5330-01-133-2778
42	Seal, Dust	3018-01339-01	5330-01-247-8438
43	Seal, Front Cover Oil	3860095	5330-00-110-8437
44	Seal, Rear Main Oil (6.2L)	23500139	5530-01-249-1629
45	Seal, Rear Main Oil (6.5L)	23503969	5330-01-378-8577
46	Seal, Rear Oil Pan	14022683	5330-01-150-7744
47	Seal, Valve	3835333	2805-00-752-0158
48	Seal, Valve Stem (Exhaust only)	23502552	5330-01-315-7047
49	Snapring	23500298	5365-01-234-0447
50	Spring	4088-35869-01	5360-01-190-6214
51	Spring, Helical	28397	5360-01-318-1893
52	Stud, Governor	23352	5307-01-188-9217
53	Valve Stop	26070	
54	Woodruff Key	106751	5315-00-012-4548
55	Capscrew, Hex-Head, 5/16-18x1-in.	B1821BH031C100N	5306-00-226-4827
56	Capscrew, Hex-Head, 3/8-16x1-in.	B1821BH038C100N	5305-00-068-0510

APPENDIX E
TOOL IDENTIFICATION LIST

Section I. INTRODUCTION

E-1. SCOPE

This appendix lists common tools, special tools, and test equipment needed to maintain the 6.2L and 6.5L diesel engines.

E-2. EXPLANATION OF COLUMNS

a. Column (1) - Item Number. This number is assigned to each entry in the listing and is referenced in the "Initial Setup" of applicable tasks under the headings of "Test Equipment," Tools," and "Special Tools."

b. Column (2) - Item Name. Name or identification of the tool or test equipment.

c. Column (3) - National Stock Number. The national stock number of the tool or test equipment.

d. Column (4) - Part Number. The manufacturer's part number.

e. Column (5) - Reference. This column lists the references used to identify and illustrate the tools and test equipment needed for maintaining the 6.2L and 6.5L diesel engines.

Section II. TOOL IDENTIFICATION LIST

(1) Item Number	(2) Item Name	(3) National Stock Number	(4) Part Number	(5) Reference
1	Tool kit, general mechanic's automotive	5180-00-177-7033		SC 5180-90-CL-N26
2	Engine repair stand	4910-00-506-0037	1725A	TM 9-2815-237-34P
3	Engine lifting sling	4910-01-193-7808	J33139	TM 9-2815-237-34P
4	Hydraulic valve lifter remover	5120-01-209-6870	J29834	TM 9-2815-234-34P
5	Injection nozzle socket	5120-01-171-5233	J29873	TM 9-2815-237-34P
6	Glow plug socket	5120-00-277-1463	FUS121	TM 9-2815-237-34P
7	Hex head driver, 6 mm	5120-01-055-1308	FAM6	TM 9-2815-237-34P
8	Hex head driver, 8 mm	5120-01-053-4159	FAM8	TM 9-2815-237-34P
9	Hex head driver, 5/16 in.	5120-00-683-8602	FA10B	TM 9-2815-237-34P
10	Dial indicator	5210-00-277-8840		SC 4910-95-CL-A02
11	Cylinder ridge reamer	5110-00-237-8598		SC 4910-95-CL-A02
12	Camshaft bearing remover and installer	5120-01-206-3818	J35178	TM 9-2815-237-34P
13	Feeler gauge	5210-01-045-3526		SC 4910-95-CL-A31
14	Cylinder bore gauge	5210-00-494-1774		SC 4910-95-A63
15	Straightedge	6675-00-224-8807		SC 4910-95-A63
16	Micrometer	5210-00-5547134		SC 4910-95-A63
17	Cylinder boring machine	4910-00-473-6361		SC 4910-95-A63
18	Cylinder honing unit	5130-00-991-0699		SC 491 0-95-A63
19	Puller	5120-00-595-9305		SC 4910-95-CL-A31
20	Piston ring expander	5120-00-857-3190		SC 4910-95-CL-A31
21	"V" blocks	3460-00-725-5076		SC 4910-95-CL-A02
22	Valve spring compressor	5120-00-239-8686		SC 4910-95-A63
23	Bench grinder	3415-00-517-7744		SC 4910-95-CL-A31
24	Rotary wheel wire brush	5130-00-473-6444		SC 4910-95-CL-A31
25	Electric drill	5130-00-889-8993		SC 4910-95-CL-A31
26	Rotary end wire brush	5130-00-293-2409		SC 4910-95-CL-A31
27	Valve spring tester	6635-00-641-7346		SC 4910-95-A63
28	Valve face grinding machine	4910-00-540-4679		SC 4910-95-A63
29	Valve seat grinding kit	4910-00-473-6437		SC 4910-95-A63

Section II. TOOL IDENTIFICATION LIST (Cont'd)

(1) Item Number	(2) Item Name	(3) National Stock Number	(4) Part Number	(5) Reference
30	Hammer, soft-head	5120-01-071-5356		SC 4910-95-A62
31	Injection nozzle tester	4910-00-255-8641		SC 4910-95-CL-A01
32	Multimeter	6625-01-139-2512	AN/PSM-45	SC 4910-95-CL-A74
33	Rear main seal installer (6.2L)	5120-01-210-8792	J33153	TM 9-2815-237-34P
34	Rear main seal installer (6.5L)	5120-01-422-0334	J39084	TM 9-2815-237-34P
35	Piston ring compressor	5120-00-250-6055		SC 4910-95-CL-A02
36	Crowfoot, 16 mm	5120-01-242-8165	J35159	TM 9-2815-237-34P
37	Crowfoot, 19 mm	5120-01-230-9421	J35160	TM 9-2815-237-34P
38	Fuel injection nozzle, remover	5120-01-171-5233	J29873	TM 9-2815-237-34P
39	Drive shaft bearing, installer	5120-01-208-7771	23805	TM 9-2815-237-34P
40	Drive shaft seal, installer	5120-01-208-7752	22727	TM 9-2815-237-34P
41	Throttle and shutoff shaft bushing kit	5120-01-189-0448	18411	TM 9-2815-237-34P
42	Torx drive, T27, socket	5120-01-078-1899	22939	TM 9-2815-237-34P
43	Governor weight support	5120-01-197-0236	16313	TM 9-2815-237-34P
44	End cap wrench	5120-01-207-5563	20548	TM 9-2815-237-34P
45	Roller-to-roller setting tool	5120-01-200-4526	19969	TM 9-2815-237-34P
46	Holding fixture	5120-01-208-7753	23615	TM 9-2815-237-34P
47	Pilot tube mandrel	5120-01-208-1767	16314	TM 9-2815-237-34P
48	Delivery valve, extractor test set, calibration	5120-00-816-7059	26081	TM 9-2815-237-34P
49	Linkage gap tool	5120-01-249-0370	23080	TM 9-2815-237-34P
50	Automatic advance indicator	5120-01-249-0368	23745	TM 9-2815-237-34P
51	Vacuum valve block gauge	5120-01-249-0369	23716	TM 9-2815-237-34P
52	Protractor	6675-01-247-2286	22089	TM 9-2815-237-34P
53	Air timing gauge	5220-01-247-0361	23715	TM 9-2815-237-34P
54	Hex-head driver, 5/32 in.	5120-00-969-0071	J35170	TM 9-2815-237-34P
55	Hex-head driver, 1/8 in.	5120-01-016-9202	J35171	TM 9-2815-237-34P
56	Hex-head driver, 1/4 in.	5120-00-596-8508	FA8A	TM 9-2815-237-34P
57	Adapter plate		67-6878	Kit 74-8033
58	Advance indicator		74-1094	Kit 74-8033

Section II. TOOL IDENTIFICATION LIST (Cont'd)

(1) Item Number	(2) Item Name	(3) National Stock Number	(4) Part Number	(5) Reference
59	Alignment gauge	4910-01-397-1616	67-4787	Kit 74-8033
60	Barring tool	4910-01-211-0879	67-2840	Kit 74-8033
61	Connector		67-5687	Kit 74-8033
62	High-pressure line		74-1250	Kit 74-8033
63	Hose assembly		74-1337	Kit 74-8033
64	Hose assembly		74-1030	Kit 74-8033
65	Male pipe fitting		03-5253	Kit 74-8033
66	Quick-connect plug		67-3456	Kit 74-8033
67	Test adapter		74-1131	Kit 74-8033
68	Test lead, black			Kit 74-8033
69	Test lead, red		77-0432	Kit 74-8033
70	Throttle arm positioner assembly		67-5649	Kit 74-8033
71	Tubing	4720-01-206-4429	03-6063	Kit 74-8033

APPENDIX F
TORQUE LIMITS

F-1. GENERAL

This section provides general torque limits for screws used on the M998 series vehicles. Special torque limits are indicated in the maintenance procedures for applicable components. The general torque limits given in this appendix shall be used when specific torque limits are not indicated in the maintenance procedure. Unless otherwise specified, standard torque tolerance shall be ± 10 percent. These general torque limits cannot be applied to screws that retain rubber components. The rubber components will be damaged before the correct torque limit is reached. If a special torque limit is not given in the maintenance instructions, tighten the screw or nut until it touches the metal bracket, then tighten it one more turn.

F-2. TORQUE LIMITS

Table F-1 lists dry torque limits. Dry torque limits are used on screws that do not have lubricants applied to the threads. Table F-2 lists wet torque limits. Wet torque limits are used on screws that have high pressure lubricants applied to the threads.

F-3. HOW TO USE TORQUE TABLE

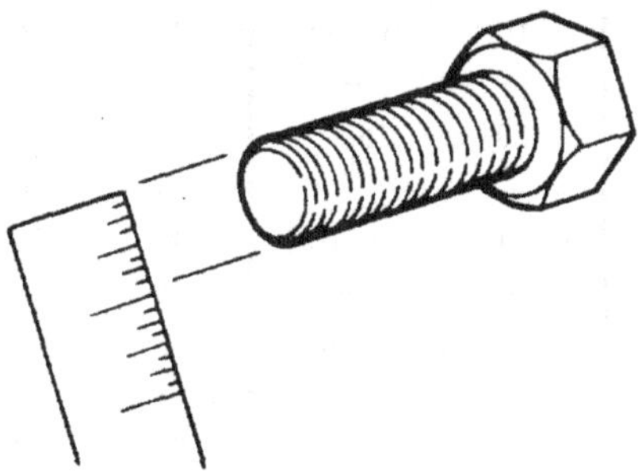

a. Measure the diameter of the screw you are installing.

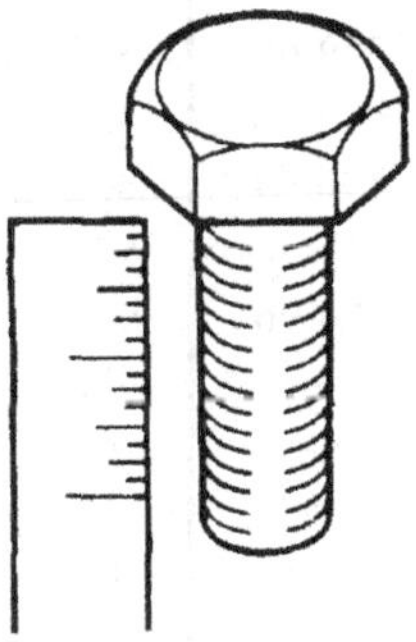

b. Count the number of threads per inch.

c. Under the heading SIZE, look down the left-hand column until you find the diameter of the screw you are installing (there will usually be two lines beginning with the same size).

d. In the second column under SIZE, find the number of threads per inch that matches the number of threads you counted in step b.

CAPSCREW HEAD MARKINGS

Manufacturer's marks may vary.
These are all SAE Grade 5
(3-line).

e. lb find the grade of screw you are installing, match he markings on the head to the correct picture of CAPSCREW HEAD MARKINGS on the torque table.

f. Look down the column under the picture you found in step e. until you find the torque limit (in lb-ft or N•m) for the diameter and threads per inch of the screw you are installing.

Table F-1. Torque Limits for Dry Fasteners

CAPSCREW HEAD MARKINGS

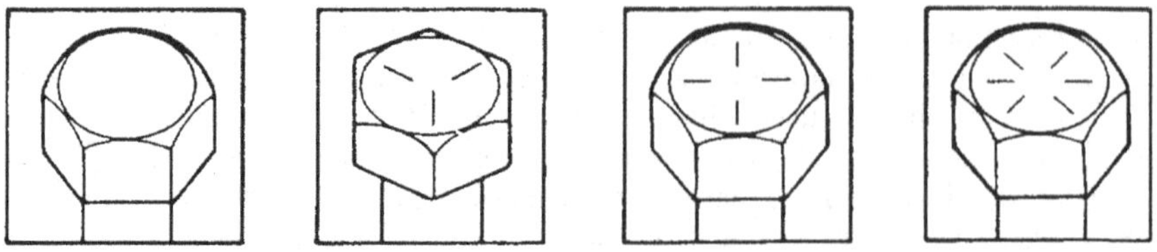

SIZE			TORQUE							
			SAE GRADE NO. 1 or 2		SAE GRADE No. 5		SAE GRADE NO .6 or 7		SAE GRADE No. 8	
DIA. INCHES	THREADS PER INCH	MILLIMETERS	POUND FEET	NEWTON MIETER5	POUND FEET	NEWTON METERS	POUND FEET	NEWTON METERS	POUND FEET	NEWTON METERS
1/4	20	6.35	5	6.78	8	10.85	10	13.56	12	16.27
/14	28	6.35	6	8.14	10	13.56	–	–	14	18.98
5/16	18	7.94	11	14.92	17	23.05	19	25.76	24	32.52
5/16	24	7.94	13	17.63	19	25.76	–	–	27	36.61
3/8	16	9.53	18	24.41	31	42.04	34	46.10	44	59.66
3/8	24	9.53	20	27.12	35	47.46	–	–	49	66.44
7/16	14	11.11	28	37.97	49	66.44	55	74.58	70	94.92
7/16	20	–	30	40.68	55	74.58	–	–	78	105.77
1/2	13	12.70	39	52.88	75	101.70	85	115.26	105	142.38
1/2	20		41	55.60	85	115.26	–	–	120	162.72
9/16	12	14.29	51	69.16	110	149.16	120	162.72	155	210.18
9/16	18	–	55	74.58	120	162.72	–	–	170	230.52
5/8	11	15.88	63	85.43	150	203.40	167	226.45	210	284.76
5/8	18	–	95	128.82	170	230.52	–	–	240	325.44
3/4	10	19.05	105	142.38	270	366.12	280	379.68	375	508.50
3/4	16	–	115	155.94	295	400.02	–	–	420	569.52
7/8	9	22.23	160	216.96	395	535.62	440	596.64	605	820.38
7/8	14	–	175	237.30	435	589.86	–	–	675	915.30
1	8	25.40	235	318.66	590	800.04	660	894.96	910	1233.96
1	14	–	250	339.00	660	894.96	–	–	990	1342.44
1-1/8	–	25.58	–	–	800- 880	1084.8- 1193.3	–	–	1280- 1440	1735.7- 1952.8
1-1/4	–	31.75	–	–	–	–	–	–	1820- 2000	2467.9- 2712.0
1-3/8	–	34.93	–	–	1460- 1680	1979.8 2278.1	–	–	2380- 2720	3227.3- 3688.3
1-1/2	–	38.10	–	–	1940- 2200	2630.6- 2983.2	–	–	3160- 3560	4285.0- 4827.4

Table F-2. Torque Limits for Wet Fasteners

CAPSCREW HEAD MARKINGS

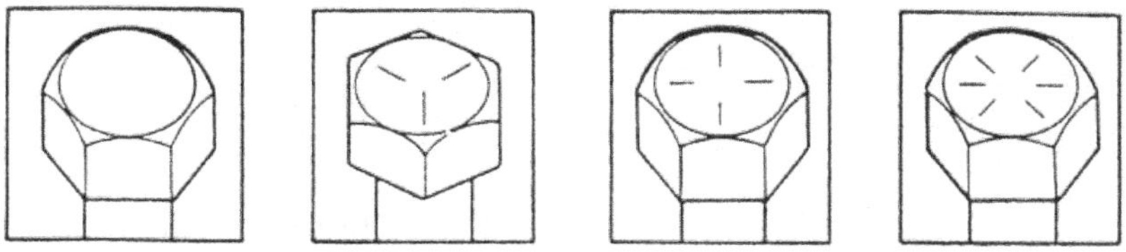

SIZE			TORQUE							
			SAE GRADE NO. 1 or 2		SAE GRADE NO. 5		SAE GRADE NO. 6 or 7		SAE GRADE NO. 8	
DIA. INCHES	THREADS PER INCH	MILLIMETERS	POUND FEET	NEWTON METERS	POUND FEET	NEWTON METERS	POUND FEET	NEWTON METERS	POUND FEET	NEWTON METERS
1/4	20	6.35	4.5	6.10	7.2	9.76	9.0	12.20	10.8	14.64
1/4	28	6.35	5.4	7.33	9.0	12.20	–	–	12.6	17.08
5/16	18	7.94	9.9	13.43	15.3	20.75	17.1	23.18	21.6	29.27
5/16	24	7.94	11.7	15.87	17.1	23.18	–	–	24.3	32.95
3/8	16	9.53	16.2	21.97	27.9	37.84	30.6	41.49	39.6	53.69
3/8	24	9.53	18.0	24.41	31.5	42.71	–	–	44.1	59.80
7/16	14	11.11	25.2	34.17	44.1	59.80	49.5	67.12	63.0	85.42
7/16	20	–	27.0	36.61	49.5	67.12	–	–	70.2	95.19
1/2	13	12.70	35.1	47.59	67.5	91.53	76.5	103.73	94.5	128.14
1/2	20	–	36.9	50.04	76.5	103.73	–	–	108.0	146.45
9/16	12	14.29	45.9	62.24	99.0	134.24	108.0	146.45	139.5	189.16
9/16	18	–	49.5	67.12	108.0	146.45	–	–	153.0	207.47
5/8	11	15.88	56.7	76.89	135.0	183.06	150.3	203.80	189.0	256.28
5/8	18	–	85.5	115.94	153.0	207.47	–	–	216.0	296.90
3/4	10	19.05	94.5	128.14	243.0	329.51	252.0	341.71	337.5	457.65
3/4	16	–	103.5	140.35	265.5	360.02	–	–	378.0	536.87
7/8	9	22.23	144.0	195.26	355.5	482.06	396.0	536.98	544.5	738.34
7/8	14	–	157.5	213.57	391.5	530.87	–	–	607.5	823.77
1	8	25.40	211.5	286.79	531.0	720.04	594.0	805.46	819.0	1110.56
1	14	–	225.0	305.10	594.0	805.46	–	–	891.0	1208.20
1-1/8	–	25.58	–	–	720.0-792.0	976.32-1073.97	–	–	1152.0-1296.0	1562.13-1757.52
1-1/4	–	31.75	–	–	–			–	1638-1800	2221.11-2440.80
1-3/8	–	34.93	–	–	1314.0-1512.0	1781.82-2050.29	–	–	2142.0-2448.0	2904.57-3319.47
1-1/2	–	38.10	–	–	1746.0-1980.0	2367.54-2684.88	–	–	2844.0-3204.0	3856.5-4344.66

INDEX

INDEX (Cont'd)

INDEX (Cont'd)

INDEX (Cont'd)

By Order of the Secretary of the Army:

DENNIS J. REIMER
General, United States Army
Chief of Staff

Official:

YVONNE M. HARRISO N
Administrative Assistant to the
Secretary of the Army
01399

By Order of the Secretary of the Air Force:

RONALD R. FOGLEMAN
General, United States Air Force
Chief of Staf f

Official:

HENRY VICCELLIO, JR.
General, United States Air Force
Commander, Air Force Materiel Command

By Order of the Marine Corps:

D. R. BLOOME R
Colonel, United States Marine Corps
Director, Program Support
Marine Corps Systems Command

Distribution:

To be distributed in accordance with DA Form 12-38-E (Block 0623) requirements for, " TM 9-2815-237-34.

☆ U.S. GOVERNMENT PRINTING OFFICE: 1996 746-014/20196

Made in the USA
Monee, IL
08 November 2026

67377480R00149